T0297194

Springer Proceedings in Mathematics & Statistics

Volume 156

More information about this series at http://www.springer.com/series/10533

Springer Proceedings in Mathematics & Statistics

This book series features volumes composed of select contributions from workshops and conferences in all areas of current research in mathematics and statistics, including OR and optimization. In addition to an overall evaluation of the interest, scientific quality, and timeliness of each proposal at the hands of the publisher, individual contributions are all refereed to the high quality standards of leading journals in the field. Thus, this series provides the research community with well-edited, authoritative reports on developments in the most exciting areas of mathematical and statistical research today.

Valery A. Kalyagin • Petr A. Koldanov
Panos M. Pardalos
Editors

Models, Algorithms and Technologies for Network Analysis

NET 2014, Nizhny Novgorod, Russia, May 2014

Editors
Valery A. Kalyagin
Laboratory of Algorithms and Technologies for Network Analysis (LATNA)
National Research University
Higher School of Economics
Nizhny Novgorod, Russia

Petr A. Koldanov
Laboratory of Algorithms and Technologies for Network Analysis (LATNA)
National Research University
Higher School of Economics
Nizhny Novgorod, Russia

Panos M. Pardalos
Department of Industrial and System Engineering
Center for Applied Optimization
University of Florida
Gainesville, FL, USA

ISSN 2194-1009 ISSN 2194-1017 (electronic)
Springer Proceedings in Mathematics & Statistics
ISBN 978-3-319-29606-7 ISBN 978-3-319-29608-1 (eBook)
DOI 10.1007/978-3-319-29608-1

Library of Congress Control Number: 2016937007

Mathematics Subject Classification (2010): 90-02, 90C31, 90C27, 90C09, 90C10, 90C11, 90C35, 90B06, 90B10, 90B15, 90B18, 90B40, 90B80, 68R01

Printed on acid-free paper

This Springer imprint is published by Springer Nature
The registered company is Springer International Publishing AG Switzerland

Preface

This volume is a collective monograph with different contributions. Many of them are related with the 4th International Conference on Network Analysis held in Nizhny Novgorod, Russia, May 11–13, 2014. The main focus of these contributions is the development of modern approaches for network analysis with different applications. The previous books which cover the similar topics are: [1–4].

According to the covered topics, the volume can be divided into three parts: optimization in networks, network data mining, and economics and other applications.

The first part deals with optimization problems in networks. In the chapter "Maximally Diverse Grouping and Clique Partitioning Problem with Skewed General Variable Neighborhood Search," a new variant of variable neighborhood search referred to as skewed general variable neighborhood search is used to solve a maximally diverse grouping problem and a clique partitioning problem. Extensive computational results show that the developed heuristic significantly outperforms its competitors.

In the chapter "Test Generation for Digital Circuits Based on Continuous Approach to Circuit Simulation Using Different Continuous Extensions of Boolean Functions," the analysis of continuous extensions of Boolean functions for test generation using continuous optimization is conducted. It represents the results of the developed software for a number of ISCAS schemes.

In the chapter "Konig Graphs for 4-Paths II: Wided Cycles," characterization of the graphs, whose each induced subgraph has the property that the maximum number of induced 4-paths is equal to the minimum cardinality of the set of vertices such as every induced 4-path contains at least one of them, is given. All such graphs obtained from simple cycles by replacing some vertices into cographs are described.

In the chapter "Optimization Algorithms for Shared Groups in Multicast Routing," the multicast group routing problem is considered. The problem requires the construction of one or more routing trees such that each destination has its demand satisfied by one or more data sources. The problem can be viewed as a generalization of the multicast routing problem with a single data source. This problem has important applications in the design of collaborative communication networks, among other uses. While the problem is NP-hard, it is possible to develop algorithms for its solution that approximate the solution in practice. Existing techniques for solving multicast group routing problems are discussed. Some fast heuristics for this problem are proposed. It is shown that computational experiments support the quality of the results achieved by these algorithms.

In the chapter "Minimizing the Fuel Consumption of a Multiobjective Vehicle Routing Problem Using the Parallel Multi-Start NSGA II Algorithm," a new multiobjective formulation of the vehicle routing problem (VRP), the multiobjective fuel consumption vehicle routing problem (MFCVRP), using two different objective functions is presented. The first objective function corresponds to the optimization of the total travel time and the second objective function is the minimization of the fuel consumption of the vehicle taking into account the travel distance, the load of the vehicle, and other route parameters. Two cases of the MFCVRP are solved.

In the chapter "Manifold Location Routing Problem with Applications in Network Theory," the problem of determining locations of facilities and the item distribution to customers from these facilities from supply chain networks is considered. Determination of the facility locations is a well-known problem in the literature, named facility location problem (FLP). The distribution of items via vehicles is known as the VRP. It is emphasized that solving VR and FL problems simultaneously can yield a robust facility location and reduced-cost distribution within the customer-supplier network.

In the chapter "A Branch and Bound Algorithm for the Cell Formation Problem," the cell formation problem (an NP-hard optimization problem) is considered for cell manufacturing systems. A branch and bound algorithm which provides exact solutions of the cell formation problem is proposed.

The second part of the book is devoted to data mining in networks. In the chapter "Hybrid Community Detection in Social Networks," several ideas to design hybrid methods for community detection are discussed.

In the chapter "Spectral Properties of Financial Correlation Matrix," the random matrix theory (RMT) is applied to investigate the cross-correlation matrix of a financial time series in four different stock markets: Russian, American, German, and Chinese. The deviations of distribution of eigenvalues of market correlation matrix from RMT global regime are investigated. Specific properties of each market are observed and discussed.

In the chapter "Statistical Uncertainty of Minimum Spanning Tree in Market Network," the procedure for the minimum spanning tree construction as a multiple decision statistical procedure is considered. The statistical uncertainty of the procedure is investigated.

In the chapter "Uncertainty of Identification of Cliques and Independent Sets in Pearson and Fechner Correlations Networks," two market network models for the analysis of NYSE stock market are used: Pearson correlation network and Fechner correlation network. The problem of estimation of statistical uncertainty of identification of maximum cliques and maximum independent sets in Pearson and Fechner correlation networks is considered. It is shown that identification of maximal cliques and maximal independent sets in Fechner correlation network is distribution-free in a certain class of distributions which is not true for Pearson correlation network.

In the chapter "Investigation of Connections Between Pearson and Sign Correlations in Market Network," the connection between the sign correlation of Fechner and the classic Pearson correlation is investigated. Testing hypothesis of the connection by real market data is conducted.

In the chapter "Testing the Stationarity of Sign Coincidence in Market Network," a hypothesis of the stationarity of observations from financial market are considered. For testing the stationarity hypotheses, the Kolmogorov–Smirnov test and multiple comparisons Bonferroni and Holm procedures are applied. The stationarity hypothesis is tested for sign coincidence of returns by binomial proportions. It is shown that the null hypothesis of stationarity is rejected for prices and not rejected for returns and their sign coincidence on some significance level.

In the chapter "Synchronization and Network Measures in a Concussion EEG Paradigm," the neurophysiologic changes in a patient who suffered a concussion during a football practice via quantitative and graph theoretic measures and to evaluate the results with respect to the preconcussion state are characterized. The deviations in the selected quantitative and graph theoretic measures partially corroborate the usual clinical characteristics of the post-concussion state, but further investigation for additional data and evaluation of alternative quantitative measures is needed.

In the chapter "Video-Based Pedestrian Detection on Mobile Phones with the Cascade Classifiers," the problem of real-time pedestrian recognition on mobile phones is discussed. A specialized procedure of data gathering and preprocessing to train cascade classifiers is proposed. Experimental results in testing under real road conditions with several mobile phones are given. It is emphasized, that sometimes it is necessary to choose faster, but less accurate, object detection algorithm, because in this case it is possible to process more number of frames in a fixed period of time. Hence, the total object detection accuracy can be increased.

In the chapter "Clustering in Financial Markets," graph partition of a particular kind of complex networks referred to as power law graphs is considered. In particular, analysis on the market graph, constructed from time series of price return on the American stock market, is conducted. Two different methods originating from clustering analysis in social networks and image segmentation are applied to obtain graph partitions, and the results are evaluated in terms of the structure and quality of the partition. It is shown that the market graph possesses a clear clustered structure only for higher correlation thresholds. Partitions for different time series are considered to study the dynamics and stability in the partition structure.

The third part contains applications of network analysis. In the chapter "Key Borrowers Detected by the Intensities of Their Short-Range Interactions," the issue of systemic importance that an individual financial institution can disturb the whole financial system is discussed. Interconnectedness is considered as one of the key drivers of systemic importance. Several measures have been proposed in the literature in order to estimate the interconnectedness of financial institutions and systems. However, they do not fully take into consideration an important dimension of this characteristic: intensities of agent interactions. This paper proposes a novel method that solves this issue. The approach is based on the power index and centrality analysis and is employed to find a key borrower in a loan market. The approach is applied at the European Union level.

In the chapter "A Semantic Solution for Seamless Data Exchange in Supply Networks," the semantic solution for data exchange in dynamically changing supply networks is proposed. Despite the dynamic nature of supply networks, there is a necessity to manage their efficiency. The first step in this direction is proper data exchange, which leads to transparency of networks and speeds up interactions of their participants. Despite plenty of data standards, there is a lack of approaches to data modeling that allowed seamless knowledge sharing within supply networks. This paper introduces a new ontology-based data metamodel of supply networks based on organizational ontology, consistent theory for data modeling, and the ontologized standard in logistics domain.

In the chapter "Langmuir Solitons in Plasma with Inhomogeneous Electron Temperature and Space Stimulated Scattering on Damping Ion-Sound Waves," an analytical solution for solitons in plasma is obtained in an approximate form. It is shown that analytical and numerical results have a good agreement.

In the chapter "Equilibria in Networks with Production and Knowledge Externalities," a game equilibrium is investigated in a network in each node of which an economy is described by the simple two-period model of endogenous growth with production and knowledge externalities. Each node of the network obtains an externality produced by the sum of knowledge in neighbor nodes. Uniqueness of the inner equilibrium is proved. Three ways of behavior of each agent are distinguished: active, passive, and hyperactive. Behavior of agents in dependence on received externalities is studied. It is shown that the equilibrium depends on the network structure.

We would like to take the opportunity to thank all the authors and referees for their efforts to contribute the chapters. In addition, we would like to thank Springer for giving us the opportunity for this work. This work is partly supported by Russian Federation Government grant N. 11.G34.31.0057.

Nizhny Novgorod, Russia — Valery A. Kalyagin
Nizhny Novgorod, Russia — Petr A. Koldanov
Gainesville, FL, USA — Panos M. Pardalos

References

1. Goldengorin, B.I., Kalyagin, V.A., Pardalos, P.M. (eds.): Models, Algorithms and Technologies for Network Analysis. In: Proceedings of the First International Conference on Network Analysis. Springer Proceedings in Mathematics and Statistics, vol. 32. Springer, Cham (2013)

2. Goldengorin, B.I., Kalyagin, V.A., Pardalos, P.M. (eds.): Models, Algorithms and Technologies for Network Analysis. In: Proceedings of the Second International Conference on Network Analysis. Springer Proceedings in Mathematics and Statistics, vol. 59. Springer, Cham (2013)
3. Batsyn, M.V., Kalyagin, V.A., Pardalos, P.M. (eds.): Models, Algorithms and Technologies for Network Analysis. In: Proceedings of Third International Conference on Network Analysis. Springer Proceedings in Mathematics and Statistics, vol. 104. Springer, Cham (2014)
4. Kalyagin, V.A., Pardalos, P.M., Rassias, T.M. (eds.): Network Models in Economics and Finance. Springer Optimization and Its Applications, vol. 100. Springer, Cham (2014)

Contents

Contributors

Habib Abdulrab INSA de Rouen, LITIS Lab., Rouen, France

Fuad Aleskerov DeCAn Lab, Institute of Control Sciences of Russian Academy of Sciences, National Research University (NRU) Higher School of Economics, Moscow, Russia

Elena Andreeva National Research University Higher School of Economics, Nizhny Novgorod, Russia

Irina Andrievskaya LIA Lab, National Research University (NRU) Higher School of Economics, Moscow, Russia

N.V. Aseeva National Research University Higher School of Economics, Nizhny Novgorod, Russia

Eduard Babkin National Research University Higher School of Economics, Nizhny Novgorod, Russia

Mikhail Batsyn Laboratory of Algorithms and Technologies for Networks Analysis, National Research University Higher School of Economics, Nizhny Novgorod, Russia

Jack Brimberg Royal Military College of Canada, Kingston, ON, Canada

Lei Cui Department of Computer Science, University of Texas at Dallas, Richardson, TX, USA

Gianluca Del Rossi University of South Florida, Tampa, FL, USA

Ding-Zhu Du Department of Computer Science, University of Texas at Dallas, Richardson, TX, USA

Hongwei Du Harbin Institute of Technology Shenzhen Graduate School, Shenzhen, China

E.M. Gromov National Research University Higher School of Economics, Nizhny Novgorod, Russia

Joseph Gutmann University of South Florida, Tampa, FL, USA

Valery A. Kalyagin National Research University Higher School of Economics, N. Novgorod, Russia

Nickolay Kascheev Faculty of Business Informatics and Applied Mathematics, National Research University Higher School of Economics, Nizhny Novgorod, Russia

Daniil Kascheev Faculty of Business Informatics and Applied Mathematics, National Research University Higher School of Economics, Nizhny Novgorod, Russia

Maxim Kazakov National Research University Higher School of Economics, N. Novgorod, Russia

Petr Koldanov National Research University Higher School of Economics, Nizhny Novgorod, Russia

Alexander P. Koldanov Lab LATNA, National Research University Higher School of Economics, Nizhny Novgorod, Russia

Anastasia Komissarova National Research University Higher School of Economics, Nizhny Novgorod, Russia

Alexei Korolev National Research University Higher School of Economics, St. Petersburg, Russia

Oleg Kremnyov National Research University Higher School of Economics, Nizhny Novgorod, Russia

Andrey Latyshev National Research University Higher School of Economics, Nizhny Novgorod, Russia

John Lloyd University of South Florida, Tampa, FL, USA

Magdalene Marinaki School of Production Engineering and Management, Technical University of Crete, Chania, Greece

Yannis Marinakis School of Production Engineering and Management, Technical University of Crete, Chania, Greece

Vladimir Matveenko National Research University Higher School of Economics, St. Petersburg, Russia

Athanasios Migdalas Department of Civil Engineering, Aristotle University of Thessalonike, Thessalonike, Greece

Industrial Logistics, Lulea University of Technology, Lulea, Sweden

Nenad Mladenović University of Valenciennes, Valenciennes, France

Mathematical Institute SANU, Belgrade, Serbia

Dmitry Mokeev Laboratory of Algorithms and Technologies for Networks Analysis, National Research University Higher School of Economics, Nizhny Novgorod, Russia

Dmitry E. Mozokhin National Research University Higher School of Economics, Nizhny Novgorod, Russia

T.V. Nasedkina National Research University Higher School of Economics, Nizhny Novgorod, Russia

Carlos A.S. Oliveira Quantitative Research Department, F-Squared Investments Inc., Ewing, NJ, USA

I.V. Onosova National Research University Higher School of Economics, Nizhny Novgorod, Russia

Ioannis Pappas University of Florida, Gainesville, FL, USA

Panos M. Pardalos Department of Industrial and Systems Engineering, University of Florida, Gainesville, FL, USA

National Research University Higher School of Economics, Nizhny Novgorod, Russia

Elena Permjakova National Research University (NRU) Higher School of Economics, Moscow, Russia

Tatiana Poletaeva National Research University Higher School of Economics, Nizhny Novgorod, Russia

Iraklis-Dimitrios Psychas School of Production Engineering and Management, Technical University of Crete, Chania, Greece

James Sackellares MD VA Hospital, Gainesville, FL, USA

Andrey V. Savchenko Laboratory of Algorithms and Technologies for Network Analysis, National Research University Higher School of Economics, Nizhny Novgorod, Russia

Ksenia G. Shipova National Research University Higher School of Economics, Nizhny Novgorod, Russia

Kristina Sörensen KTH Royal Institute of Technology, Stockholm, Sweden

Emre Tokgoz Department of Industrial Engineering, Quinnipiac University, Hamden, CT, USA

Theodore B. Trafalis Department of Industrial and System Engineering, University of Oklahoma, Norman, OK, USA

V.V. Tyutin National Research University Higher School of Economics, Nizhny Novgorod, Russia

Dragan Urošević Mathematical Institute SANU, Belgrade, Serbia

Irina Utkina Laboratory of Algorithms and Technologies for Networks Analysis, National Research University Higher School of Economics, Nizhny Novgorod, Russia

Weili Wu Department of Computer Science, University of Texas at Dallas, Richardson, TX, USA

Part I
Optimization in Networks

Maximally Diverse Grouping and Clique Partitioning Problems with Skewed General Variable Neighborhood Search

Jack Brimberg, Nenad Mladenović, and Dragan Urošević

Abstract The maximally diverse grouping problem (MDGP) requires finding a partition of a given set of elements into a fixed number of mutually disjoint subsets (or groups) in order to maximize the overall diversity between elements of the same group. The clique partitioning problem (CPP) has a similar form as the MDGP, but is defined as the minimization of dissimilarity of elements in an unknown number of groups. In this paper a new variant of variable neighborhood search referred to as skewed general variable neighborhood search (SGVNS) is used to solve both problems. Extensive computational results show that the developed heuristic significantly outperforms its competitors. This demonstrates the usefulness of a combined approach of diversification afforded with *skewed VNS* and intensification afforded with the local search in *general* VNS.

Keywords Diverse grouping • Clique partitioning • Variable neighborhood search

1 Introduction

The maximally diverse grouping problem (MDGP) requires finding a partition of a given set of elements into a fixed number of mutually disjoint subsets (or groups) in order to maximize the overall diversity between elements of the same group.

J. Brimberg
Royal Military College of Canada, Kingston, ON, Canada
e-mail: jack.brimberg@rmc.ca

N. Mladenović (✉)
University of Valenciennes, Valenciennes, France

Mathematical Institute SANU, Belgrade, Serbia
e-mail: nenad.mladenovic12@gmail.com

D. Urošević
Mathematical Institute SANU, Knez Mihailova 36, Belgrade, Serbia
e-mail: draganu@mi.sanu.ac.rs

V.A. Kalyagin et al. (eds.), *Models, Algorithms and Technologies for Network Analysis*, Springer Proceedings in Mathematics & Statistics 156,
DOI 10.1007/978-3-319-29608-1_1

The clique partitioning problem (CPP) is similar. Instead of maximizing diversity among groups, one wants to find an unknown number of groups such that the sum of edge weights over the induced subgraphs (or cliques) is minimized.

MDGP Formulation. The MDGP may be formulated as follows (see, e.g., Fan et al. [11]): $P = \{p_i : i = 1, \ldots, N\}$ denotes the set of elements of interest, and p_{ik}, $k \in \{1, \ldots, K\}$, the attribute values associated with each element p_i. The measure of diversity between any pair of elements p_i and p_j may be given by some distance function, e.g.,

$$d_{ij} = \sqrt{\sum_{k=1}^{K} (p_{ik} - p_{jk})^2}$$

which calculates the Euclidean distance between corresponding points in the attribute space. The objective is the partition of a set P into G disjoint groups so that the sum of diversities over the individual groups is maximized.

Let $x_{ig} = 1$ if p_i is assigned to the group g, and 0 otherwise, $i = 1, 2, \ldots, N$, $g = 1, 2, \ldots, G$. The model may be written as the following quadratic binary integer programming problem:

$$\max \sum_{g=1}^{G} \sum_{i=1}^{N-1} \sum_{j=i+1}^{N} d_{ij} x_{ig} x_{jg} \tag{1}$$

$$s.t.$$

$$\sum_{g=1}^{G} x_{ig} = 1, \; i = 1, 2, \ldots, N \tag{2}$$

$$\sum_{i=1}^{N} x_{ig} \geq a_g, \; g = 1, 2, \ldots, G \tag{3}$$

$$\sum_{i=1}^{N} x_{ig} \leq b_g, \; g = 1, 2, \ldots, G \tag{4}$$

$$x_{ig} \in \{0, 1\}, i = 1, 2, \ldots, N, g = 1, 2, \ldots, G. \tag{5}$$

The constraints (2) ensure that each element is assigned to exactly one group. Constraints (3) and (4) impose minimum and maximum group sizes, respectively. In most studies of MDGP, it is assumed that group sizes must be equal so that $N = mG$, where m is an integer, and $a_g = b_g = m$ for all groups $g = 1, 2, \ldots, G$. Then constraints (3) and (4) in the model above can be replaced with

$$\sum_{i=1}^{N} x_{ig} = m, \; g = 1, 2, \ldots, G. \tag{6}$$

The above model allows groups to differ in size, although in practical cases such as study groups, the variation in size should be kept relatively small.

CPP Formulation. The CPP can be described as follows. Given a weighted, non-oriented, and complete graph $G = (V, E, w)$, we wish to find a partition of the node set into an unknown number of nonempty, disjoint subsets $V_1, V_2, \ldots, V_k$ such that the sum of edge weights over the k induced subgraphs (or cliques) $G_1, G_2, \ldots, G_k$ is minimized. Let c_{ij} denote the weight (or cost) of edge (i,j), for all pairs of nodes (i,j), $1 \leqslant i < j \leqslant N$, where N equals the number of nodes in graph G similarly as in the MDGP. Note, however, that unlike the MDGP where diversities are specified by edge distances $d_{ij} \geq 0$, the costs c_{ij} can now take on negative values. A mathematical formulation of the problem is given by (see [6]):

$$
\begin{aligned}
&\min \sum_{(i,j)\in E} c_{ij}x_{ij} \\
&\text{s.t.} \\
&x_{ij} + x_{jr} - x_{ir} \leq 1, && \forall 1 \leqslant i < j < r \leqslant N \\
&x_{ij} - x_{jr} + x_{ir} \leq 1, && \forall 1 \leqslant i < j < r \leqslant N \\
&-x_{ij} + x_{jr} + x_{ir} \leq 1, && \forall 1 \leqslant i < j < r \leqslant N \\
&x_{ij} \in \{0, 1\}, && \forall 1 \leqslant i < j \leqslant N.
\end{aligned} \tag{CPP}
$$

If $x_{ij} = 1$, then nodes i and j are in the same cluster, and the cost c_{ij} of edge (i,j) is added in the objective function; otherwise, it is not. The constraint set ensures that all edges in the complete subgraph G_t, $t = 1, 2, \ldots, k$, are included in the solution.

The CPP may be extended to incomplete graphs by inserting fictitious edges with large positive weights between pairs of nodes wherever edges are missing in the original graph. The large positive weight ensures that each subgraph G_t, $t = 1, 2, \ldots, k$, is a clique that contains no fictitious edges. If all edges in a complete graph G have negative (or zero) weights, the problem becomes trivial, with the optimal solution given by $k = 1$ and $G_1 = G$. Thus, the CPP becomes interesting only when some of the edge weights have positive values.

MDGP Applications. One of the earliest applications of MDGP was in the formation of student work groups. For example, in MBA programs it is very important to divide a class into diverse study groups in order to enhance the learning environment (Weitz and Lakshminarayanan [27], Desrosiers et al. [9]). Another application concerns the formation of peer review groups to evaluate research proposals. Again, the objective is to form diverse groups in order to ensure that projects are evaluated from several different points of view (see Hettich and Pazzani [16]). For other applications cited in the literature see, e.g., Lotfi and Cerveny [18] for exam scheduling, Weitz and Lakshminarayanan [27] for very large scale integration (VLSI) design, and Kral [17] for the storage of large programs onto paged memory.

Problems similar to MDGP have been considered by Bhadury et al. [2], Baker and Benn [1], and Desrosiers et al. [9]. A simplified model is proposed in [2], where a high degree of intra-team similarity and/or inter-team dissimilarity is requested.

A network flow problem, known as the dining problem, is used to assign MBA students to different projects. The case study presented in [1] consists of assigning 235 students to eight tutor groups. The model used belongs to the mixed linear goal programming class and is equivalent to the min-sum formulation of Desrosiers et al. [9], where the ℓ_1-norm is used to measure dissimilarities between students. In the proposed local search, swaps between the two worst teams are used to improve the current solution. Desrosiers et al. [9] use a centroid to represent each group of entities, and examine two objectives: *min-sum*—minimize the sum of distances between group centroids and the general centroid of all data; *min-max*—minimize the maximum of such distances. The authors also apply their method to partition 120 MBA students in groups of 5 at HEC, Montreal. If the ℓ_1-norm is used, the model becomes linear. However, if Euclidean distance is considered, the corresponding model is more similar to MDGP, and belongs to the class of quadratic 0–1 programs. Exact and heuristic solution techniques are adapted for solving the problem.

CPP Applications. The CPP has important applications in the social sciences (e.g., see [6], and the references therein). Wang et al. [26] demonstrate that the CPP compares favorably with K-means and latent class analysis for recovery of cluster structure in real data sets. An advantage of the CPP model is that we do not need to know the number of clusters beforehand. One of the most cited applications of CPP is in the aggregation of binary equivalence relations. In this context, the c_{ij} parameters represent the number of attributes on which nodes i and j disagree minus the number on which they agree. For example, if there are 10 attributes in all being compared and given nodes i and j have the same measurement on 8 of them, $c_{ij} = 2 - 8 = -6$, and there is a strong tendency to place these two nodes in the same cluster. If on the other hand $c_{ij} = 8 - 2 = 6$, there would be a tendency to place them in separate clusters.

Recent Heuristics for Solving MDGP. The MDGP is known to be NP-hard [12], and therefore, heuristics have been developed to solve it. Fan et al. [11] propose a hybrid genetic algorithm to solve the problem. Gallego et al. [13] suggest tabu search with strategic oscillation. The artificial bee colony optimization (ABCO) method proposed by Rodriguez et al. [23] may be considered as a state-of-the-art heuristic for solving MDGP. A competitive heuristic based on general variable neighborhood search (GVNS) is found in [24]. GVNS is a variant where a deterministic local search that uses several neighborhoods (called variable neighborhood descent, or VND for short) is applied in the local search step in place of the single neighborhood used in basic VNS.

Recent Heuristics for Solving CPP. The earliest methods to solve CPP use a simple relocation procedure (analogous to the Insertion move for MDGP) that interchanges nodes across clusters until a local optimum is reached (e.g., [22]). Marcotorchino and Michaud [19] expanded the search to include merging of clusters and exchanging cluster memberships for pairs of nodes. Simulated annealing and tabu search heuristics are proposed by De Amorim et al. [8], based on the simple relocation neighborhood above. However, this also includes the possibility of relocating a node to the empty set, thereby increasing the number of clusters by 1.

Charon and Hudry [7] investigate various "noising" procedures that distort the data in order to allow uphill moves. Brusco and K ohn [6] develop a state-of-the-art "neighborhood search heuristic," following the ideas of the variable neighborhood search (VNS) metaheuristic, where the shaking (or perturbation) step is carried out in a random fashion instead of the systematic approach of VNS. The local search employs the single node relocation neighborhood.

Outline. In this paper we present two related algorithms recently developed in [4, 5] for solving the MDGP and CPP, respectively. The methodology is an extension of the GVNS based heuristic proposed in [24]. In brief, the new algorithms consider a skewed version of GVNS that allows moves to inferior points located in promising regions of the solution space. In the next section the relationship between the two problems is analyzed. The new heuristic is described in detail in Sect. 3, as applied to the MDGP including a description of the data structure used for efficient implementation of the VND local search. In Sect. 4 the implementation differences for solving the CPP are summarized. A summary of computational results on a wide range of test problems taken from the literature is given is Sects. 5 and 6 for MDGP and CPP, respectively. For almost all problem instances examined, significantly better solutions are obtained than those of the best competitors.

Concluding remarks are given in Sect. 7. This work unifies the results of two journal articles [4, 5].

2 Relation Between CPP and MDGP

In the MDGP the number of groups (or cliques) is fixed, and typically these groups must have the same size, or nearly the same size. However, in the CPP we don't know the number of groups, and there are no limitations on group size. To get around this difficulty, we simply set the number of groups initially to an upper limit of N (the number of nodes in G), and allow some of these groups to be empty. No limitations on group size are imposed, so that the number of groups is effectively treated as a variable in the model formulation.

Now let $y_{ig} = 1$, if node i belongs to group g, and 0 otherwise. We reformulate CPP as the following equivalent quadratic binary integer program:

$$\begin{aligned} & \max -\sum_{g=1}^{N}\sum_{i=1}^{N-1}\sum_{j=i+1}^{N} c_{ij}y_{ig}y_{jg} \\ & \text{s.t.} \\ & \sum_{g=1}^{N} y_{ig} = 1, \qquad i = 1,\ldots,N \\ & y_{ig} \in \{0,1\}, \qquad i = 1,\ldots,N, g = 1,\ldots,N \end{aligned} \tag{CPP$_e$}$$

This formulation is equivalent to the general MDGP except that (a) all constraints on group size have been deleted and (b) the number of groups is no longer specified. Note that the nonlinear formulation in (CPP$_e$) allows a reduction in the number of constraints compared to the original linear formulation (CPP) from $O(N^3)$ to N. Meanwhile the number of binary variables remains close to the same in both formulations.

A similar $0/1$ quadratic program as (CPP$_e$) may be found in Wang et al. [25] as an alternate formulation of the CPP. However, in [25], a maximum number of cliques ($k_{\max} < N$) are assumed to be known, which is not necessary here. Also, those authors use the quadratic formulation to solve the Group Technology Problem, whereas we are showing here that the CPP may be viewed as a relaxed form of the MDGP.

By recasting the model in the form of an MDGP, we are now able to borrow any solution method for MDGP and apply it (after some small modifications) to solve our original CPP. We will take advantage of this fact by adapting our new VNS based heuristic that was originally applied on the MDGP, and is described next.

3 Solving MDGP with Variable Neighborhood Search

VNS is a well-known metaheuristic, or framework for building heuristics, whose basic idea is a systematic change of neighborhood structures during a search for a better solution (Mladenović and Hansen [20] and Brimberg et al. [3]). The inner loop of basic VNS contains three steps: (a) shaking; (b) local search; and (c) neighborhood change. A successful VNS variant, called general VNS, uses a mechanism of changing neighborhoods not only in the diversification or shaking step, but also in the intensification or deterministic local search step (Hansen et al. [15], Mladenović et al. [21]). Skewed VNS is another VNS variant that modifies the "neighborhood change" step. A current solution is allowed to move to an inferior solution only if the latter is very far and of similar quality.

This section gives the details of the VNS implementation for solving MDGP. The main idea is to combine the concepts of general VNS (GVNS) and skewed VNS (SVNS), by allowing skewed moves within GVNS; that is, we combine the intensification of the local search in GVNS with the diversification provided by SVNS. The resulting framework is called skewed general variable neighborhood search (SGVNS), as proposed in Brimberg et al. [4].

3.1 Solution Space of MDGP

The solution space includes all possible feasible divisions of elements into groups. A division is feasible if and only if each created group g contains at least a_g and at most b_g elements. A current solution is represented by a vector x^c of length N such that $x^c[i]$ is the label of the group containing the element i ($i = 1, 2, \ldots, N$). In order to speed up the local search, we also maintain matrix sd^c such that $sd^c[i][g]$ is the sum of diversities between the element i and all the elements assigned to the group g in the current solution:

$$sd^c[i][g] = \sum_{j=1,2,\ldots,N;x^c[j]=g} d_{ij}.$$

Note that for the current solution, matrix sd^c can be computed in $O(N^2)$.

For the solution represented by the vector x^c (and by the corresponding matrix sd^c) it is possible to calculate the objective function value given in formula (1) in the following way:

$$f^c = f(x^c) = \sum_{i=1}^{N-1} \sum_{j=i+1}^{N} d_{ij}\chi(x^c, i, j)$$

where

$$\chi(x_c, i, j) = \begin{cases} 1 & \text{if } x^c[i] = x^c[j] \\ 0, & \text{if } x^c[i] \neq x^c[j] \end{cases}.$$

Note also that the same objective value can be calculated by using the previously calculated matrix sd^c:

$$f(x^c) = \frac{1}{2} \sum_{i=1}^{N} sd^c[i][x^c[i]]$$

3.2 VND Local Search

The local search is implemented using VND, for which the following neighborhoods are designed (see [24]): Insertion, Swap, and 3-Chain.

Insertion. Neighborhood Insertion contains solutions obtained by moving only one element from its current group to another group. By using the previously described matrix sd^c, it is possible to compute efficiently for each feasible move the change in value of the objective function f. Denote with x^n a solution obtained from solution x^c by moving the element i from its current group g_1 to a group g_2. In this case, the sum of diversities in all groups except groups g_1 and g_2 is unchanged. The element i is removed from the group g_1 and because of that, the sum of diversities in the group g_1 decreases by the sum of diversities between i and all other elements belonging to the group g_1. The element i is inserted into g_2; hence, the sum of diversities in g_2 increases by the sum of all diversities between i and the elements belonging to the group g_2. It is easy to conclude that the difference between objective function values for solutions x^c and x^n is

$$\Delta f = f(x^n) - f(x^c) = sd^c[i][g_2] - sd^c[i][g_1].$$

If $\Delta f > 0$, the element i is moved from the group g_1 to the group g_2, ($x^c[i] = g_2$); then, groups g_1 and g_2 are modified and values $sd^c[j][g_1]$ and $sd^c[j][g_2]$ must be updated in the following way:

$$sd^c[j][g_1] = sd^c[j][g_1] - d_{ji}$$

and

$$sd^c[j][g_2] = sd^c[j][g_2] + d_{ji}.$$

Since the updating must be performed for each element j, updating of matrix sd^c after performing an Insertion move has complexity $O(N)$. The number of solutions in the Insertion neighborhood of x^c is $O(GN)$. See Algorithm 1 for details of the search. Note that the subroutine `Updatesd` is used to denote in general the updating of matrix sd^c after a move is made to a neighboring solution (Insertion, Swap, or 3-Chain).

Swap. Neighborhood Swap contains solutions obtained by swapping a single pair of elements belonging to different groups. Let element i be in group g_i and element j in group g_j of the current solution x^c. Denote with x^n the solution obtained after moving the element i into group g_j and the element j into group g_i. Since the element i is removed from the group g_i, the diversities between element i and the elements remaining in the group g_i do not contribute to the objective function value of the

```
Function LSIns(x, sd, f);
rez ← false;
for v ← 1 to N do
    for g ← 1 to G do
        if x[v] ≠ g then
            df ← sd[v][g] − sd[v][x[v]];
            if df > 0 then
                x[v] ← g;
                f ← f + df;
                Updatesd(x, sd, v, g);
                rez ← true
            end
        end
    end
end
return rez
```

Algorithm 1: VND implementation of local search in Insertion neighborhood

```
Function LSSwap(x, sd, f);
rez ← false;
for v ← 1 to N − 1 do
    for u ← v + 1 to N do
        if x[v] ≠ x[u] then
            df ← sd[v][x[u]] + sd[u][x[v]] − sd[v][x[v]] − sd[u][x[u]] − 2d_{v,u};
            if df > 0 then
                Swap(x, v, u);
                f ← f + df;
                Updatesd(x, sd, v, u);
                rez ← true;
                return rez
            end
        end
    end
end
return rez
```

Algorithm 2: VND implementation of local search in Swap neighborhood

new solution. But, because the element i is inserted in the group g_j, all diversities between i and the elements belonging to the group g_j contribute to the objective function value of the new solution. Similar facts are true for the element j. So, we can finally calculate the difference between the objective values of the current and neighboring solutions:

$$\Delta f = f(x^n) - f(x^c) = (sd^c[i][g_j] - sd^c[i][g_i]) + (sd^c[j][g_i] - sd^c[j][g_j]) - 2d_{ij}.$$

It is obvious that the change of the objective value for each solution from neighborhood Swap is done in $O(1)$, while the cardinality of Swap is $O(N^2)$. After performing a Swap move, it is necessary to update matrix sd^c, and the complexity of this update is $O(N)$. (We can consider Swap or 2-opt as two successive Insertions.)

3-Chain. A 3-Chain move is determined by three elements, i, j, and k, belonging to three different groups, g_i, g_j, and g_k, respectively, by moving the element i to the group g_j, the element j to the group g_k, and the element k to the group g_i. Neighborhood 3-Chain consists of all solutions obtained by performing a single 3-Chain move.

We can calculate the difference between objective function values as follows:

$$\begin{aligned}\Delta f = f(x^n) - f(x^c) =& (sd^c[i][g_j] - sd^c[i][g_i]) + (sd^c[j][g_k] - sd^c[j][g_j]) + \\ & (sd^c[k][g_i] - sd^c[k][g_k]) - (d_{ij} + d_{jk} + d_{ki}).\end{aligned}$$

So, the complexity of solution checking is $O(1)$, but the cardinality of the whole neighborhood is $O(N^3)$. Updating of the matrix sd^c after a 3-Chain move is made is performed in a straightforward way by combining three successive Insertion moves.

```
Function VND-2(x, sd,f);
end ← false;
while not end do
    repeat
        rez ← LSIns(x, sd,f);
    until rez = false;
    if LSSwap(x, sd,f) = false then end ← true
end
```

Algorithm 3: VND-2

Two VND Variants. We propose two variants of variable neighborhood descent:

- VND-2—use neighborhoods Insertion and Swap in this order and
- VND-3—use all three developed neighborhoods in the following order: Insertion, Swap, and 3-Chain.

If no improvement is found in a given neighborhood of the current solution, the search resumes in the next neighborhood in the sequence. If an improvement is found, the local search always resumes in the Insertion neighborhood of the new current solution, i.e., the first neighborhood in the sequence. If no improvement can be found in any of the listed neighborhoods of x^c, the local search is terminated. See Algorithms 1–3 for a standard implementation of VND-2. Our implementation is slightly different in that all pairs of elements are examined once in LSSwap, and hence, a few Swap moves may occur before exiting the loop and returning to the Insertion neighborhood. Also note that VND-2 is the only variant attempted in the computational experiments due to limitations placed on the computing time.

3.3 Initial Solution

The calculation of the initial solution is done in two phases. In the first phase, we ensure that each group g contains at least a_g elements by inserting the chosen elements. In the second phase, we distribute the remaining elements so that each group g contains at most b_g elements. At the beginning, we select G elements at random and insert them into different groups.

Denote with E_g a set of elements currently assigned to a group g. During the first phase, we maintain the set of groups that have fewer elements than the desired minimum:

$$G' = \{g : |E_g| < a_g\}.$$

Each iteration of the first phase consists of selecting at random one unassigned element, denoted with i, and assigning it to a selected group. The selected element must be inserted into one of the groups from the set G'. So, for each group $g \in G'$, we calculate the average distance between the selected element i and all the elements belonging to the group g, and select the group whose average distance value is the

```
Procedure InitialSolution(x, sd, f);
AV ← {1, 2, ..., N}; AG ← {1, 2, ..., G};
for g ← 1 to G do c_g ← 0 end
while AG ≠ ∅ do
  | g ← RandomElem(AG); v ← RandomElem(AV);
  | x[v] ← g; c_g ← c_g + 1;
  | AV ← AV \ {v};
  | if c_g = a_g then AG ← AG \ {g} end
end
AG ← ∅;
for g ← 1 to G do
  | if c_g < b_g then AG ← AG ∪ {g} end
end
while AV ≠ ∅ do
  | g ← RandomElem(AG); v ← RandomElem(AV);
  | x[v] ← g; c_g ← c_g + 1;
  | AV ← AV \ {v};
  | if c_g = b_g then AG ← AG \ {g} end
end
Computesd(x, sd);
f ← Computeobj(x)
```

Algorithm 4: Initial solution

largest. The selected group g is the one for which the expression D_{ig} defined as

$$D_{ig} = \frac{\sum_{j \in E_g} d_{ij}}{|E_g|}$$

is maximized.

The first phase ends when the set G' becomes empty. During the second phase, we maintain a set of groups which has fewer elements than the desired maximum:

$$G' = \left\{g : |E_g| < b_g\right\}.$$

All other steps of the second phase are the same. The second phase finishes when all elements are assigned.

The initial solution can also be created at random: instead of assigning a selected element to a group with maximal average distance, we assign it to a randomly selected group.

3.4 Shaking

In order to perform a perturbation of the current solution, we define the kth neighborhood as the set of solutions obtained by k consecutive Swap moves. Thus, in the shaking step, we generate a random solution from the kth neighborhood of

```
Procedure Shake(k, x, sd, f);
while k > 0 do
    (u, v) ←(RandVert, RandVert);
    if x[u] ≠ x[v] then
        Swap(x, u, v);
        k ← k − 1
    end
end
Computesd(x, sd);
f ← Computeobj(x)
```

Algorithm 5: Shake procedure

```
Function SGVNS(k_min, k_step, k_max, t_max);
InitialSolution(x^c, sd^c, f^c);
VND-2(x^c, sd^c, f^c);
(x^b, sd^b, f^b) ← (x^c, sd^c, f^c);
k ← k_min;
while CPUTime() ≤ t_max do
    (x^n, sd^n, f^n) ← (x^c, sd^c, f^c);
    Shake(k, x^n, sd^n, f^n);
    VND-2(x^n, sd^n, f^n);
    if f(x^n)/f(x^c) + αρ(x^c, x^n) > 1 and f(x^n)/f(x^b) + αρ(x^b, x^n) > 1 then
        (x^c, sd^c, f^c) ← (x^n, sd^n, f^n);
        if f(x^c) > f(x^b) then (x^b, sd^b, f^b) ← (x^c, sd^c, f^c) end;
        k ← k_min
    else
        k ← k + k_step;
        if k > k_max then k ← k_min
        end
    end
end
return x^b
```

Algorithm 6: SGVNS

the current solution by performing k random Swap moves. In each of the moves, we randomly select two elements that belong to two different groups.

3.5 Skewed General Variable Neighborhood Search

Unlike classical VNS, the idea is to allow skewed moves to inferior solutions in promising regions of the solution space. In our implementation this means that after shaking and local search, we move to the new solution x^n if the following conditions are satisfied:

$$\frac{f(x^n)}{f(x^c)} + \alpha\rho(x^c, x^n) > 1 \quad \text{and} \quad \frac{f(x^n)}{f(x^b)} + \alpha\rho(x^b, x^n) > 1, \tag{7}$$

where x^b is the best found solution, and x^c the current solution. We use $\rho(x^{(1)}, x^{(2)})$ to denote the distance between solutions $x^{(1)}$ and $x^{(2)}$. This distance is defined in the following way:

$$\rho(x^{(1)}, x^{(2)}) = \frac{\left|x^{(1)} \Delta x^{(2)}\right|}{N^2/G}, \tag{8}$$

where

$$\begin{aligned} x^{(1)} \Delta x^{(2)} = \{(i,j) : &((x^{(1)}[i] = x^{(1)}[j]) \wedge (x^{(2)}[i] \neq x^{(2)}[j])) \vee \\ &((x^{(1)}[i] \neq x^{(1)}[j]) \wedge (x^{(2)}[i] = x^{(2)}[j]))\}, \end{aligned} \tag{9}$$

which estimates the "fraction" of pairs belonging to the same group in one solution, but not to the same group in the other solution. Note that the average number of elements in any of the groups is approximately $\frac{N}{G}$, and thus there are $\binom{\frac{N}{G}}{2} \approx \frac{N^2}{2G^2}$ pairs of elements in each group. Because there are G groups, the total number of pairs is $\frac{N^2}{2G}$ in one solution, and $2 \times \frac{N^2}{2G} = \frac{N^2}{G}$ elements in both solutions. The parameter α is assigned a value of 0.05 (selected after detailed testing).

Combining the different parts described above leads to the implementation of SGVNS that is summarized in Algorithm 6.

4 Implementation Differences for Solving CPP with SGVNS

To avoid duplication in this section, we only summarize the differences between the SGVNS implementation for CPP and the one for MDGP. Further details on the CPP implementation are given in [5]. Computational results in Sect. 6 demonstrate that the new heuristic provides a powerful solution approach just as its counterpart for the MDGP.

The solution space is constructed in the same way as for the MDGP. Thus, the current solution is represented by vector x^c of length N ($x^c = (x_1^c, x_2^c, \ldots, x_N^c)$) such that x_i^c is the label of the group (clique) containing element i ($i = 1, 2, \ldots, N$). In order to speed up the local search, we also maintain matrix sd^c such that $sd^c[i][g]$ is the sum of edge weights between element i and all elements assigned to the group g in the current solution.

The number of groups (active columns in sd^c) is initially set to N with one element in each group. This, in fact, gives the initial solution. Unlike MDGP with a constant number of groups (G), the number of groups ($g_{\max}$) in CPP will be allowed to change during the solution process. To accomplish this, we always maintain an empty group in the list of groups (last column in sd^c).

The local search is implemented using the same VND as for MDGP, including the same nested structure of Insertion and Swap neighborhoods. If the Insertion step is performed, we change the current solution, and then it becomes necessary to update matrix sd^c. For example, if the element i moves from the group g_1 to the group g_2, then groups g_1 and g_2 are modified and values $sd^c[j][g_1]$ and $sd^c[j][g_2]$ must be updated in the following way:

$$sd^c[j][g_1] = sd^c[j][g_1] - c_{ji}$$

and

$$sd^c[j][g_2] = sd^c[j][g_2] + c_{ji}.$$

Since this updating must be performed for each element j, updating of matrix sd^c after performing an Insertion move has complexity $O(N)$. On the other hand, the cardinality of the Insertion neighborhood is now $O(g_{\max}N)$, where $g_{\max}$ is the current number of groups (including the empty group). Note that if a group becomes empty as the result of an Insertion move, it is eliminated, the remaining groups are renumbered accordingly, and $g_{\max}$ is reduced by 1. If the Insertion move results in group $g_{\max}$ (the empty one) becoming a singleton, $g_{\max}$ is increased by 1.

Neighborhood Swap contains solutions obtained by swapping a single pair of elements belonging to different groups (elements exchange the group they are currently assigned to). As for MDGP, the change in objective value for each solution from neighborhood Swap is done in $O(1)$, while the cardinality of Swap is $O(N^2)$. After performing a Swap move, it is necessary to update the matrix sd^c, and the complexity of this update is $O(N)$ again.

As in MDGP, the VND proceeds first in neighborhood Insertion until a local optimum is obtained. The search then proceeds to neighborhood Swap. If after one cycle, at least one Swap move is made, the search resumes in neighborhood Insertion; otherwise, the VND stops. The distance between two solutions x^n and x^b is now defined in the following way:

$$d(x^n, x^b) = \frac{\left|\{(i,j) | 1 \leq i < j \leq n, ((x_i^b = x_j^b) \wedge (x_i^n \neq x_j^n)) \vee ((x_i^b \neq x_j^b) \wedge (x_i^n = x_j^n))\}\right|}{\sum_{g=1}^{g_b} \binom{c_g^b}{2}},$$

where g_b is the number of groups in the best solution x^b, and c_g^b is the number of elements in group g in the best solution. Intuitively, the expression in the numerator is the number of pairs of elements belonging to the same cluster (or group) in one solution, but not belonging to the same group in the other solution. On the other hand the denominator is equal to the number of edges participating in cliques in the best solution. The expression $d(x^c, x^n)$ denotes the distance between solutions x^c and x^n and calculates in a similar way. The same criteria as in MDGP are applied to allow skewed moves.

5 Computational Results for MDGP

The purpose of this section is to compare our heuristic with other state-of-the-art methods for solving MDGP. We first describe five data sets usually used in the literature; then, we analyze the performance of SGVNS in order to estimate the best values of its parameters. Note that the test instances are all randomly generated as explained in the corresponding references given below. Finally, we perform an extensive computational comparison. Also note that all computer runs by us for both MDGP and CPP are conducted on an Intel x64 based machine with 3.4 GHZ CPU (with 4 cores) and 4GB of RAM. The SGVNS heuristic is coded in C++ computer language.

There are five sets of large instances used for the experiments. Set I contains three groups of instances: *RanReal, RanInt*, and *Geo* (random instances proposed in [11, 13] and [14], respectively).

- The set *RanReal* consists of 160 $N \times N$ matrices in which the distance values d_{ij} are real numbers from the interval $[0, 100]$, generated by using a uniform distribution. The number of elements N, the number of groups G, and bounds for group size are given in Table 1. There are 20 instances for each combination of parameters (i.e., each row in Table 1), 10 instances with equal group size (EGS) and 10 with different group size (DGS). This data set was introduced by Fan et al. [11] with N ranging from 10 to 240. Galego et al. [13] generated larger instances with $N = 480$ and $N = 960$. In our tests, we use only instances with $N \in \{120, 240, 480, 960\}$.
- The *RanInt* set has the same structure and size as *RanReal* (shown in Table 1), but the distances are random integers from the interval [0,100], generated by a uniform distribution.
- The set *Geo* was introduced by Glover et al. [14], and has the same structure and size as the previous two. However, the d_{ij} are calculated as Euclidean distances between pairs of points with coordinates randomly generated from [0,10]. The number of coordinates for each point is generated randomly in the range [2,21].

Table 1 Group sizes for *RanReal, RanInt*, and *Geo* instances

		DGS				EGS
N	G	a_g^{min}	a_g^{max}	b_g^{min}	b_g^{max}	$a_g = b_g$
10	2	3	5	5	7	5
12	4	2	3	3	5	3
30	5	5	6	6	10	6
60	6	7	10	10	14	10
120	10	8	12	12	16	12
240	12	15	20	20	25	20
480	20	18	24	24	30	24
960	24	32	40	40	48	40

Table 2 Group sizes for instances with $N = 2000$ elements (Set II)

	DGS		EGS
G	a_g	b_g	$a_g = b_g$
10	173	227	200
25	51	109	80
50	26	54	40
100	13	27	20
200	6	14	10

Table 3 Group sizes for instances with $N = 3000$ elements (Set IV)

	DGS		EGS
G	a_g	b_g	$a_g = b_g$
25	80	160	120
50	40	80	60
100	20	40	30
200	10	20	15

Set II contains 20 instances labelled *Type*1_22, as presented by Duarte and Martí [10]. All instances have $N = 2000$ elements, while diversities d_{ij} are randomly selected from the interval $[1, 10]$. In this case, we made tests with different numbers of groups. These numbers as well as the bounds for group sizes are given in Table 2 (same as Rodriguez et al. [23]):

Set III contains 20 instances with $N = 2000$ elements. Diversities between elements are integers randomly selected from the interval $[0, 10]$. We conducted tests here only with $G = 50$. In the first series of tests (different group sizes), the lower bound for group size was set to 32 and upper bound was set to 48; in the second series, both lower and upper bounds were set to 40.

Set IV contains five instances with $N = 3000$ elements. These instances differ in the number of pairs of elements connected by edges (density). Edge lengths are random integers from the interval [1,10] representing the diversity (d_{ij}) between corresponding elements (pairs of elements that are not connected have diversity equal to 0). In this case, we made tests with different numbers of groups. These numbers as well as bounds for group sizes are given in Table 3.

Set V contains five instances with $N = 5000$ elements. The corresponding graphs are constructed in the same manner as for Set IV. Again, we made tests with different numbers of groups as summarized in Table 4.

Table 4 Group sizes for instances with $N = 5000$ elements (Set V)

	DGS		EGS
G	a_g	b_g	$a_g = b_g$
25	133	267	200
50	67	133	100
100	33	67	50
200	17	33	25

5.1 VNS Parameter Values

Three SGVNS parameters need to be specified in the given implementation: the limit on CPU time t_{max}, the maximum neighborhood size k_{max}, and parameter α for MDGP (for skewed moves). The values of t_{max} are the same as those used in the artificial bee colony optimization (ABCO) algorithm [23], and are given in the last column of each table. The value of k_{max} is chosen to be 60 for all instances, while α is set to 0.05 as explained below.

In order to choose a proper value for α, we ran our SGVNS on ten instances from Set III with parameter $\alpha \in [0.01, 0.1]$. In Fig. 1 are plotted average objective function values (left) and the corresponding CPU times spent on these ten instances (right). From this figure we see that the best results are obtained for $\alpha = 0.05$, since the maximum average objective function value and the smallest average CPU time used to find the best solution are both obtained with $\alpha = 0.05$.

5.2 Comparison of Average Results

In this subsection we compare the results obtained by our Skewed GVNS (SGVNS) with four heuristics mentioned in Sect. 1: the hybrid genetic algorithm (HGA) by Fan et al. [11]; tabu search with strategic oscillation (TS-SO) by Gallego et al. [13]; artificial bee colony optimization (ABCO) by Rodriguez et al. [23], and general variable neighborhood search (GVNS) by Urošević [24].

Comparison on Set I Instances. In Tables 5, 6, and 7, we summarize the results on the three different groups making up Set I: *RanReal*, *RanInt*, and *Geo*. The first two columns identify the number of elements and the type of problem instance: DGS for different group size, and EGS for equal group size. As noted above, ten instances are randomly generated for each problem type and size. Columns 3, 4, 5, 6, and 7 report average values on those ten instances obtained by HGA, TS-SO, ABCO, GVNS, and SGVNS, respectively. The detailed results for each instance may be found in Appendices 6.1, 6.2, and 6.3 of [4], as well as on our web site http://www.mi.sanu.ac.rs/~nenad/mdgp.

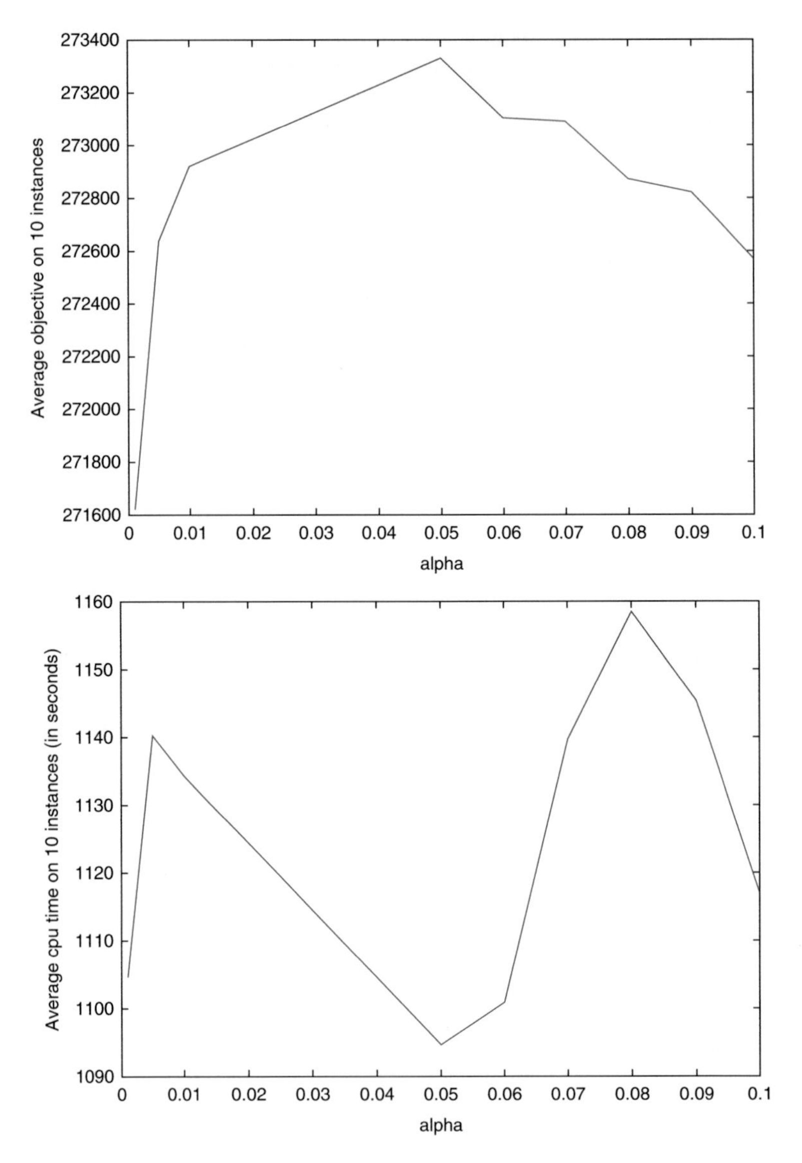

Fig. 1 Average objective value on ten instances obtained by executing SGVNS with different values of parameter α, and average CPU times spent on these ten instances

Table 5 Summarized results for *RanReal* instances

		Objective value					Time			
N	Type	HGA	TS-SO	ABCO	GVNS	SGVNS	ABCO	GVNS	SGVNS	t_{max}
120	DGS	48,495.45	49,643.36	49,825.42	50,086.90	**50,239.97**	1.64	1.61	1.77	3.00
240	DGS	155,818.84	157,830.33	158,388.42	159,516.23	**159,921.56**	10.82	10.03	14.50	20.00
480	DGS	374,330.63	382,315.62	383,749.35	387,260.29	**388,836.28**	81.55	64.09	97.73	120.00
960	DGS	1,194,663.91	1,216,301.76	1,222,625.17	1,233,855.03	**1,235,290.34**	433.50	524.75	516.19	600.00
120	EGS	46,783.77	46,981.43	47,120.62	47,282.53	**47,322.66**	1.50	1.46	1.58	3.00
240	EGS	152,450.44	153,619.77	154,101.95	155,121.55	**155,466.78**	12.25	10.48	13.56	20.00
480	EGS	367,839.17	373,108.00	373,272.88	376,843.17	**378,509.23**	71.33	62.25	88.02	120.00
960	EGS	1,183,111.23	1,199,584.41	1,201,610.10	1,213,863.54	**1,213,920.02**	483.51	499.21	533.31	600.00
Average		440,436.68	447,423.09	448,836.74	452,978.66	**453,750.86**	137.01	146.74	158.33	185.75

Table 6 Summarized results for *RanInt* instances

N	Type	Objective value					Time			
		HGA	TS-SO	ABCO	GVNS	SGVNS	ABCO	GVNS	SGVNS	t_{max}
120	DGS	48,947	49,937	50,006	50,266	**50,428**	1.51	1.41	1.78	3.00
240	DGS	155,134	158,367	158,686	159,833	**160,369**	10.98	11.35	14.52	20.00
480	DGS	375,062	383,334	385,097	388,019	**389,803**	79.26	72.36	94.17	120.00
960	DGS	1,197,698	1,217,330	1,224,767	1,234,575	**1,238,025**	445.39	423.34	500.30	600.00
120	EGS	46,838	47,070	47,239.5	47,402	**47,468**	1.50	1.48	1.49	3.00
240	EGS	152,517	154,048	154,393	155,523	**155,841**	9.79	10.70	13.95	20.00
480	EGS	368,888	374,255	374,637	377,920	**379,370**	71.26	71.75	94.57	120.00
960	EGS	1,184,368	1,202,757	1,208,386	1,215,223	**1,217,030**	480.19	521.50	513.63	600.00
Average		441,182	448,387	450,401	453,595	**454,792**	137.48	139.23	154.30	185.75

Table 7 Summarized results for *Geo* instances

		Objective value					Time			
N	Type	HGA	TS-SO	ABCO	GVNS	SGVNS	ABCO	GVNS	SGVNS	t_{max}
120	DGS	91,605.01	91,811.12	92,263.31	92,253.37	**92,461.40**	1.76	1.47	1.56	3.00
240	DGS	291,615.40	293,331.81	293,903.02	293,772.35	**294,524.70**	12.96	9.82	9.80	20.00
480	DGS	814,724.99	819,978.28	822,618.18	822,467.24	**824,446.92**	95.59	60.90	57.38	120.00
960	DGS	2,486,329.65	2,498,360.73	2,508,775.18	2,507,772.88	**2,513,906.88**	501.06	327.07	355.99	600.00
120	EGS	84,077.87	84,109.75	84,126.82	84,120.25	**84,304.03**	1.63	1.41	1.42	3.00
240	EGS	277,664.99	279,326.04	279,305.20	279,305.00	**279,831.13**	12.79	8.43	9.25	20.00
480	EGS	783,497.55	783,567.86	783,728.49	783,754.73	**785,192.81**	96.60	59.15	61.67	120.00
960	EGS	2,430,953.72	2,431,140.90	2,431,001.65	2,431,145.33	**2,435,081.64**	496.22	314.23	288.05	600.00
Average		907,558.65	910,203.31	911,965.23	911,823.89	**913,718.69**	152.33	97.81	98.14	185.75

The next three columns report the average CPU times (sec) used to find the best values by the three heuristics (ABCO, GVNS, and SGVNS), within the maximum time (sec) allocated for each run (t_{max}) given in the last column. HGA and TS-SO also used the same t_{max} values, but actual running times were not presented in the corresponding papers.
We observe:

(1) In terms of solution quality, SGVNS outperforms all four heuristics on average in all instance sets and for each N. Moreover, the detailed results in Appendices 6.1, 6.2, and 6.3 of [4] show that the best known results are improved by SGVNS on almost all 240 instances. There are just a few ties with GVNS.
(2) The performance of the four other heuristics are ordered (in decreasing quality) as follows: GVNS, ABCO, TS-SO, and HGA.
(3) The same order of methods with respect to solution quality is maintained for both types of problems: DGS and EGS.
(4) SGVNS superiority over the other heuristics is larger on *RanReal* and *RanInt* instances than on *Geo* instances; largest average improvements of 3.73 %, 3.78 %, and 1.18 % were obtained by SGVNS for *RanReal*, *RanInt*, and *Geo*, respectively.
(5) The CPU times to obtain the best solution (within t_{max}) are similar for each heuristic.

Table 8 summarizes average % improvements of SGVNS over the other four methods on all three instance types of Set I. For example, the value 3.47 in the first line and third column of Table 8 reports that 3.47 % average improvement is obtained by SGVNS when compared with HGA on ten random *RanReal* instances; input parameters of all those instances are $N = 120$ and DGS. The detailed results on all instances may be found on our web site http://www.mi.sanu.ac.rs/~nenad/mdgp. From Table 8 one can easily see that SGVNS outperforms on average all heuristics on all three instance types of Set I. The only exception applies to the EGS *RanReal* instances with $N = 960$, where GVNS and SGVNS exhibit the same average performances.

Comparison on Set II Instances. The number of elements is fixed here at $N = 2000$. This data set is originally used in [10] for testing heuristics for the maximum diversity problem. The maximum running time of each heuristic is set to 1200 s. In this case, 20 instances of the same type are used to report average results on each line of Table 9, where only a comparison with ABCO and GVNS is presented. The last two columns of Table 9 give the average % improvement obtained by SGVNS over ABCO and GVNS. Observe that the largest average improvement of 2.2 % was obtained for the largest instance (200 non-equal groups). In addition, improvements of the best known values are reported for all 200 instances (see Appendices 6.4 and 6.5 in [4]).

From the last columns of Table 9, one can observe that ABCO and GVNS have similar solution quality, but ABCO is slightly faster.

Table 8 Percent improvement by SGVNS

		RanReal				RanInt				Geo			
N	Type	HGA	TS-SO	ABCO	GVNS	HGA	TS-SO	ABCO	GVNS	HGA	TS-SO	ABCO	GVNS
120	DGS	3.47	1.19	0.83	0.30	2.94	0.97	0.84	0.32	0.93	0.70	0.21	0.22
240	DGS	2.57	1.31	0.96	0.25	3.26	1.25	1.05	0.33	0.99	0.41	0.21	0.26
480	DGS	3.73	1.68	1.31	0.41	3.78	1.66	1.21	0.46	1.18	0.54	0.22	0.24
960	DGS	3.29	1.54	1.03	0.12	3.26	1.67	1.07	0.28	1.10	0.62	0.20	0.24
120	EGS	1.14	0.72	0.43	0.08	1.33	0.84	0.48	0.14	0.27	0.23	0.21	0.22
240	EGS	1.94	1.19	0.88	0.22	2.13	1.15	0.93	0.20	0.77	0.18	0.19	0.19
480	EGS	2.82	1.43	1.38	0.44	2.76	1.35	1.25	0.38	0.22	0.21	0.19	0.18
960	EGS	2.54	1.18	1.01	**0.00**	2.68	1.17	0.71	0.15	0.17	0.16	0.17	0.16
Average		2.69	1.28	0.98	0.23	2.77	1.26	0.94	0.28	0.70	0.38	0.20	0.21

Table 9 Comparison on Set II (*Type1_22*) instances, within $t_{max} = 1200$ s

	ABCO		GVNS		SGVNS		% impr.	
G, Type	Obj.	Time	Obj.	Time	Obj.	Time	ABCO	GVNS
10, DGS	1,130,150	879.37	1,129,955	1199.28	**1,132,013**	1030.93	0.16	0.18
25, DGS	535,988	896.34	536,227	1195.33	**539,500**	1107.20	0.65	0.61
50, DGS	288,096	1020.73	288,429	1197.03	**292,035**	1123.02	1.35	1.23
100, DGS	156,519	997.22	156,738	1191.27	**159,652**	1059.54	1.96	1.83
200, DGS	86,639	1026.13	86,812	1190.73	**88,592**	1064.84	2.20	2.01
10, EGS	1,111,470	896.42	1,111,256	1198.17	**1,113,326**	949.86	0.17	0.19
25, EGS	482,647	1001.53	482,812	1196.79	**486,063**	1057.12	0.70	0.67
50, EGS	261,148	959.03	261,291	1193.84	**264,768**	1011.17	1.37	1.31
100, EGS	141,992	994.53	142,202	1184.53	**144,672**	1039.15	1.85	1.71
200, EGS	76,266	1069.54	75,935	1186.80	**76,555**	1114.88	0.38	0.81
Average	427,092	974.08	427,166	1193.38	**429,718**	1055.77	1.08	1.06

Table 10 Comparison of ABCO, GVNS, and SGVNS on Set III—DGS

	ABCO		GVNS		SGVNS		% improvement of SGVNS over	
Name	Obj.	Time	Obj.	Time	Obj.	Time	ABCO	GVNS
MDG-a_21	270,135	1181.33	270,483	1200.07	**273,098**	1073.74	1.08	0.96
MDG-a_22	270,037	1183.55	270,127	1200.07	**273,207**	1017.50	1.16	1.13
MDG-a_23	269,754	1177.73	270,007	1200.14	**273,559**	998.02	1.39	1.30
MDG-a_24	269,975	1151.14	270,205	1200.43	**273,198**	1088.31	1.18	1.10
MDG-a_25	269,590	1183.83	270,018	1198.21	**273,344**	1164.90	1.37	1.22
MDG-a_26	269,693	1170.51	269,951	1200.13	**273,569**	1173.18	1.42	1.32
MDG-a_27	269,608	1142.44	270,331	1195.33	**273,225**	1173.30	1.32	1.06
MDG-a_28	269,357	1154.58	270,139	1179.86	**273,242**	1044.67	1.42	1.14
MDG-a_29	269,566	1141.00	270,102	1196.20	**273,571**	1048.53	1.46	1.27
MDG-a_30	269,456	1179.67	270,096	1200.11	**273,293**	1180.24	1.40	1.17
MDG-a_31	269,492	1188.32	270,525	1200.04	**273,887**	1094.56	1.60	1.23
MDG-a_32	269,648	1141.74	270,132	1200.05	**273,634**	1195.78	1.46	1.28
MDG-a_33	269,347	1163.62	270,213	1199.08	**273,376**	1189.03	1.47	1.16
MDG-a_34	269,778	1195.74	270,200	1199.96	**273,533**	1123.70	1.37	1.22
MDG-a_35	269,669	1161.94	270,037	1200.01	**273,301**	1167.06	1.33	1.19
MDG-a_36	269,451	1195.89	270,364	1199.78	**273,567**	1123.29	1.50	1.17
MDG-a_37	269,355	1187.64	270,272	1200.04	**273,289**	1175.44	1.44	1.10
MDG-a_38	270,048	1191.83	270,221	1199.78	**273,548**	1180.45	1.28	1.22
MDG-a_39	269,453	1159.42	270,104	1200.04	**273,633**	954.10	1.53	1.29
MDG-a_40	269,802	1157.89	270,489	1199.88	**273,424**	1178.65	1.32	1.07
Avg	269,661	1170.49	270,201	1198.46	**273,425**	1117.22	1.38	1.18

Comparison on Set III Instances. The number of elements (N) in each problem instance is equal to 2000. These instances were originally proposed in [10].

The format of Tables 10 and 11 is the same as that of Table 9. SGVNS outperforms ABCO and GVNS on all instances with improvement of more than 1 % in each case. It also appears that GVNS is on average slightly more effective than ABCO.

Comparison on Sets IV and V. In each of the last two data sets, five problem instances (in3000_1, in3000_2, in3000_3, in3000_4, in3000_5 and in5000_1, in5000_2, in5000_3, in5000_4, in5000_5) are run for 4 different numbers of groups. Results are reported in Tables 12, 13, 14 and 15. In Set IV, there are 3000 elements while instances in Set V contain 5000 elements. Results on each data set are divided in two tables, according to non-equal and equal number of elements in each group. Values of the SGVNS parameters are not changed, i.e., they are set to $k_{max} = 60$ and $\alpha = 0.05$. The running times are given in the second column under each method in the tables.

Analyzing the results from Tables 12, 13, 14, and 15, one can make the following observations:

Table 11 Comparison of ABCO, GVNS, and SGVNS on Set III—EGS

Name	ABCO		GVNS		SGVNS		% improvement of SGVNS over	
	Obj.	Time	Obj.	Time	Obj.	Time	ABCO	GVNS
MDG-a_21	261,982	1162.61	261,403	1997.58	**264,517**	1170.87	0.96	1.18
MDG-a_22	262,199	1072.44	261,385	1996.58	**264,762**	1051.83	0.97	1.28
MDG-a_23	262,019	895.30	261,269	2000.26	**264,273**	1113.63	0.85	1.14
MDG-a_24	262,552	727.74	261,692	1199.93	**264,597**	1057.89	0.77	1.10
MDG-a_25	262,159	1176.78	261,124	1199.58	**264,559**	1176.99	0.91	1.30
MDG-a_26	261,976	1060.61	261,439	1200.17	**264,931**	1155.92	1.12	1.32
MDG-a_27	262,152	1116.08	261,066	1189.11	**264,458**	1198.19	0.87	1.28
MDG-a_28	262,032	1031.30	261,279	1197.37	**264,367**	1170.52	0.88	1.17
MDG-a_29	262,301	1199.36	261,797	1198.28	**264,660**	1144.33	0.89	1.08
MDG-a_30	262,407	1195.95	261,390	1195.80	**264,426**	1164.86	0.76	1.15
MDG-a_31	262,406	1170.77	261,370	1186.90	**264,756**	1099.82	0.89	1.28
MDG-a_32	261,979	1093.85	261,230	1199.96	**264,907**	1135.58	1.11	1.39
MDG-a_33	262,112	904.70	261,475	1199.89	**264,736**	985.59	0.99	1.23
MDG-a_34	262,160	1160.30	261,315	1200.06	**264,778**	1150.74	0.99	1.31
MDG-a_35	262,076	1168.83	261,554	1200.00	**264,482**	1196.73	0.91	1.11
MDG-a_36	262,087	1152.89	261,324	1200.13	**264,287**	1146.73	0.83	1.12
MDG-a_37	262,324	1159.34	261,415	1186.34	**264,371**	1060.53	0.77	1.12
MDG-a_38	262,460	901.72	262,017	1179.38	**265,107**	1147.56	1.00	1.17
MDG-a_39	262,402	817.17	261,363	1183.26	**264,428**	1162.56	0.77	1.16
MDG-a_40	262,296	1194.68	261,397	1187.83	**264,426**	1179.89	0.81	1.15
Average	262,204.05	1068.12	261,415.20	1314.92	**264,591.40**	1133.54	0.90	1.20

Table 12 Results for Set IV, different group size

		ABCO		GVNS		SGVNS		% impr.	
G	Name	Obj.	Time	Obj.	Time	Obj.	Time	ABCO	GVNS
25	in3000_1	2,081,408	4999.92	2,068,602	2998.42	**2,104,929**	2967.82	1.12	1.73
	in3000_2	4,576,718	4918.75	4,564,659	2997.66	**4,608,599**	2813.95	0.69	0.95
	in3000_3	6,773,996	4966.69	6,767,840	2964.36	**6,817,205**	2781.44	0.63	0.72
	in3000_4	9,740,824	4914.86	9,731,506	2999.88	**9,795,296**	2901.59	0.56	0.65
	in3000_5	11,509,076	4916.46	11,501,445	2987.76	**11,554,861**	2734.88	0.40	0.46
50	in3000_1	1,387,969	4968.28	1,380,451	2976.47	**1,412,436**	2953.62	1.73	2.26
	in3000_2	2,754,949	4933.49	2,745,609	2838.58	**2,810,695**	2963.15	1.98	2.32
	in3000_3	3,883,091	4929.64	3,887,036	3000.8	**3,948,025**	2655.85	1.64	1.54
	in3000_4	5,328,649	4985.30	5,323,850	2995.08	**5,391,550**	2981.76	1.17	1.26
	in3000_5	6,140,589	4974.03	6,141,923	2999.51	**6,196,772**	2949.51	0.91	0.89
100	in3000_1	972,851	4932.49	958,500	3001.23	**989,227**	2964.43	1.66	3.11
	in3000_2	1,713,158	4907.45	1,709,197	2971.47	**1,755,292**	2963.30	2.40	2.63
	in3000_3	2,279,018	4968.87	2,284,311	3002.64	**2,333,689**	2999.49	2.34	2.12
	in3000_4	2,949,313	4967.83	2,955,104	2942.81	**3,008,777**	2747.19	1.98	1.78
	in3000_5	3,297,539	4924.33	3,304,546	2999.84	**3,350,941**	2340.22	1.59	1.38
200	in3000_1	697,926	4979.23	700,488	2990.07	**729,274**	2838.70	4.30	3.95
	in3000_2	1,090,814	4904.01	1,097,610	2907.4	**1,129,015**	2813.64	3.38	2.78
	in3000_3	1,358,310	4984.13	1,359,374	2908.29	**1,399,287**	2931.42	2.93	2.85
	in3000_4	1,642,375	4947.49	1,648,337	3002.18	**1,688,298**	2887.60	2.72	2.37
	in3000_5	1,778,995	4907.05	1,779,837	3001.14	**1,815,266**	2304.51	2.00	1.95
Average		3,597,878	4946.51	3,595,511	2974.28	**3,641,972**	2824.70	1.81	1.88

Table 13 Results for Set IV, equal group size

G	Name	ABCO		GVNS		SGVNS		% impr.	
		Obj.	Time	Obj.	Time	Obj.	Time	ABCO	GVNS
25	in3000_1	1,962,677	2873.06	1,946,244	2999.26	**1,987,940**	2958.42	1.27	2.10
	in3000_2	4,245,425	2865.05	4,232,917	2996.18	**4,292,166**	2874.76	1.09	1.38
	in3000_3	6,241,880	2911.18	6,234,529	3000.79	**6,296,800**	2705.09	0.87	0.99
	in3000_4	8,912,946	2920.60	8,913,960	3000.72	**8,964,748**	2886.29	0.58	0.57
	in3000_5	10,500,149	2995.81	10,489,960	2998.77	**10,545,933**	2973.27	0.43	0.53
50	in3000_1	1,325,165	2986.32	1,311,677	2986.47	**1,345,757**	2847.59	1.53	2.53
	in3000_2	2,571,189	2987.36	2,563,902	3001.44	**2,618,042**	2975.28	1.79	2.07
	in3000_3	3,591,028	2958.15	3,583,529	3000.95	**3,656,776**	2982.3	1.80	2.00
	in3000_4	4,880,018	2882.77	4,870,578	2983.77	**4,942,527**	2961.43	1.26	1.46
	in3000_5	5,601,185	2909.66	5,599,074	2981.06	**5,648,618**	2879.66	0.84	0.88
100	in3000_1	923,369	2870.16	925,412	2986.45	**959,521**	2687.5	3.77	3.55
	in3000_2	1,607,960	2991.39	1,608,454	2949.27	**1,655,723**	2632.08	2.88	2.85
	in3000_3	2,122,943	2888.37	2,114,157	2998.88	**2,175,690**	2998.89	2.42	2.83
	in3000_4	2,716,075	2860.90	2,707,751	2991.83	**2,764,835**	2994.7	1.76	2.06
	in3000_5	3,021,445	2949.48	3,018,386	2988.83	**3,066,391**	2868.07	1.47	1.57
200	in3000_1	670,498	2963.90	668,612	2953.48	**695,609**	2993.39	3.61	3.88
	in3000_2	1,031,953	2913.15	1,029,343	2970.58	**1,062,813**	2956.62	2.90	3.15
	in3000_3	1,269,850	2888.33	1,265,304	3000.90	**1,311,845**	2878.6	3.20	3.55
	in3000_4	1,509,208	2953.54	1,515,664	2937.68	**1,542,180**	2798.8	2.14	1.72
	in3000_5	1,632,235	2904.26	1,624,954	2989.85	**1,650,303**	1693.61	1.09	1.54
Average		3,316,860	2923.67	3,311,220	2985.86	**3,359,211**	2827.32	1.84	2.06

Table 14 Results for Set V, different group size

G	Name	ABCO		GVNS		SGVNS		% impr.	
		Obj.	Time	Obj.	Time	Obj.	Time	ABCO	GVNS
25	in5000_1	4,989,105	4950.79	5,021,077	4899.84	**5,069,622**	4941.69	1.59	0.96
	in5000_2	11,679,751	4971.94	11,722,266	5001.23	**11,794,507**	4971.95	0.97	0.61
	in5000_3	17,665,939	4903.92	17,736,691	4999.40	**17,822,899**	4961.27	0.88	0.48
	in5000_4	25,965,407	4904.73	26,023,736	5000.12	**26,136,965**	4819.15	0.66	0.43
	in5000_5	31,074,974	4979.76	31,130,997	4986.50	**31,213,238**	4971.43	0.44	0.26
50	in5000_1	3,221,257	4979.17	3,226,908	5000.1	**3,298,080**	4972.73	2.33	2.16
	in5000_2	6,809,271	4932.09	6,841,586	5000.4	**6,947,557**	4741.19	1.99	1.53
	in5000_3	9,897,659	4976.61	9,933,108	5002.09	**10,050,394**	4950.87	1.52	1.17
	in5000_4	13,964,167	4998.87	14,011,062	5001.07	**14,121,832**	4933.78	1.12	0.78
	in5000_5	16,367,506	4946.00	16,423,667	4996.74	**16,499,098**	4944.36	0.80	0.46
100	in5000_1	2,180,367	4917.99	2,178,504	4973.79	**2,245,837**	4454.31	2.92	3.00
	in5000_2	4,151,676	4926.17	4,167,293	4913.14	**4,263,886**	4199.93	2.63	2.27
	in5000_3	5,718,563	4993.27	5,750,498	4997.5	**5,847,628**	4751.82	2.21	1.66
	in5000_4	7,699,561	4957.89	7,724,695	4914.08	**7,818,875**	4861.49	1.53	1.20
	in5000_5	8,782,963	4939.48	8,810,609	4998.73	**8,891,786**	4799.7	1.22	0.91
200	in5000_1	1,535,396	4955.32	1,536,851	4996.28	**1,577,453**	4290.67	2.67	2.57
	in5000_2	2,594,797	4952.85	2,585,083	4934.04	**2,660,807**	4902.03	2.48	2.85
	in5000_3	3,353,962	4926.08	3,378,785	4980.39	**3,444,632**	4895.61	2.63	1.91
	in5000_4	4,240,057	4950.18	4,252,457	4959.05	**4,326,483**	4986.66	2.00	1.71
	in5000_5	4,692,472	4971.13	4,708,897	5000.12	**4,763,069**	4421.97	1.48	1.14
Average		9,329,242	4951.71	9,358,238	4977.73	**9,439,732**	4788.63	1.70	1.40

Table 15 Results for Set V, equal group size

		ABCO		GVNS		SGVNS		% impr.	
G	Name	Obj.	Time	Obj.	Time	Obj.	Time	ABCO	GVNS
25	in5000_1	4,680,966	4923.31	4,715,807	5001.9	**4,765,328**	4906.36	1.77	1.04
	in5000_2	10,766,978	4981.41	10,836,919	5004.02	**10,903,109**	4976.35	1.25	0.61
	in5000_3	16,213,193	4942.36	16,280,534	5000.56	**16,369,676**	4680.3	0.96	0.54
	in5000_4	23,738,950	4903.52	23,758,910	4935.9	**23,856,059**	4932.02	0.49	0.41
	in5000_5	28,262,166	4971.16	28,313,109	4975.28	**28,401,883**	4934.04	0.49	0.31
50	in5000_1	3,032,986	4942.96	3,047,114	4933	**3,118,260**	4887.01	2.73	2.28
	in5000_2	6,330,923	4976.58	6,350,779	4997.8	**6,468,129**	4975.9	2.12	1.81
	in5000_3	9,122,569	4978.54	9,163,073	4865.49	**9,265,860**	4972.03	1.55	1.11
	in5000_4	12,781,294	4981.91	12,841,963	4971.41	**12,931,152**	4924.33	1.16	0.69
	in5000_5	14,931,100	4913.28	14,959,736	4990.61	**15,055,032**	4991.1	0.82	0.63
100	in5000_1	2,076,330	4910.96	2,070,039	4932.43	**2,134,605**	4831.58	2.73	3.02
	in5000_2	3,883,432	4942.52	3,890,450	5001.98	**3,967,141**	4062.47	2.11	1.93
	in5000_3	5,297,314	4968.03	5,301,296	4989.81	**5,401,804**	4974.37	1.93	1.86
	in5000_4	7,049,769	4922.36	7,056,998	4802.16	**7,154,405**	4952.22	1.46	1.36
	in5000_5	7,999,911	4943.48	8,015,062	4972.45	**8,090,974**	4829.48	1.13	0.94
200	in5000_1	1,484,270	4903.24	1,466,900	5001.52	**1,523,690**	4418.2	2.59	3.73
	in5000_2	2,466,164	4980.91	2,446,930	4919.12	**2,519,226**	4711.67	2.11	2.87
	in5000_3	3,163,535	4938.68	3,147,499	5000.38	**3,230,835**	4840.55	2.08	2.58
	in5000_4	3,951,614	4914.46	3,936,666	4980.05	**4,007,218**	4646.46	1.39	1.76
	in5000_5	4,338,852	4915.96	4,329,628	4997.6	**4,390,118**	4437.95	1.17	1.38
Average		8,031,241	4942.78	8,596,471	4963.67	**8,677,725**	4794.22	1.602	1.544

(1) Regarding solution quality, SGVNS is better than ABCO and GVNS in all instances, with average % improvement varying from 1.4 to 2.1 % across the tables;
(2) The average CPU time of the three methods is similar, except in Table 12, where GVNS and SGVNS results are obtained in half of the ABCO time (compare average times of 2974.28, 2824.70, and 4946.51 s used by GVNS, SGVNS, and ABCO, respectively); and
(3) The largest improvement of 4.3 % of SGVNS over ABCO is obtained on the instance in3000_1 with $G = 200$ (see Table 12). Computing times of 2839 and 4979 s are used here by SGVNS and ABCO, respectively.

6 Computational Results for CPP

Thirteen Benchmark Instances. The first set of test instances consists of seven benchmark instances originally considered by Charon and Hudry [7] (rand100-100, rand300-100, rand500-100, rand300-5, zahn300, sym300-50, and regnier300-50) and six instances generated by Brusco and Köhn [6] (rand200-100, rand400-100, rand100-5, rand200-5, rand400-5, and rand500-5).

Thirty Large New Instances. In order to compare heuristics on large problems, we introduce new CPP instances. They consist of sets with 1000, 1500, and 2000 vertices, each having ten instances. The edge lengths are generated at random as integers uniformly distributed in the range $[-100, 100]$.

Preliminary Testing and Parameter Values. The SGVNS heuristic for CPP has five parameters: α, k_{min}, k_{max}, k_{step}, and t_{max}. Values of the three neighborhood parameters are set as follows: $k_{max} = \max\{100, N/5\}$ and $k_{min} = k_{step} = \max\{1, k_{max}/50\}$.

To obtain the value of parameter α (used in the skewed phase of SGVNS), we run our heuristic ten times on an instance with $N = 1000$ and with different values of α, i.e., $\alpha \in \{0.01, 0.02, 0.03, 0.04, 0.05, 0.06, 0.08, 0.10, 0.15, 0.20\}$. Results are given in Table 16. From the table we can conclude that the best value for parameter α is 0.02, and we use this value in the remaining tests. A limit on CPU time set at $t_{max} = N$ (in seconds) is used as the stopping condition in all runs.

We also compare our SGVNS heuristic with an exact method. To this end we selected a CPLEX-LP solver for the linear formulation of the CPP, and a CPLEX quadratic solver for the equivalent quadratic program (see (CPP_e) above). The quadratic solver required excessive computer time, and thus, was limited to very small test instances ($N \leq 15$). The SGVNS heuristic obtained the optimal solution in all cases tested ($N = 10, 15, 20, 25, 30, 35$). For example, for the test instance with $N = 35$ nodes, the CPLEX-LP solver took 4127 s to find the optimal solution, whereas SGVNS obtained the same solution in 0.03 s!

Comparison with State-of-the-Art Heuristics. We now compare results obtained by our SGVNS with those from the literature. For this purpose, the state of the art comprises the "neighborhood search" heuristics referred to as NS-R and

Table 16 SGVNS results on large test instance using ten restarts and different values of α

α	f_{best}	f_{avg}	f_{worst}	Time
0.01	−875,776	−870,699.00	−861,370	989.96
0.02	−878,331	−877,307.50	−875,902	931.99
0.03	−878,153	−876,886.70	−875,075	878.77
0.04	−878,163	−877,169.60	−875,565	977.02
0.05	−877,970	−877,017.70	−874,985	993.88
0.06	−877,970	−877,208.70	−874,985	869.10
0.08	−877,881	−877,032.30	−875,561	996.54
0.10	−877,870	−877,097.00	−875,365	995.44
0.15	−877,713	−876,793.70	−875,263	981.28
0.20	−877,857	−876,935.50	−875,263	912.92

Table 17 Comparison of NS-R, NS-TS, and SGVNS on small instances; SGVNS is restarted ten times with time limit $t_{max} = N$

				SGVNS		
Name	Best	NS-R	NS-TS	f_{best}	f_{avg}	Time
rand100-5	−1407	−1407	−1407	−1407	−1407	0.33
rand100-100	−24,296	−24,296	−24,296	−24,296	−24,296	1.42
rand200-5	−4079	−4079	−4079	−4079	−4079	26.59
rand200-100	−74,924	−74,924	−74,924	−74,924	−74,924	12.56
rand300-5	−7732	−7723	−7729	−7732	−7728	87.12
rand300-100	−152,709	−152,709	−152,709	−152,709	−152,709	24.81
sym300-50	−17,592	−17,592	−17,592	−17,592	−17,592	143.45
regnier300-50	−32,164	−32,164	−32,164	−32,164	−32,164	3.24
zahn300	−2504	−2503	−2504	−2504	−2504	29.4
rand400-5	−12,133	−12,096	−12,120	−12,133	−12,123	206
rand400-100	−222,757	−222,647	−222,374	−222,757	−222,735	212.65
rand500-5	−17,127	−17,008	−17,086	−17,127	−17,095.5	255.29
rand500-100	−309,125	−308,620	−308,341	−309,107	−308,754	291.00

NS-TS in [6]. In Table 17, we compare SGVNS with these two heuristics on the 13 small test instances. The second column reports the best known values obtained by our SGVNS. Objective function values obtained by NS-R and NS-TS are given in columns 3 and 4, respectively. The next 3 columns give results obtained by SGVNS: the best objective value in 10 restarts, the average value and the average CPU time spent before the best solutions were found. Note that the running times of the other two heuristics were set to 500 s in [6]. It appears that results of equivalent quality are obtained by all three heuristics for these smaller instances, although SGVNS is slightly better on some. However, we shall see that for the larger instances, SGVNS reports significantly better results.

In Table 18 the same methods are compared on our new large size test instances. This time the best in ten, average in ten, and average CPU times for all three

Table 18 Results on large instances, based on ten restarts for all methods and with the time limit set to number of vertices ($t_{max} = N$)

Name	NS-R			NS-TS			SGVNS			%Im over	
	f_{best}	f_{avg}	Time	f_{best}	f_{avg}	Time	f_{best}	f_{avg}	Time	NS-R	NS-TS
r1000_01	−867,120	−864,600.33	1054.32	−866,105	−865,510.67	1033.85	−878,331.00	−877,307.50	874.28	1.28	1.39
r1000_02	−867,506	−866,265.33	1048.33	−872,131	−867,497.67	1113.92	−881,844.00	−88,0551.90	931.99	1.63	1.10
r1000_03	−869,632	−866,454.33	1047.74	−867,505	−862,686.33	1086.74	−879,577.00	−878,174.10	1037.07	1.13	1.37
r1000_04	−863,587	−860,479.33	1059.32	−864,306	−860,894.67	1080.27	−874,888.00	−873,902.30	1067.97	1.29	1.21
r1000_05	−871,291	−869,459.33	1055.17	−873,648	−872,259.67	1092.70	−882,815.00	−881,738.80	987.96	1.31	1.04
r1000_06	−883,498	−880,247.60	1043.91	−888,777	−883,757.30	1070.99	−886,735.00	−886,129.30	916.05	0.37	−0.23
r1000_07	−874,515	−873,598.33	1069.01	−874,095	−870,816.00	1056.59	−889,906.00	−887,235.50	1131.97	1.73	1.78
r1000_08	−877,350	−873,365.00	1043.38	−878,923	−874,766.67	1081.24	−879,916.00	−878,897.60	802.65	0.29	0.11
r1000_09	−884,951	−883,422.33	1060.23	−888,638	−886,901.00	1023.65	−893,210.00	−891,992.40	697.56	0.92	0.51
r1000_10	−868,132	−865,049.00	1050.94	−866,735	−864,425.67	1088.51	−878,342.00	−876,374.20	1082.99	1.16	1.32
Avg.	−872,758.2	−870,294.09	1053.24	−87,4086.3	−870,951.56	1072.85	−882,556.40	−881,230.36	953.05	1.11	0.96
r1500_01	−1,600,474	−1,594,384.33	1596.51	−1,592,399	−1,591,123.67	1782.53	−1,613,347.00	−1,610,805.22	1185.06	0.80	1.30
r1500_02	−1,583,069	−1,577,906.33	1541.10	−1,584,404	−1,580,197.33	1649.66	−1,617,944.00	−1,614,664.20	1231.76	2.16	2.07
r1500_03	−1,583,597	−1,581,930.00	1563.01	−1,591,449	−1,583,654.00	1657.47	−1,613,265.00	−1,610,329.10	1292.90	1.84	1.35
r1500_04	−1,578,713	−1,575,948.33	1541.04	−1,590,351	−1,580,704.67	1713.55	−1,612,603.00	−1,609,826.30	1123.63	2.10	1.38
r1500_05	−1,585,727	−1,582,063.33	1553.26	−1,586,615	−1,584,059.33	1601.34	−1,614,855.00	−1,611,385.70	1240.69	1.80	1.75

(continued)

Table 18 (continued)

Name	NS-R			NS-TS			SGVNS			%Im over	
	f_{best}	f_{avg}	Time	f_{best}	f_{avg}	Time	f_{best}	f_{avg}	Time	NS-R	NS-TS
r1500_06	−1,582,878	−1,578,623.33	1541.54	−1,588,589	−1,580,136.33	1670.01	−1,614,049.00	−1,610,638.50	1326.36	1.93	1.58
r1500_07	−1,609,452	−1,601,790.00	1536.98	−1,617,723	−1,607,697.00	1656.17	−1,636,085.00	−1,633,918.60	1253.96	1.63	1.12
r1500_08	−1,565,381	−1,562,638.00	1545.45	−1,571,224	−1,563,231.67	1591.34	−1,586,554.00	−1,579,111.10	1286.64	1.33	0.97
r1500_09	−1,600,449	−1,598,608.33	1557.15	−1,607,941	−1,600,318.33	1642.45	−1,625,854.00	−1,612,040.60	1251.96	1.56	1.10
r1500_10	−1,593,909	−1,588,740.33	1552.03	−1,585,602	−1,582,132.33	1644.01	−1,607,360.00	−1,595,823.20	1042.75	0.84	1.35
Avg.	−1,588,364.9	−1,584,263.23	1552.81	−1,591,629.7	−1,585,325.47	1660.85	−1,614,191.60	−1,608,854.25	1223.57	1.60	1.40
r2000_01	−2,467,372	−2,460,571.60	2081.64	−2,476,734	−2,459,899.60	2237.85	−2,493,143.00	−2,485,535.50	1596.02	1.03	0.66
r2000_02	−2,463,804	−2,451,838.00	2062.19	−2,467,493	−2,447,006.10	2176.77	−2,484,001.00	−2,473,849.60	1451.65	0.81	0.66
r2000_03	−2,462,756	−2,458,149.90	2062.58	−2,477,969	−2,460,834.60	2152.34	−2,490,379.00	−2,481,812.60	1569.89	1.11	0.50
r2000_04	−2,469,300	−2,459,721.00	2051.13	−2,478,567	−2,461,141.70	2129.76	−2,497,504.00	−2,485,469.90	1633.03	1.13	0.76
r2000_05	−2,445,974	−2,435,498.50	2107.07	−2,445,534	−2,436,632.40	2365.45	−2,472,230.00	−2,459,713.90	1641.98	1.06	1.08
r2000_06	−2,451,637	−2,436,883.30	2085.50	−2,449,319	−2,436,018.60	2132.28	−2,473,995.00	−2,462,179.60	1684.78	0.90	1.00
r2000_07	−2,459,744	−2,449,809.70	2072.86	−2,456,281	−2,448,892.00	2290.89	−2,483,346.00	−2,472,279.20	1760.67	0.95	1.09
r2000_08	−2,440,670	−2,431,941.10	2065.87	−2,439,846	−2,423,618.60	1115.77	−2,467,568.00	−2,451,094.40	1818.02	1.09	1.12
r2000_09	−2,468,137	−2,458,719.80	1062.17	−2,461,927	−2,454,540.10	1124.97	−2,495,666.00	−2,479,213.90	1740.09	1.10	1.35
r2000_10	−2,415,438	−2,409,663.80	1059.33	−2,422,183	−2,411,409.50	1173.93	−2,453,464.00	−2,437,303.70	1663.52	1.55	1.27
Avg.	−2,454,483.2	−2,445,279.67	1871.03	−2,457,585.3	−2,443,999.32	1890.00	−2,481,129.60	−2,468,845.23	1655.96	1.07	0.95

heuristics are reported. In the last two columns we give the % improvement of SGVNS over the other two state-of-the-art heuristics.

It appears that in 29 of the 30 large instances, SGVNS solutions were of better quality and obtained in less overall CPU time. In most cases the improvement in quality was impressive (≥ 1 %). Therefore, our new heuristic may be viewed as state-of-the-art for solving the CPP.

7 Conclusions

In this paper, we implement a new variant of VNS to solve the MDGP. This variant combines general VNS (where several neighborhood structures are used in the local search) and skewed VNS (where a move to a worse solution is allowed if it is slightly inferior and relatively far). Therefore, it is referred to as skewed general VNS (SGVNS). Based on extensive computational tests, we show that the SGVNS heuristic outperforms the current state of the art significantly. Moreover, the best known solutions have been improved on 531 out of 540 test instances taken from the literature.

The SGVNS heuristic with some modifications is then applied to the related CPP. Similar success is obtained, with significant improvement of solution quality reported on several new large scale instances of CPP.

Future research directions include the design and testing of different neighborhoods within VND, and the application of SGVNS to other combinatorial problems.

Acknowledgements The work of Nenad Mladenović was conducted at National Research University Higher School of Economics and supported by RSF (Russian Federation) grant 14-41-00039.

References

1. Baker, B.M., Benn, C.: Assigning pupils to tutor groups in a comprehensive school. J. Oper. Res. Soc. **52**, 623–629 (2001)
2. Bhadury, J., Mighty, E.J., Damar, H.: Maximizing workforce diversity in project teams: a network flow approach. Omega **28**, 143–153 (2000)
3. Brimberg, J., Hansen, P., Mladenović, N.: Attraction probabilities in variable neighborhood search. 4OR **8**(2), 181–194 (2010)
4. Brimberg, J., Mladenović, N., Urošević, D.: Solving the maximally diverse grouping problem by skewed general variable neighborhood search. Inf. Sci. **295**, 650–675 (2015)
5. Brimberg, J., Janićijević, S., Mladenović, N., Urošević, D.: Solving the clique partitioning problem as a maximally diverse grouping problem. Optim. Lett. (accepted for publication). doi:10.1007/s11590-015-0869-4
6. Brusco, M.J., Köhn, H.F.: Clustering qualitative data based on binary equivalence relations: neighborhood search heuristics for the clique partitioning problem. Psychometrika. **74**(4), 685–703 (2009)

7. Charon, I., Hudry, O.: Noising methods for a clique partitioning problem. Discret. Appl. Math. **154**(5), 754–769 (2006)
8. De Amorim, S.G., Barthélemy, J.P., Ribeiro, C.C.: Clustering and clique partitioning: simulated annealing and tabu search approaches. J. Classif. **9**, 17–41 (1992)
9. Desrosiers, J, Mladenović, N., Villeneuve, D.: Design of balanced MBA student teams. J. Oper. Res. Soc. **56**, 60–66 (2005)
10. Duarte, A., Martí, R.: Tabu search and grasp for the maximum diversity problem. Eur. J. Oper. Res. **178**, 71–84 (2007)
11. Fan, Z.P., Chen, Y., Ma, J., Zeng, S.: A hybrid genetic algorithmic approach to the maximally diverse grouping problem. J. Oper. Res. Soc. **62**, 92–99 (2011)
12. Feo, T., Khellaf, M.: A class of bounded approximation algorithms for graph partitioning. Networks **20**, 181–195 (1990)
13. Gallego, M., Laguna, M., Martí, R., Duarte, A.: Tabu search with strategic oscillation for the maximally diverse grouping problem. J. Oper. Res. Soc. **64**, 724–734 (2013)
14. Glover, F., Kuo, C.C., Dhir, K.S.: Heuristic algorithms for the maximum diversity problem. J. Inf. Optim. Sci. **19**(1), 109–132 (1998)
15. Hansen, P., Mladenović, N., Moreno Pérez, J.A.: Variable neighborhood search: methods and applications. Ann. Oper. Res. **175**(1), 367–407 (2010)
16. Hettich, S., Pazzani, M.J.: Mining for element reviewers: lessons learned at the national science foundation. In: Proceedings of the KDD'06, pp. 862–871. ACM, New York (2006)
17. Kral, J.: To the problem of segmentation of a program. Inform. Process. Mach. **2**, 116–127 (1965)
18. Lotfi, V., Cerveny, R.: A final exam scheduling package. J. Oper. Res. Soc. **42**, 205–216 (1991)
19. Marcotorchino, F., Michaud, P.: Heuristic approach to the similarity aggregation problem. Methods Oper. Res. **43**, 395–404 (1981)
20. Mladenović, N., Hansen, P.: Variable neighborhood search. Comput. Oper. Res. **24**, 1097–1100 (1997)
21. Mladenović, N., Todosijević, R., Urošević, D.: An efficient general variable neighborhood search for large traveling salesman problem with time windows. Yugosl. J. Oper. Res. **23**(1), 19–30 (2012)
22. Régnier, S.: Sur quelques aspects mathématiques des problèmes de classification automatique. I.C.C. Bull. **4**, 175–191 (1965)
23. Rodriguez, F.J., Lozano, M., García-Martínez, C., González-Barrera, J.: An artificial bee colony algorithm for the maximally diverse grouping problem. Inf. Sci. **230**, 183–196 (2013)
24. Urošević, D.: Variable neighborhood search for maximum diverse grouping problem. Yugosl. J. Oper. Res. **24**(1), 21–33 (2014)
25. Wang, H., Alidaee, B., Glover, F., Kochenberger, G.: Solving group technology problems via clique partitioning. Int. J. Flex. Manuf. Syst. **18**, 77–97 (2006)
26. Wang, H., Obremski, T., Alidaee, B., Kochenberger, G.: Clique partitioning for clustering: a comparison with k-means and latent class analysis. Commun. Stat. Simul. Comput. **37**, 1–13 (2008)
27. Weitz, R.R., Lakshminarayanan, S.: An empirical comparison of heuristic methods for creating maximally diverse groups. J. Oper. Res. Soc. **49**(6), 635–646 (1998)

Test Generation for Digital Circuits Based on Continuous Approach to Circuit Simulation Using Different Continuous Extensions of Boolean Functions

Nickolay Kascheev and Daniil Kascheev

Abstract This paper provides the analysis of continuous extensions of Boolean functions for test generation using continuous optimization. It represents the results of the developed software for a number of ISCAS schemes.

Keywords Boolean function • Testing • Test generation • Continuous model

1 Introduction

Most classical methods for tests construction are based on the stuck-at faults model and combinatorial algorithms for tests generation. The algorithms are based on optimization techniques that use variations of the branch-and-bound methods. Complexity of such methods is well-known.

Nowadays industry requires development of effective methods of test generation. In this paper, we propose a method for test generation based on continuous optimization. We describe main ideas and features of the solution. We present the results of the software running on a number of ISCAS circuits. Such an approach offers the possibility to develop efficient algorithms of tests generation for a wide set of faults.

2 Continuous Approach of Digital Circuits Simulation

Continuous approach for the simulation of circuit behavior is based on the concept of continuous analogue of Boolean function. This paper describes the method which is to replace Boolean functions of circuit elements by their continuous analogues.

N. Kascheev (✉) • D. Kascheev
Faculty of Business Informatics and Applied Mathematics, National Research University, Higher School of Economics, B. Pecherskaya str. 25, Nizhny Novgorod 603600, Russia
e-mail: nikolay.kascheev@gmail.com; danko.kas@gmail.com

V.A. Kalyagin et al. (eds.), *Models, Algorithms and Technologies for Network Analysis*, Springer Proceedings in Mathematics & Statistics 156,
DOI 10.1007/978-3-319-29608-1_2

Table 1 Continuous models of Boolean functions

Logic function	Continuous analogue
$Y = \bar{X}$	$\tilde{y} = 1 - x$
$Y = X_1 \cap X_2$	$\tilde{y} = x_1 x_2$
$Y = X_1 \cup X_2$	$\tilde{y} = x_1 + x_2 - x_1 x_2$
$Y = X_1 \oplus X_2$	$\tilde{y} = x_1 + x_2 - 2x_1 x_2$

This idea was first proposed by Kano [1] and continued in [5]. We extend the scope of usage of this approach.

Let's use the following notation: $x = x_1, \ldots, x_n$ coordinates of a point in n-dimensional space. $T^n = (x|0 \leq x_i \leq 1, i = 1, \ldots, n)$. So T^n is a unit hypercube, a set elements $V^n \subset T^n$ are the vertices of the hypercube $T^n \cdot V^n = (x|x_i \in \{0, 1\}, i = 1, \ldots, n)$.

Let's consider $\lambda(x)$ some n-place Boolean function that can be defined as the mapping of vertices of unit n-dimensional hypercube in the set $\{0, 1\}$. So the following expression is valid $\lambda(x) : V_n \rightarrow V1$. Continuous model of Boolean functions $\lambda(x)$ is any function $f(x)$ that maps an n-dimensional arithmetic space R_n in the set of real numbers $R1$, and coincides with the set V_n.

$f(x) : R^n \rightarrow R^1, f(x) = \lambda(x), \forall x \in V^n$. As we can see the only significant condition imposed of the function $f(x)$ is a requirement to match the values $f(x)$ and $\lambda(x)$ in the corners of the unit n-dimensional hypercube.

Let's consider conjunctive Boolean basis. We define the following continuous models of the basic functions: $\tilde{y} = x_1 x_2$ for the conjunction; $\tilde{y} = 1 - x$ for the denial. It is easy to understand that values of continuous models really coincide with the values of the corresponding Boolean functions on the set of vertices of the unit hypercube. Let's define continuous models for all major functions of the Boolean algebra of logic (Table 1). So, the values 0 and 1 of continuous model are matched only on the edges and vertices of the unit hypercube T_n.

Continuous model of the combinational circuit is called a continuous network. Continuous model of a digital circuit is constructed by means of replacement of the gates logic functions with corresponding continuous analogues.

Obviously, any logic function can be represented by an infinite number of continuous models. For this work we have used a set of continuous models of Boolean functions shown (see Sect. 4).

3 Test Pattern Generation Algorithm

The problem of finding the fault test can be reduced to the search of the input set that identifies the difference between a working and a faulty circuit (see Fig. 1). Thus, the system which checks the equivalence of the two schemes: serviceable and unserviceable, helps us to state the objective function in the following general form:

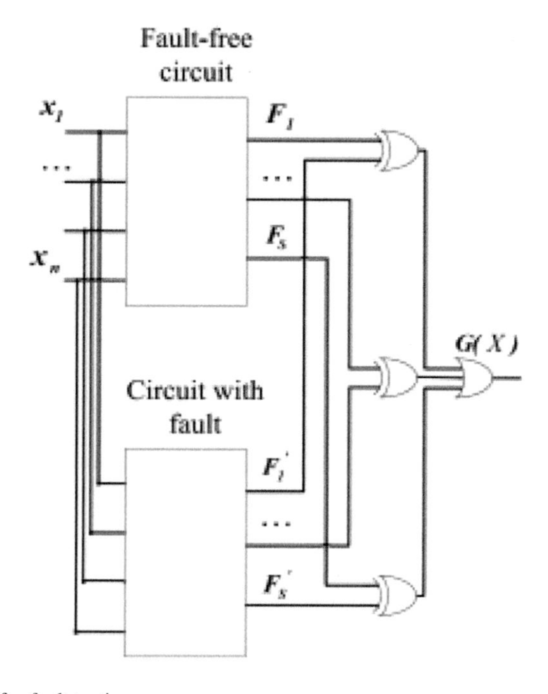

Fig. 1 Structure for fault testing

$$G(x) = \bigcup_{k=1}^{S} (F_k \oplus F_k'(x)) \tag{1}$$

As we can see from Fig. 1, the value of G is equal to 1 if the outputs of fault free and faulty circuits are different, and zero—otherwise [3].

Most test generation algorithms use combinatorial methods, but if we create a continuous model of the function (1), we can generate a circuit test using continuous methods [2–4]. Test generation task, which is based on a continuous approach, can be reduced to find the global maximum of the objective function. So it becomes possible to use a global optimization algorithm to resolve the task of test generation for logic circuits [2–4].

The objective function needs to achieve its maximum value equal to 1 only if its argument is a test vector. So, to generate a test vector it is necessary:

1. To obtain a continuous model of the scheme;
2. To bring in a fault to the circuit and to obtain a continuous model of the circuit with this fault;
3. To construct a continuous objective function;
4. To find an input vector that maximizes the objective function;
5. To verify the resulting test vector for the considered fault.

Since we are interested in the global maximum of the objective function on the entire domain of the particular function so finding the maximum of the objective function—it is the task of global optimization. Note that although the global maxima can be several, value of the objective function is equal to 1 at all points of extremes. Maximum is already known, it simplifies the solution of the problem. Each coordinate of the global maximum will be one of the possible test vectors.

We propose the following algorithm to find the maximum of the objective function, based on the method of coordinate-wise ascent [2–4]:

1. The center of the unit hypercube is the point of an initial approximation of the test suite.

$$X^1 = (x_1^1, x_2^1, \ldots, x_n^1) = (0.5, 0.5, \ldots, 0.5)$$

2. Next step is to calculate estimate of the derivative

$$\frac{dG(x)}{dx_i}, \; i = 1, 2, \ldots, n.$$

   with respect to each coordinate, which value is not equal to 0 or 1, concerning to the current point. The estimate is determined as the absolute difference between the objective function value at the current point and the value at the point with the increment for one of the coordinates.
3. Then coordinate has to be chosen that has maximum estimate. The calculated difference is used as criteria for determining values for the selected coordinates. If the value of the difference is non-negative, it is set to 1, negative values set to 0.
4. If the newly obtained point function attains a value 1, the test set is found, and the algorithm terminates. Otherwise, next iteration should be started at point 2. The test vector is the set of 1, 0, and 0.5. The value 0.5 corresponds a coordinate which can be 1 or 0.

We artificially limit the number of iterations. If for a given number of iterations the solution is not found, such a fault is considered undetectable. Obviously, the restriction imposed on the number of iterations is a compromise between the speed of the algorithm and the percentage of fault coverage.

4 Investigation of Various Continuous Models of Boolean Functions for Test Generation

Efficiency of the test generation algorithm depends on the method of the objective function optimization and the method of the objective function construction. The purpose of this paper is to consider the dependence of the efficiency of the test suites generation on the type of continuous models (Tables 2, 3, 4, and 5).

Table 2 Continuous models option No. 2

$\bar{X}$	$\tilde{y} = \frac{1-x}{1+x}$
$X_1 \cup X_2$	$\tilde{y} = \frac{(1+x_1)(1+x_2)-(1-x_1)(1-x_2)}{(1+x_1)(1+x_2)+(1-x_1)(1-x_2)}$
$X_1 \oplus X_2$	$\tilde{y} = \frac{x_1(1+x_1)(1+x_2)+(1-x_1)(1-x_2)}{(1+x_1)(1+x_2)+x_1x_2(1-x_1)(1-x_2)}$

Table 3 Continuous models option No. 3

$\bar{X}$	$\tilde{y} = 1 - \sin(\frac{\Pi x}{2})$
$X_1 \cup X_2$	$\tilde{y} = 1 - \sin((1 - \sin(\frac{\Pi x_1}{2}) - \sin((\frac{\Pi x_2}{2}) + \sin(\frac{\Pi x_1}{2})\sin(\frac{\Pi x_2}{2}))\frac{\Pi}{2})$
$X_1 \oplus X_2$	$\tilde{y} = 1 - \sin(\frac{\Pi}{2}(\sin(\frac{\Pi x_1}{2}) + \sin(\frac{\Pi x_2}{2}) - \sin(\frac{\Pi x_1}{2})\sin(\frac{\Pi x_2}{2}))(1 - \sin(\frac{\Pi x_1 x_2}{2}))$

Table 4 Continuous models option No. 4

$\bar{X}$	$\tilde{y} = \cos(\frac{\Pi}{2}x)$
$X_1 \cup X_2$	$\tilde{y} = \cos(\frac{\Pi}{2}(\cos(\frac{\Pi}{2}x_1)(\cos(\frac{\Pi}{2}x_2))$
$X_1 \oplus X_2$	$\tilde{y} = \cos(\frac{\Pi}{2}(\cos(\frac{\Pi}{2}x_1)(\cos(\frac{\Pi}{2}x_2)(\cos(\frac{\Pi}{2}x_1x_2))$

Table 5 Continuous models option No. 5

$\bar{X}$	$\tilde{y} = 1 - \log_{10}(9x + 1)$
$X_1 \cup X_2$	$\tilde{y} = 1 - \log_{10}(9(1 - \log_{10}((9x_1 + 1)(9x_2 + 1)) + \log_{10}(9x_1 + 1)\log_{10}(9x_2 + 1)) + 1)$
$X_1 \oplus X_2$	$\tilde{y} = (1 - \log_{10}(1 - \log_{10}((9x_1 + 1)(9x_2 + 1)) + \log_{10}(9x_1 + 1)\log_{10}(9x_2 + 1)$ $(1 - \log_{10}(9x_1x_2 + 1))$

Evaluating the effectiveness was made on a set of ISCAS circuits faults. Time execution optimizing was not a part of this analysis.

5 Results

Different sets of continuous models have been tested on circuits from a set of ISCAS in the software for generation test sets for stuck-at faults. Table 6 shows the software running results. The column Detectable shows the number of detectable faults. The column Coverage shows the number of detected faults by the use of our approach with different continuous extensions of Boolean functions.

6 Conclusion

In this paper we have proposed various types of continuous models of Boolean functions to implement a continuous approach to modeling of discrete devices and fault test generation. The results of testing of stuck-at faults ISCAS circuits show the

Table 6 The Software Running Results

Circuit	Coverage					Detectable
	opt.1	opt.2	opt.3	opt.4	opt.5	
S27	32	32	32	32	32	32
S349	335	332	319	319	335	335
S713	458	434	414	414	470	476
S1196	1239	1230	1140	1140	1240	1242
S1238	1283	1279	1209	1209	1283	1283
Total	3347	3307	3114	3114	3360	3368

effectiveness of the logarithmic function usage. The next work is to use the obtained results to optimize the test generation algorithms based on the continuous approach to test generation for discrete circuits.

References

1. Kano, H.: Test pattern generation for logic networks by real number logic simulation. In: AUTOTESTCONF'79, pp. 168–178 (1979)
2. Kascheev, N., Mindrov, A.: The use of continuous model of combinational circuit for test generation. In: Simulation and CAD Systems, pp. 47–50 (1989)
3. Kascheev, N., Ryabkov, Y., Danilov, S.: Test generation for synchronous digital circuits based on continuous approach to circuit modeling. In: Proceedings of East-West Design and Test Workshop - EWDTW'04, Crimea (2004)
4. Podyablonsky, F., Kascheev, N.: Generalized faulty block model for automatic test pattern generation. In: Proceedings of IEEE East-West design and Test Symposium, Sept 2009, pp. 81–83 (2009)
5. Rivin, I., Chakradhar, S.T.: Discrete test generation by continuous methods. In: Proceedings of 12th IEEE VLSI Test Symposium, April 1994, pp.100–105 (1994)

König Graphs for 4-Paths: Widened Cycles

Dmitry Mokeev

Abstract We characterize the graphs whose induced subgraphs all have the following property: The maximum number of induced 4-paths is equal to the minimum cardinality of the set of vertices such that every induced 4-path contains at least one of them. In this chapter we describe all such graphs obtained from simple cycles by replacing some vertices with cographs.

Keywords König graphs • Widened cycles • 4-paths • Duality

1 Introduction

Let $\mathscr{X}$ be a set of graphs. A set of disjoint induced subgraphs of a graph G isomorphic to graphs in $\mathscr{X}$ is an $\mathscr{X}$-matching of G. The $\mathscr{X}$-matching problem is to find a maximum $\mathscr{X}$-matching in a graph. A subset of vertices of graph G that covers all induced subgraphs of G isomorphic to graphs in $\mathscr{X}$ is its vertex cover of G with respect to $\mathscr{X}$, or simply its $\mathscr{X}$-cover. The $\mathscr{X}$-cover problem is to find a minimum $\mathscr{X}$-cover in a graph. A König graph for $\mathscr{X}$ is a graph in which every induced subgraph has the property that the maximum cardinality of its $\mathscr{X}$-matching is equal to the maximum cardinality of its $\mathscr{X}$-cover [1]. The class of all König graphs for set $\mathscr{X}$ is denoted as $\mathscr{K}(\mathscr{X})$. If $\mathscr{X}$ consists of a single graph H, then we will discuss H-matchings, H-covers, and König graphs for H.

One can find some similar terms in the literature: a König–Egervary graph [2], a graph with the König property [3], a König graph [4]. They all have the same meaning, which is a graph in which the cardinalities of maximum matching and minimum vertex cover are equal. Note that the definition of a König graph in this chapter is not a generalization of the concept. This definition and usage of the term are motivated by the fact that the class of bipartite graphs, referred to in König's theorem, is exactly the class of all graphs whose cardinalities of maximum matching

D. Mokeev (✉)
Laboratory of Algorithms and Technologies for Networks Analysis, National Research University Higher School of Economics, 136, Rodionova Str., Nizhny Novgorod, Russia
e-mail: MokeevDB@gmail.com

V.A. Kalyagin et al. (eds.), *Models, Algorithms and Technologies for Network Analysis*, Springer Proceedings in Mathematics & Statistics 156,
DOI 10.1007/978-3-319-29608-1_3

and minimum vertex cover are equal not only for the graph but also for all its induced subgraphs. Thus, the class of bipartite graphs coincides with the class of all König graphs for P_2 in this sense.

Several papers have been devoted to the $\mathscr{X}$-matching problem, especially its algorithmic aspects (see, e.g., [5, 6]). It is known that the H-matching problem is nondeterministic polynomial time complete for any graph H having a connected component with three or more vertices. The complexity theory of extremal problems on graphs was developed by V. E. Alekseev and D. S. Malyshev (see, e.g., [1]).

Formulated as integer linear programming problems, $\mathscr{X}$-matching and $\mathscr{X}$-cover form a pair of dual problems. Thus, König graphs are graphs such that for any induced subgraph there is no duality gap. In this regard König graphs are similar to perfect graphs, having the same property with respect to another pair of dual problems (vertex coloring and maximum clique), which helps to efficiently solve these problems on perfect graphs [7].

Class $\mathscr{K}(\mathscr{X})$ is hereditary for any $\mathscr{X}$ and therefore can be described by a set of forbidden graphs (minimal graphs by the relation "to be an induced subgraph" not belonging to $\mathscr{X}$). Such a characterization for P_2 is given by König's theorem with a known criterion for bipartite graphs. In addition to this classical theorem, the following results are known for this type of simple graph: In [8] all forbidden subgraphs are described for the class $\mathscr{K}(\mathscr{C})$, where $\mathscr{C}$ is the set of all simple cycles. In [1] several families of forbidden graphs for $\mathscr{K}(P_3)$ are found, and it is conjectured that the set of these families forms a complete set of forbidden graphs for this class.

Graph G is P_4-connected if both G and $\overline{G}$ are connected and each vertex of G is a member of at least one induced P_4.

In [9] we give two ways of characterizing the class $\mathscr{K}(P_4)$. One of them is the standard description of hereditary class by forbidden subgraphs. We prove that the set of forbidden subgraphs includes 10 infinite families and 62 individual graphs. In the other approach, we show how to construct a graph of the given class by widening the subdivision from an arbitrary P_4-connected bipartite graph except a simple cycle.

Theorem 1. *Any graph obtained from a widened subdivision from an arbitrary P_4-connected bipartite graph except a simple cycle is König for P_4.*

The aim of this work is to show which graphs obtained by a widened subdivision from simple cycles are König. In Chaps. 2 and 3, we give some definitions and auxiliary propositions and in Chap. 4 we formulate and prove the main theorem, which describes all König widened subdivisions of simple cycles./marginparComp: Should we change "Chaps." and "Chap." in this paragraph to "Sects." and "Sect." since the author is referring to sections within this chapter?

In what follows, a "König graph" means a König graph for P_4. The maximum number of subgraphs in a P_4-matching of G is denoted as $\mu_{P_4}(G)$, and the minimum number of vertices in its P_4-covering is denoted as $\beta_{P_4}(G)$.

The induced subgraph P_4 is called a quartet. We denote by (v_1, v_2, v_3, v_4) a quartet that consists of vertices v_1, v_2, v_3, v_4.

We denote by $|G|$ a number of vertices in G.

Considering cycle C_n, assume that the vertices are labeled along the cycle as $0, 1, \ldots, n-1$. The arithmetic operations with the vertex labels are performed modulo n. Each residue class of vertices modulo 4 is called a 4-class.

2 Some Definitions

Definition 1. A graph without quartets is called a cograph.

Definition 2. We say that a cograph is trivial if it consists of one vertex and is nontrivial otherwise.

Definition 3. The operation of replacing vertex x with a cograph consists of the following steps:

1. The vertex is removed from the graph.
2. Several new vertices are added to it. New vertices are interconnected to form a cograph.
3. Each new vertex is connected by an edge to each vertex adjacent to x in the original graph.

Definition 4. We call an induced path of a graph terminal if one of its vertices is terminal (has degree 1) and the others have degree at most 2. We call a contact vertex of a terminal path one that is adjacent to one of the vertices of the path but does not belong to it (if it exists).

Definition 5. The operation of replacing a terminal path of two vertices with a cograph consists of the following steps:

1. Vertices of this path are removed from the graph.
2. New vertices $k_1, k_2, \ldots, k_p$ are added to the graph. New vertices are connected to each other and connected to the contact vertex.
3. New vertices $l_1, l_2, \ldots, l_{p-1}$ and possibly vertex l_p are added to the graph; $N(l_i) = \{k_1, k_2, \ldots, k_i\}$.
4. Each vertex of set $\{k_1, k_2, \ldots, k_p, l_1, l_2, \ldots, l_{p-1}\}$ and the contact vertex can be replaced with an arbitrary cograph.

Definition 6. The operation of replacing a terminal path of three vertices with a cograph consists of the following steps:

1. Vertices of this path are removed from the graph.
2. Several new vertices are added to the graph. New vertices are interconnected to form a cograph.
3. Let y be a contact vertex of the path. New vertices are connected to vertex y so that a maximum induced path that contains y and the added vertices has length 3.

Definition 7. Let H be a bipartite graph. The operation of widening a subdivision of H consists of the following steps:

1. Each cyclic edge (an edge that belongs to a cycle) of this graph is subdivided by one vertex.
2. Each vertex added at the subdivision and each vertex of degree 1 or 2 not belonging to the cycle is replaced with an arbitrary cograph.
3. Some of the old vertices belonging to cycles are also replaced with arbitrary cographs. They are vertices of degree 2 except under the following conditions:
   (a) For each cycle, if there is a vertex v adjacent to three or more vertices of degree more than 1, then any vertex of a 4-class containing v cannot be replaced with a cograph.
   (b) For each cycle, if a there is a vertex v of degree 3 or more in the cycle, then a vertex of the 4-class containing v and a vertex of another 4-class consisting of old vertices cannot simultaneously be replaced with cographs.
4. Some terminal paths of two and three vertices in the obtained graph can be replaced with arbitrary cographs.

We call the obtained graph a widened subdivision of the original bipartite graph H.

The notion of P_4-structure was invented by Chvatal [10] in 1984.

Definition 8. For a given graph $G = (V, E)$, its P_4-structure is defined as the 4-uniform hypergraph on $V(G)$ whose edges are all the 4-element sets that induce a quartet in G.

One can associate the P_4-matching and P_4-cover of the graph with the matching and vertex covering of its P_4-structure. It motivates us to use this term to formulate some useful definitions.

Definition 9. We say that a graph G is P_4-connected if its P_4-structure is the connected hypergraph.

Obviously, if there is a vertex in a graph that does not belong to any quartet, then it is an isolated vertex in the P_4-structure. Thus, each P_4-connected graph does not contain such vertices.

Since P_4 is a self-complementary graph, graphs G and $\overline{G}$ have the same P_4-structure. It means that each P_4-connected graph G is a connected graph and a complement of a connected graph.

Proposition 1. *A connected graph G is P_4-connected if and only if its complement is connected and each of its vertices belongs to at least one induced 4-path.*

Definition 10. We say that a subgraph H of graph G is its P_4-connected component if H is a maximal P_4-connected subgraph in G.

Obviously, P_4-matching of the graph is the sum of P_4-matchings of all its P_4-connected components. Similarly, the P_4-cover of the graph is the sum of P_4-covers of all its P_4-connected components.

In what follows we consider only P_4-connected graphs. If the graph is not P_4-connected, one can consider each of its P_4-connected components.

3 Widened Subdivisions of Even Cycles

Here we consider widened subdivisions of even cycles and show when such graphs are König. Note that all degrees of the simple cycle equal 2. Therefore, it doesn't contain terminal paths. It means that each widened subdivision of an even cycle can be obtained from a cycle of $4k$ vertices, where $k \geq 2$, by replacing each vertex with an arbitrary (possibly trivial) cograph.

Definition 11. We replace a vertex labeled i in a cycle with an arbitrary cograph. We call such a cograph a section and denote it as S_i. We say that a section is trivial if it consists of one vertex and otherwise is nontrivial.

In [9] we prove the following lemma:

Lemma 1. *Let A be a section in the graph G. A vertex $v \in A$ belongs to the minimum covering of graph G if and only if all the vertices of this section also belong to this minimum covering.*

Thus, each minimum covering of widened subdivisions of an even cycle consists of whole sections.

In [9] we prove the following lemma:

Lemma 2. *Each widened subdivision of the tree belongs to $\mathscr{K}(P_4)$.*

In [9] we describe infinite families of forbidden subgraphs for König graphs. One of those classes is denoted as $\mathscr{D}$.

Definition 12. Denote by $D(k_1, k_2, k_3, k_4)$ the set of graphs obtained from a cycle of length $n = k_1 + k_2 + k_3 + k_4$ by replacing four vertices numbered $0,\ k_1, k_1 + k_2,\ k_1 + k_2 + k_3$ with cographs of two vertices. This set always consists of exactly 16 graphs that differ by the structure of their cograph that replaced the vertex (K_2 or O_2).

Denote

$$\mathscr{D} = \bigcup_{\substack{k_1 \equiv k_2 \equiv k_3 \equiv \\ \equiv k_4 \equiv 1\,(mod\,4), \\ k_i \geq 5,\ i=2,3,4, \\ or \\ k_1 \equiv 1\,(mod\,4),\, k_1 \geq 5, \\ k_2 \equiv k_4 \equiv 2\,(mod\,4), \\ k_3 \equiv 3\,(mod\,4)}} \left\{ G \mid G \in D(k_1, k_2, k_3, k_4)\ or\, \overline{G} \in D(k_1, k_2, k_3, k_4) \right\}.$$

Now we define the shift-of-section operation for the cover of a widened subdivision of a cycle.

Definition 13. Let G be obtained from a cycle or path of m vertices by replacing each of its vertices with an arbitrary (possibly trivial) cograph. Denote all of its sections as $S_1, \ldots, S_m$ along the cycle (path). Let C be a P_4-cover of G. Let $n \leq 3$, $S_i \in C$, $\forall 1 \leq k \leq n[S_{i+k} \notin C]$ and $\exists 1 \leq l \leq 4 - n[S_{i-l} \in C]$. The operation of right-shifting section S_i in C into n consists of replacing section S_i by S_{i+n}. In other words, the result of this operation is a P_4-cover $C' = C \setminus S_i \cup S_{i+n}$. The operation of left-shift is defined similarly. The result of this operation is a P_4-cover $C'' = C \setminus S_i \cup S_{i-n}$.

It's easy to see that the following proposition is true:

Proposition 2. *Let C be a minimum P_4-cover of G and C' be a P_4-cover obtained from C by shifting one of its sections into the trivial section. Then C' is a minimum P_4-cover of G.*

We prove some propositions before formulating the main theorem.

Proposition 3. *Let G be obtained from a cycle of $4n$ vertices (where $n > 2$) by replacing each vertex with an arbitrary cograph, and let i be such that sections S_i, S_{i+1}, S_{i+2} are nontrivial and sections S_{i-1}, S_{i+3} are trivial. Then there exist a P_4-cover C and a P_4-matching M of G such that $|M| = |C|$.*

Proof. Consider graph G' obtained from G by deleting sections $S_{i-1}, S_i, S_{i+1}, S_{i+2}, S_{i+3}$. By Lemma 2, a P_4-cover C' and a P_4-matching M' exist in G' such that $\left|M'\right| = \left|C'\right|$. Then $C = C' \cup S_{i-1} \cup S_{i+3}$ is a P_4-cover of G and $M = M' \cup \{(x_{i-1}, x_i, x_{i+1}, x_{i+2}) \cup (x_i, x_{i+1}, x_{i+2}, x_{i+3})\}$, where $x_j \in S_j, y_j \in S_j, x_j \neq y_j, j \in \{i-1, \ldots, i+3\}$ is its P_4-matching. Since sections S_{i-1}, S_{i+3} are trivial, $|M| = \left|M'\right| + 2 = \left|C'\right| + 2 = |C|$. □

Proposition 4. *Let graph $G \in Free(\mathcal{D})$ be obtained from a cycle of $4n$ vertices (where $n > 2$) by replacing each vertex with an arbitrary cograph in which three nontrivial sections do not exist in a row and the numeration of such sections is such that for some $q \geq 0$, sections S_0, S_{4q+3} are nontrivial and sections $S_1, S_2, \ldots, S_{4q+2}$ are trivial. Then there exist a P_4-cover C and a P_4-matching M of G such that $|M| = |C|$.*

Proof. Consider graph G' obtained from G by deleting sections $S_1, S_2, \ldots, S_{4q+2}$ and deleting one vertex each from sections S_0 and S_{4q+3}. By Lemma 2, a P_4-cover C' and a P_4-matching M' exist in G' such that $\left|M'\right| = \left|C'\right|$. Denote $S'_1 = S_{4q+3} \setminus \{x_{4q+3}\}, S'_{4(n-q-1)+2} = S_0 \setminus \{x_0\}$, where x_{4q+3} and x_0 are deleted vertices. For all $i \in \{2, 3, \ldots, 4(n-q-1)+1\}$, denote $S'_i = s_{4q+2+i}$. Note that $S'_1, \ldots, S'_{4(n-q-1)+2}$ are all sections in G'. Obviously, if C' contains at least one of sections S'_2, S'_3, then $M = M' \cup \{(x_0, x_1, x_2, x_3), \ldots, (x_{4q}, x_{4q+1}, x_{4q+2}, x_{4q+3})\}$, where $S_j = \{x_j\}$ for all $1 \leq 4q + 2$, is the P_4-matching of G and $C = C' \cup S_1 \cup S_5 \cup \ldots \cup S_{4q+1}$ is its P_4-cover. It's easy to see that $|M| = |C|$. Now we prove that such a cover always exists. Note then that if C' contains at least one of sections $S'_{4(n-q-1)}, S'_{4(n-q-1)+1}$, then we can reduce such a case to a previously used numeration of sections. Since three nontrivial sections are not in a row in G, at least one of S'_2, S'_3 is trivial. It means that any minimum P_4-cover containing S'_1 can be reduced to a minimum P_4-cover containing S'_2 or S'_3 by a right-shift of S'_1. Now suppose that C' does not contain S'_1, S'_2, S'_3, and $S'_{4(n-q-1)}, S'_{4(n-q-1)+1}, S'_{4(n-q-1)+2}$. It means that it contains S'_4 and $S'_{4(n-q-2)+3}$. Let i be a minimum number such that i is not divisible by 4 and $S'_i \in C'$. Three cases are possible:

1. $i = 4k + 3$. Then there exists $j < k$ such that section S'_{4j+3} is nontrivial. Otherwise, the consequent left-shift of $S'_{4k}, S'_{4k-4}, \ldots, S'_4$ into 1 reduces the minimum P_4-cover containing S'_3. Thus, in G sections $S_0, S_{4q+3}, S_{4(q+j+1)+1}$ are nontrivial. It means that $S'_{4(j+1+l)} = S_{4(q+j+1+l)+2}$ is trivial for all $l \in \{0, 1, \ldots, n-j-q-3\}$. Otherwise, G contains a forbidden subgraph from the set $D(4l+1, 4(n-q-j-l)+2, 4q+3, 4j+2)$. Thus, $S'_4 \cup S'_8 \cup \ldots \cup S'_{4j} \cup S'_{4(j+1)} \cup \ldots \cup S'_{4(n-q+1)}$ is the minimum P_4-cover containing $S'_{4(n-q+1)}$.
2. $i = 4k + 2$. Then section $S'_{4(k-1)+3}$ is nontrivial. Otherwise, using the left-shift of S'_{4k}, we can reduce this case to the case 1. Thus, in the same way as in case 1 we can say that $S'_4 \cup S'_8 \cup \ldots \cup S'_{4j} \cup S'_{4(j+1)} \cup \ldots \cup S'_{4(n-q+1)}$ is the minimum P_4-cover containing $S'_{4(n-q+1)}$.
3. $i = 4k + 1$. Note that for all $j \leq k$ at least one of $S'_{4j-1}, S'_{4j-2}, S'_{4j-3}$ is trivial. Also, at least one of S'_2, S'_3 is trivial. Let j^* be the minimum number such that for all $j > j^*$ sections S'_{4j-2}, S'_{4j-3} are both nontrivial and one of S'_{4j^*-2} and S'_{4j^*-3} is trivial. Then using the consequent left-shift of $S'_{4k}, S'_{4(k-1)}, \ldots, S'_{4(j^*+1)}$ into 3 and then the left-shift of S'_{4j^*} into 1 or 2, we can reduce this case to the case 1 or 2. Note that if $j^* = 1$, then such shifts reduce the minimum P_4-cover containing S'_2 or S'_3.

□

Proposition 5. *Let graph $G \in Free(\mathscr{D})$ be obtained from a cycle of $4n$ vertices (where $n \geq 2$) by replacing each vertex with an arbitrary cograph in which three nontrivial sections do not exist in a row and G contains an induced subgraph from the set $D(k_1, k_2, k_3, k_4)$, where $k_1 \equiv k_2 \equiv k_3 \equiv k_4 \equiv 1(mod 4)$. Then there exist a P_4-cover C and a P_4-matching M of G such that $|M| = |C|$.*

Proof. Since $G \in Free(\mathscr{D}$ and three nontrivial sections do not exist in a row in G, two of k_1, k_2, k_3, k_4 equal 1. If, for example, $k_1 = k_2 = 1$, then G satisfies the condition of Proposition 3. Thus, there exist a P_4-cover C and a P_4-matching M of G such that $|M| = |C|$. It means that we must consider only case $k_2 = k_4 = 1$, $k_1 \geq 5, k_3 \geq 5$ (the other cases are symmetric).

Now let $k_1 = 4l+1$; then $k_3 = n-4l-3$. We number sections of G in such a way that sections $S_0, S_1, S_{4l+2}, S_{4l+3}$ are nontrivial. Then for $1 \leq i \leq l$, sections S_{4i+1} and S_{4i-2} are trivial. Otherwise, G contains an induced subgraph from $\mathscr{D}$. Similarly, for $l+1 \leq j \leq n-1$, sections S_{4j} and S_{4j+3} are trivial.

Consider the P_4-cover $C = S_2 \cup S_6 \cup \ldots \cup S_{4l-2} \cup S_{4l+1} \cup S_{4(l+1)} \cup S_{4(l+2)} \cup \ldots \cup S_{4(n-1)} \cup S_{4n-1}$ and the P_4-matching $M = \{(v_0, v_1, v_2, v_3), \ldots, (v_{4l}, v_{4l+1}, v_{4l+2}, v_{4l+3})\} \cup \{(u_{4l+2}, u_{4l+3}, u_{4(l+1)}, u_{4(l+1)+1}), \ldots, (u_{4(n-1)+2}, u_{4(n-1)+3}, u_0, u_1)\}$, where $u_i \in s_i, v_i \in s_i, u_i \neq v_i$ for all $i \in \{0, \ldots 4n-1\}$. It's easy to see that $|M| = |C| = n+1$. □

4 The Main Theorem

Now we can formulate and prove the main theorem of this chapter.

Theorem 2. *A widened subdivision of an even cycle is König if and only if it does not contain induced subgraphs of the set $\mathscr{D}$.*

Proof. Let graph $G \in Free(\mathscr{D}$ be a widened subdivision of a cycle of $2n$ vertices. One can say that G was obtained from a cycle of $4n$ vertices by replacing each vertex with an arbitrary cograph. Now we prove that there exist a P_4-cover C and a P_4-matching M of G such that $|M| = |C|$. Consider the following cases:

1. There exist four nontrivial sections in a row in G. Let section S_0 be trivial and S_1, S_2, S_3, S_4 be nontrivial. We select vertices $v_i \in S_i$ for all $i \in \{1,2,3,4\}$. Let $G^{'}$ be the graph obtained by removing the selected vertices. If G contains no subgraphs of $\mathscr{D}$, then $G^{'}$ does not contain them either. This allows us to use induction on the number of vertices. By the induction hypothesis, there exist a P_4-matching $M^{'}$ and a P_4-cover $C^{'}$ of $G^{'}$ such that $|M^{'}| = |C^{'}|$. Note that $|C^{'}|$ includes at least one of the sections $S_i \setminus \{v_i\}, i \in \{1,2,3,4\}$, but not more than two. If there are two such sections, then S_0 is not part of $C^{'}$ (otherwise, $C^{'}$ is not at a minimum) and at least one of $S_i \setminus \{v_i\}, i \in \{1,2,3\}$ is part of $C^{'}$. Then, using a left-shift into the appropriate number, we obtain the minimum P_4-cover that includes exactly one of the sections $S_i \setminus \{v_i\}, i \in \{1,2,3,4\}$. Adding to the $C^{'}$ the deleted vertex from the corresponding section, and adding to $M^{'}$ the quartet (v_1, v_2, v_3, v_4), we obtain the P_4-cover and P_4-matching of G having equal cardinality.
2. There are not four nontrivial sections in a row, but there are three nontrivial sections in a row in G. Then by Proposition 3 there exist a P_4-cover C and a P_4-matching M of G such that $|M| = |C|$.
3. There are not three nontrivial sections in a row in G. Let $k_1, k_2, \ldots, k_s$ be distances between nontrivial sections along the cycle. Note that $\sum_{i=1}^{s} k_i = 4n$. Consider the following cases:

(a) There exists i such that $k_i \equiv 3(mod4)$. Then by Proposition 4 there exist a P_4-cover C and a P_4-matching M of G such that $|M| = |C|$.

(b) There is no i such that $k_i \equiv 3(mod4)$, but there are j_1, j_2 such that $k_{j_1} \equiv k_{j_2} \equiv 1(mod4)$ and the other k_js are even. Then there exist an odd number of js such that $k_j \equiv 2(mod4)$. Number sections of a graph in such a way that sections S_0, S_{4p+1}, S_{4q+3} are nontrivial for some $0 < p \leq q$. Sections $S_2, S_6, \ldots, S_{4n-2}$ are trivial. Otherwise, G contains an induced subgraph of $\mathcal{D}$. Then there is a P_4-cover $C = S_2 \cup S_6 \cup \ldots \cup S_{4n-2}$ and the cardinality of C equals n. Obviously, there is a P_4-matching of G of the same cardinality.

(c) There is an i such that $k_i \equiv 3(mod4)$, but there are j_1, j_2, j_3, j_4 such that $k_{j_1} \equiv k_{j_2} \equiv k_{j_3} \equiv k_{j_4} \equiv 1(mod4)$. Then there exist p, q, r such that

$$l_1 = k_1 + k_2 + \ldots + k_p \equiv 1(mod4)$$

$$l_2 = k_{p+1} + k_{p+2} + \ldots + k_q \equiv 1(mod4)$$

$$l_3 = k_{q+1} + k_{q+2} + \ldots + k_r \equiv 1(mod4)$$

$$l_4 = k_{r+1} + k_{r+2} + \ldots + k_s \equiv 1(mod4).$$

Then by Proposition 2 there exist a P_4-cover C and a P_4-matching M of G such that $|M| = |C|$.

(d) All k_is are even. Then all sections with odd numbers are trivial. Both 4-classes containing these sections are P_4-covers of G of cardinality n. Obviously, G has a P_4-matching of the same cardinality.

Thus, there exist a P_4-cover C and a P_4-matching M of G such that $|M| = |C|$. But any subgraph H also satisfies the conditions of the theorem. Therefore, $\mu_{P_4}(H) = \beta_{P_4}(H)$, and hence $G \in \mathcal{K}(P_4)$. □

Acknowledgements This research is partly supported by LATNA Laboratory, NRU HSE, RF government grant 11.G34.31.0057.

References

1. Alekseev, V.E., Mokeev, D.B.: König graphs with respect to 3-paths. Diskretnyi Analiz i Issledovanie Operatsiy. **19**, 3–14 (2012)
2. Deming, R.W.: Independence numbers of graphs—an extension of the König–Egervary theorem. Discret. Math. **27**, 23–33 (1979)
3. Lovasz, L., Plummer, M.D.: Matching theory. Akadémiai Kiadó, Budapest (1986)
4. Mishra, S., Raman, V., Saurabh, S., Sikdar, S., Subramanian, C.R.: The complexity of Konig subgraph problems and above-guarantee vertex cover. Algorithmica **61**, 857–881 (2011)
5. Hell, P.: Graph packing. Electron Notes Discrete Math. **5**, 170–173 (2000)
6. Yuster, R.: Combinatorial and computational aspects of graph packing and graph decomposition. Comput. Sci. Rev. **1**, 12–26 (2007)
7. Grötschel, M., Lovasz, L., Schrijver, A.: Geometric algorithms and combinatorial optimization. Springer, Heidelberg (1993)

8. Ding, G., Xu, Z., Zang, W.: Packing cycles in graphs II. J. Comb. Theory. Ser. B. **87**, 244–253 (2003)
9. Mokeev, D.: König graphs for 4-paths. In: Batsyn, M.V., Kalyagin, V.A., Pardalos, P.M. (eds.), Models, algorithms and technologies for network analysis, pp. 93–103. Springer, Berlin (2014)
10. Chvatal, V.: A semi-strong perfect graph conjecture. Ann. Discrete Math. **21**, 279–280 (1984)

Optimization Algorithms for Shared Groups in Multicast Routing

Carlos A.S. Oliveira and Panos M. Pardalos

Abstract Multicast group routing is a combinatorial optimization problem occurring in the field of communication networks. Given a graph $G = (V, E)$, a set of data sources $S \subset V$ and destinations $D \subset V$, the problem requires the construction of one or more routing trees such that each destination has its demand satisfied by one or more data sources. The MGR can be viewed as a generalization of the multicast routing problem with a single data source. This problem has important applications in the design of collaborative communication networks, among other uses. While the MGR problem is NP-hard, it is possible to determine algorithms for its solution that approximate the result in practice. In this paper, we discuss existing techniques for solving MGR. We also propose some fast heuristics for this problem and show that computational experiments support the quality of the results achieved by these algorithms.

Keywords Multicast group routing • Communication network • Combinatorial optimization • NP-hard problem • Heuristics

1 Introduction

Multicast routing is the general problem of creating routes in a communication network so that data packets can be correctly sent from sources to a set of destinations [12, 13]. Multicast routing has become an active area of research due to its multiple applications, which include work collaboration, video conferencing, groupware, virtual reality, and cache placement [15]. Multicast routing also provides

C.A.S. Oliveira (✉)
Quantitative Research Department, F-Squared Investments Inc., Ewing, NJ, USA
e-mail: oliveira@ufl.edu

P.M. Pardalos
Department of Industrial and Systems Engineering, University of Florida, 303 Weil Hall, Gainesville, FL 32608, USA

National Research University Higher School of Economics, Nizhny Novgorod, Russia
e-mail: pardalos@ufl.edu

V.A. Kalyagin et al. (eds.), *Models, Algorithms and Technologies for Network Analysis*, Springer Proceedings in Mathematics & Statistics 156,
DOI 10.1007/978-3-319-29608-1_4

multiple opportunities for the application of combinatorial optimization techniques. Many of the problems occurring in this area can be modeled as combinatorial problems [2, 9, 10, 14, 17, 18].

Despite the importance of multicast routing, several of the problems occurring in this area are NP-hard, and require the development of intelligent algorithms capable of exploring the unique combinatorial properties of the target problems. In this paper we deal with the determination of optimal routes in multicast groups. First, we provide some definitions that will be necessary to define multicast routing problems. Later, we describe the multicast group routing (MGR) problem and its applications.

1.1 Problem Formulation

A multicast network is represented by a directed graph $G = (V, E)$, where V is the set of nodes in the communications network, and E is the set of direct links between nodes. A multicast group is a set $R = S \cup D$, for $S \subset V$ and $D \subset V$ with $S \cap D = \emptyset$, where S is a set of data sources and D is a set of destinations. The objective of the problem is to send data packets from the set of sources to the set of destinations, so that all destinations have their aggregate demand satisfied. For this purpose, a route (normally represented by an induced tree $T \subseteq E$) needs to be found, while minimizing a particular objective function. Different versions of the problem arise depending on the type of data required by each destination.

If each destination node can be served only by a particular source, then we have a fixed version of the problem. In this version of the problem, links can be shared and used by two or more multicast groups. If a source can serve any destination, however, we have a non-fixed version of the problem where all routes may be shared. Since routes are allowed to share links, both problems can be viewed as a specialized version of the Steiner tree problem, which is well known to be NP-hard [7, 11]. The MGR is also a generalized version of the single source multicast routing problem, which has been studied by several people [14].

A multicast group is a set of nodes that are grouped together in order to share a particular type of data. For example, the set of nodes interested in receiving updates in a certain financial instrument may form a multicast group to consume this information. Therefore, a multicast group M is composed of at least one source and a set of destinations. Given the set of multicast groups M_i, for $i \in \{1, \dots, k\}$, we can define a multicast problem on a network $G = (V, E)$, a set of source nodes $S = \bigcup_i \sigma(M_i)$ and destinations $D = \bigcup_i \delta(M_i)$, where $\sigma : 2^V \to 2^V$ is the set of sources in a multicast group M_i, and $\delta : 2^V \to 2^V$ is the set of destinations in M_i.

In this paper, we consider algorithms for the MGR problem, which can be used to model the operation of multicast groups as described above. We assume a generalized multicast routing problem, where one or more sources in a set S are used to satisfy $|D|$ destinations. The routing algorithms considered here aim at generating a routing tree (or forest) connecting the sources to destinations. In the following sections, we explore different strategies used in the literature to solve this problem. We also introduce a few algorithms that attempt to provide near-optimal solutions for the MGR problem.

This paper is organized as follows. In Sect. 2 we discuss a few of the center-based approaches for the MGR problem, where the goal becomes finding a central node, from which one or more trees can be used to reach each of the destination nodes in the multicast group. These approaches include the topological center, which uses a measure of proximity based on the shortest paths between the central node and destination nodes. Another approach is finding the median node, which uses a weighted strategy to locate the central node. Also, the centroid strategy is presented, a technique where a few mathematical properties are used to determine the location of one or more central nodes. In Sect. 3 we discuss a few heuristics for the MGR problem. In Sect. 4 we discuss the computational experiments performed with the algorithms presented in the previous sections. Finally, we provide concluding remarks on Sect. 5.

2 Literature Review

In this section, we review some of the approaches used in the literature for the solution of MGR problems. We divide our coverage by type of general strategy used to solve the problem, starting with center-based approaches, where a particular node is used as the central point for the route construction process. Then we review mathematical programming formulations for the problem and how such MIP formulations have been used to compute solutions for the MGR problem. We also discuss other algorithms for the MGR in the last part of this section.

2.1 Center-Based Approaches

A common strategy for solving the MGR problem is the use of center-based approaches. The basic idea of such methods is to label one or more nodes of the network as center nodes for the purpose of creating routes that are shared among members of the multicast group. Using center nodes, one can more easily connect new routing subtrees and allow new groups to be connected to existing routes.

The great challenge of center-based approaches is the computation of high-quality center nodes. Depending on how such nodes are defined, it can become computationally expensive to calculate the best solution to the problem. Three main strategies have been devised to compute center nodes. The first strategy involves the calculation of the *topological center* of the network. The second approach involves the use of a *median node*. The third approach employs the concept of the *centroid*, as described in the remainder of this section.

First, let's consider the necessary notation. Let $G = (V, E)$ be the underlying network for the given multicast groups. Define as $c : E \rightarrow R$ the cost function on the arc set E. We want to select a node $v \in V$ such that the distance to every other node w in the routing group is minimized according to some measure of distance. Based on these concepts, we define the main center-based approaches as follows.

Problem 1 (Topological Center). The topological center problem requires the determination of a node v^* that minimizes the maximum distance between v and all other nodes in the network. In mathematical terms, this can be described as

$$v^* = \arg\min_{v \in V} \max_{w \neq v} d(v, w).$$

This problem can be solved in polynomial time when the underlying graph is a tree. However, for general graphs and routing trees, the problem is known to be NP-hard as shown in [2]. Beyond multicast routing, several applications for the topological center have been identified [21].

Problem 2 (Median). The median node is a slight modification of the topological center approach, where the desired center node is determined using a weighted function W. In this case, let the function $H : V \rightarrow R$ be defined as

$$H(v) = \sum_{u \in V} W(u) d(v, u).$$

In this definition, we denote by $W : V \rightarrow R$ the node weight function, while $d(u, v)$ is the shortest distance between nodes u and v. Then, the median node for a graph $G = (V, E)$ is defined as the node v^* that minimizes $H(v)$, i.e.,

$$v^* = \arg\min_{v \in V} H(v).$$

Problem 3 (Centroid). The third center-based strategy used with multicast groups employs the concept of centroid. The centroid can be defined in the following way: given an induced tree T on a graph G, the node v is a centroid of G if it is also a median of T. That is, by creating an auxiliary tree we can simplify the computation of the centroid, even if we lose the uniqueness of the solution. This type of center node is important in applications because it has a number of convenient properties, which allow the implementation of efficient algorithms. Here is one of the important properties shared by centroids:

Theorem 1. *Let be given a network $G = (V, E)$, a node $v \in V$, and an induced tree T of G. Then v is a centroid if and only if*

$$F(v) = \max_i w(T_{vi}) \leq 1/2 \sum_{u \in V} W(u).$$

This property can be used as a verification of the properties of a centroid. The following theorems are also used in the processing of centroids. They guarantee that if a centroid v is found in a network, then any other centroids are also located in the neighborhood of v.

Theorem 2. *Given a network $G = (V, E)$ and a vertex $v \in V$, then at most one subtree rooted at v can contain a centroid.*

Theorem 3. *Given a network $G = (V, E)$ and a tree T_G induced by G, then there are at most two centroids in T_G. If more than one centroid exist, they are adjacent.*

Finally, the following proposition can be used to explore centroids using a decomposition algorithm:

Theorem 4. *In a rooted tree with v as root, let there be k subtrees containing v. If $y_1, \ldots, y_k$ are the resulting subtrees, then there is a centroid on subtree y_j only if*

$$n_j > -n_j + \sum_{i=1,k} n_i$$

where n_j is the number of nodes in subtree j.

Using such properties, it is possible to find centroids for any graph G in polynomial time [8].

2.2 Mathematical Programming Formulation

Another way of formalizing the MGR problem is to use mathematical programming formulations. Such MIP formulations can be used to provide either one or more feasible solutions as well as defining lower bounds for the solution of the more complex models. Over the last years, a number of such MIP models have been proposed in the literature, as we will see in this section.

The model for group routing we consider first is referred to as multicast packing model [20]. The goal of the problem is to determine the best way to pack the data transferred by a set of multicast groups using a single routing tree. The objective function tries to minimize the maximum congestion resulting from the use of a candidate multicast route.

Let λ_e be a variable that represents the amount of flow passing through a particular edge of the network. This can be calculated as

$$\lambda_e = \sum_{k=1}^{K} t^k x_e^k,$$

where t^k is the amount of traffic generated by multicast group k, K is the number of multicast groups, and x_e^k is a decision variable that is 1 only when the arc e is being used by group k.

This results in the following MIP model, which minimizes the maximum congestion variable λ:

$$\min \lambda$$

subject to

$$\sum_{k=1}^{K} t^k x_e^k \leq \lambda \qquad \text{for all } e \in E$$

$$x_e^k \in \{0,1\} \qquad \text{for } k \in \{1,\ldots,K\}$$

This model has been successfully solved for instances of the multicast group routing problem, as demonstrated in [16].

Another set of formulations for the MGR problem follow the more traditional technique of modeling minimum tree problems, such as the Steiner problem. One particular version of these models has been proposed and solved in [3] and can be described as follows.

Let x be a vector of decision variables v_e, for $e \in E(G)$. Let $N : V \to 2^V$ be a function returning the neighbors of node v in G. To simplify notation, we also extend this function to $N : V^2 \to 2^V$ so that $N(S)$ is the union of all $N(v)$ such that $v \in S$. Then the problem becomes

$$\min c^T x$$

subject to

$$\sum_{e \in N(v)} x_e = 2 \qquad \text{for all } v \in M$$

$$\sum_{e \in N(v)} x_e \leq 2 \qquad \text{for all } v \in V(G) \setminus M$$

$$\sum_{e \in N(S)} x_e \geq 2 \qquad \text{for all } S \subset V(G) \text{ s.t. } u \in S \text{ and } M \not\subset S$$

$$x \in \{0,1\}^{|E|}$$

Where $M \subset V(G)$ is the set of nodes participating in the multicast groups. Since the number of possible subsets S is exponential, the model above is also exponential in the number of constraints. Despite this, it is possible to use branch-and-bound techniques to efficiently find a solution to this problem, as demonstrated in [3].

2.3 Distributed Algorithms for MGR

In this section, we consider distributed algorithms for routing applied to multicast groups. Although this is a combinatorial problem that can be solved using standard optimization techniques, many of the algorithms proposed and implemented for the MGR are in the format of distributed protocols. The reason is that such algorithms need to be readily implemented in computer network routers, where each node cooperates with its neighbors to achieve its routing objectives. As a result of this design characteristic, many of these algorithms for MGR problem lack guarantees of approximation or solution quality. They are targeted at providing fast, feasible solutions with little overhead, while at the same time being amenable to distributed implementations.

Among the most important distributed algorithms for multicast groups is the core-based tree (CBT) protocol [2]. The CBT operates by electing a single node to become the core of the multicast tree, using some of the techniques discussed in Sect. 2.1. Once a core node is elected, other nodes can be added to routing tree by proximity, implicitly using a distributed version of Dijkstra's algorithm. Finally, nodes are also able to process "disconnect" messages sent from destinations that want to terminate their membership in the group.

The protocol independent multicast (PIM) protocol [5] acts as a method to dynamically add and remove nodes from multicast groups. Its main design goal is to designate certain nodes as rendezvous points, while allowing them to maintain current data about the network topology. Like many such distributed algorithms, the objective is to create feasible solutions quickly, instead of trying to achieve optimality. Another proposal is the scalable multicast protocol (SCAMP) [19], which was optimized for the management of large multicast groups occurring in dynamic applications such as video conference.

3 Heuristics for Multicast Group Routing

In this section, we present a few heuristics that can be used to provide fast solutions for the multicast group routing problem. While these heuristics by their nature cannot guarantee optimality of results, nonetheless we have observed that we can achieve high-quality solutions when using these strategies. The local optimality of these strategies is generally determined by the properties of the shortest paths or spanning trees, combined with local search techniques.

3.1 Tree Connection Heuristic

The first heuristic we applied to the MGR is based on the use of a spanning tree as a starting point of the optimization process. You can see a quick listing of the procedure in Algorithm in Fig. 1. In the algorithm, the arguments are the graph

```
1  Algorithm Tree Connection
2  T ← ∅
3  R ← D ∪ S { set of Steiner nodes }
4  Compute the minimum spanning tree T for G
5  complete ← false
6  while not complete do
7     complete ← true
8     for v ∈ V do
9        if deg_T(v) = 1 and v ∉ R then
10          T ← T \ v
11          complete ← false
12       end
13    end
14 end
15 return T
```

Fig. 1 Tree construction heuristic

$G = (V, E)$, and the sets of sources S and destinations D. We denote by $deg_G : V \to N$ the degree function, which gives the number of nodes adjacent to v in $V(G)$.

The first step of this algorithm is to compute a minimum spanning tree for the graph G. For this step, we can use any of the standard algorithms for MST, which run in polynomial time. Once a spanning tree is computed, we can use it as the basis for the complete solution of the problem. Notice that for nodes that are not sources and destinations, there are two possibilities: either it is a necessary node in a path between sources and destinations or it is not part of a routing solution. To remove such nodes, the algorithm employs a routine that checks if each node needs to be part of the candidate solution.

The removal procedure starts on line 5 of Algorithm in Fig. 1. For each node $v \in G$, the procedure checks the degree $deg_G(v)$. If the degree is 1 and $v \notin R$, then it is clearly not necessary as part of the candidate solution, since it is a leaf node of the tree. Consequently, the node is removed at line 9. This process is repeated as long as an improvement can be performed by the removing additional nodes. When that is not possible, the resulting solution T is returned on line 14.

The complexity of this algorithm is dominated by the time spent in the construction of the spanning tree. Using a standard algorithm such as Kruskal's method, this can be performed in time $O(m \log n)$ [1]. This remaining of the heuristic can run on time $O(n^2)$, since the loop is executed n times, each repetition executing on constant time, and the number of repetitions of the external loop cannot be worse than $O(n)$.

```
1   Algorithm Combined SP
2   S ← ∅
3   for v ∈ D do
4       Compute the shortest path P between v and a destination d ∈ D
5       S ← S ∪ P
6   end
7   while not complete do
8       complete ← true
9       for v ∈ V do
10          if deg_S(v) = 1 and v ∉ R then
11              S ← S \ v
12              complete ← false
13          end
14      end
15  end
16  return S
```

Fig. 2 Combined shortest path heuristic

3.2 Combined Shortest Path Heuristic

The second heuristic we studied for the MGR problem is based on the use of the shortest paths between source and destination nodes in the multicast groups. The heuristic employs the tactic of combining different shortest paths into a single source tree, which is then locally optimized using a gradient descending algorithm. The details of this strategy are quickly presented in Algorithm in Fig. 2. The input to the algorithm is a graph $G = (V, E)$, a set of sources S, and a set of destinations D.

The algorithm starts by computing the shortest paths from each of the destinations towards the closest source node. This can be done using one of the existing algorithms for the shortest paths such as Dijkstra [6] or Bellman–Ford [4]. The paths are combined in line 4, so that a new path extends the existing solution $\mathcal{S}$. This can be done by adding just the part of the path $\mathcal{P}$ from s to d that is not contained in the existing tree. Once all destinations $d \in D$ have been satisfied in this way, we can start the pruning process, which is similar to what we discussed in the previous algorithm.

The complexity of the algorithm is dominated by the discovery of the shortest paths in the first part. A traditional algorithm for the shortest paths, such as Dijkstra's, can be executed in time $O(|E| + |V| \log |V|)$. Since this is performed $|D|$ times, the resulting complexity is $O(|D|(|E| + |V| \log |V|)$. The resulting parts of the algorithm can be executed in $O(n^2)$ time, as previously explained.

4 Computational Results

In this section we present preliminary results of computational testing performed with the heuristics described in Sect. 3. We have implemented these algorithms using a library of graph operations crafted by the authors using the C programming language.

All tests were performed in a 64-bit machine using the Intel Core i5, 2.4 GHz processor. The machine had 8 GB of available memory, although memory was not a limiting factor for the tests described in this section. The code was compiled using the GCC compiler, with the `O3` level of optimization.

In all tests we used a set of 60 input files that simulate realistic instances of the MGR. The number of nodes ranged from 20 to 90, while the number of edges ranged from 100 to 380. The number of sources varied from 2 to 8. Finally, the number of destinations ranged from 4 to 12.

In the first test, we decide to compare the performance of Algorithm in Fig. 1 (Algorithm Tree Connection) against a simple construction routing, which creates a solution containing random paths. The random solution can be viewed as a baseline for the improved algorithms we discussed earlier.

Table 1 shows the results of running these algorithms against 30 instances as described above. The table columns can be read as follows: the first column is an identifier for the MGR instance. The next columns give the size for the sets V, E, S, and D, respectively. The next column $\mathcal{S}_r$ gives the value achieved by the random tree generation algorithm. The column $\mathcal{S}_1^*$ displays the optimum value achieved by Algorithm in Fig. 1. Column $\mathcal{S}_2^*$ displays the optimum value achieved by Algorithm in Fig. 2. Column t displays the total running time for all iterations of the algorithm, in seconds.

We observed that Algorithm in Fig. 1 gives results that are consistently better than the baseline, with some results that are twice as good as the results achieved by the baseline implementation. The data for Algorithm in Fig. 2 is also included in Table 1. Results for this algorithm are very close to results for the first, which indicates that they are both converging to similar solutions. Algorithm 2 is able to improve the results for a few instances, but Algorithm 1 dominates the results for most of the tests.

Table 2 summarizes the results for a second set of instances, with the same number of nodes, sources, and destinations, but with different number of arcs. The table includes results for both algorithms as discussed above.

5 Concluding Remarks

In this paper we considered the minimum routing problem in the context of shared multicast groups. The MGR is a practical problem in telecommunications networks, and it has received attention due to its numerous applications in the implementation

Table 1 Results for Algorithms in Figs. 1 and 2

#	$\|V\|$	$\|E\|$	$\|S\|$	$\|D\|$	$\mathcal{S}_r$	$\mathcal{S}_1^*$	$\mathcal{S}_2^*$	t
1	20	100	2	4	168.038	102.586	111.8191	4.656
2	20	120	2	5	78.8766	67.822	68.497	4.127
3	20	140	2	6	156.62	122.528	121.886	8.442
4	20	160	3	6	159.688	79.2063	78.5152	7.184
5	20	180	3	7	182.329	127.51	127.477	5.3331
6	30	120	3	5	139.5103	104.857	105.928	7.952
7	30	140	3	6	167.0446	98.7166	99.162	7.0414
8	30	160	3	7	153.646	94.555	93.459	5.05
9	30	180	4	7	115.5549	58.5034	59.859	10.507
10	30	200	4	8	110.794	154.272	152.446	13.545
11	40	140	4	6	95.4794	65.349	65.294	10.321
12	40	160	4	7	185.774	153.983	152.523	7.85607
13	40	180	4	8	92.9569	70.0457	72.562	8.5275
14	40	200	5	8	202.68	178.306	179.782	16.367
15	40	220	5	9	90.446	56.6629	57.08418	16.162
16	50	160	5	7	183.239	90.4917	91.561	14.19
17	50	180	5	8	127.142	101.545	102.648	6.8587
18	50	200	5	9	143.459	83.805	85.075	14.707
19	50	220	6	9	58.3205	13.9511	13.654	12.996
20	50	240	6	10	121.394	89.865	88.208	7.8097
21	70	200	6	8	139.261	105.199	105.887	13.572
22	70	220	6	9	48.5774	17.2112	16.762	13.1948
23	70	240	6	10	127.4463	78.4428	79.673	18.0209
24	70	260	7	10	160.835	132.645	133.719	17.015
25	70	280	7	11	131.3358	78.047	78.376	16.84
26	90	300	7	9	80.1889	69.7786	68.4517	16.0674
27	90	320	7	10	25.9581	12.8343	12.7936	25.7936
28	90	340	7	11	121.7618	84.0041	85.002	28.674
29	90	360	8	11	199.785	131.543	132.79	27.072
30	90	380	8	12	143.6514	112.6192	114.3989	37.334

of collaborative systems. We discussed the existing strategies for MGR that are mainly based on the concept of center-based routing, along with some of their advantages and disadvantages. Our discussion includes graph-theoretical techniques as well as MIP-based formulations for the MGR.

We have also introduced two heuristic solutions for the MGR multicast routing problem in the presence of shared groups. Preliminary tests have been performed on a set of representative instances. Our algorithm has demonstrated to be able to find efficient solutions for the MGR problem in very little time, when compared to exact strategies. In future work, we intend to expand the testing of our algorithm into more types of problem instances. We also plan to compare the algorithm to LP and MIP-based strategies for the problem.

Table 2 Results for Algorithm in Fig. 2 (combined SP)

#	$\|V\|$	$\|E\|$	$\|S\|$	$\|D\|$	$\mathcal{S}_r$	$\mathcal{S}_1^*$	$\mathcal{S}_2^*$	t
31	20	120	2	4	191.294	143.6342	144.941	2.892
32	20	140	2	5	177.728	125.2151	127.6088	1.947
33	20	160	2	6	131.461	99.0888	102.896	1.691
34	20	180	3	6	171.735	152.095	154.8274	2.905
35	20	200	3	7	187.912	136.95	136.124	4.042
36	30	140	3	5	184.794	147.001	146.3734	5.664
37	30	160	3	6	179.683	136.889	136.7554	3.983
38	30	180	3	7	122.953	76.6377	75.2856	4.931
39	30	200	4	7	136.844	119.954	122.8322	9.241
40	30	220	4	8	182.194	173.7327	174.614	6.935
41	40	160	4	6	96.4981	58.8321	57.7313	14.664
42	40	180	4	7	63.165	46.4523	46.226	15.116
43	40	200	4	8	190.05	116.898	119.137	10.378
44	40	220	5	8	184.026	148.8825	147.293	11.614
45	40	240	5	9	39.532	30.478	31.0084	19.999
46	50	180	5	7	176.212	146.43	149.221	10.857
47	50	200	5	8	128.216	125.095	124.854	17.991
48	50	220	5	9	86.3907	58.694	59.6561	15.037
49	50	240	6	9	103.919	72.8204	72.2967	19.576
50	50	260	6	10	56.2119	49.014	49.508	14.815
51	70	220	6	8	157.2	114.906	116.864	14.108
52	70	240	6	9	161.4916	140.021	141.1835	14.458
53	70	260	6	10	89.4067	60.5059	60.2702	16.832
54	70	280	7	10	145.2213	104.313	103.537	18.614
55	70	300	7	11	37.5066	35.2421	35.9027	18.778
56	90	320	7	9	55.2469	48.0125	49.292	16.045
57	90	340	7	10	111.289	99.56	98.9886	22.6263
58	90	360	7	11	183.3003	146.531	147.696	28.338
59	90	380	8	11	33.9214	31.313	32.649	38.605
60	90	400	8	12	181.361	153.481	151.7783	36.939

References

1. Ahuja, R.K., Magnanti, T.L., Orlin, J.B.: Network Flows: Theory, Algorithms, and Applications. Prentice-Hall, Upper Saddle River (1993)
2. Ballardie, A., Francis, P., Crowcroft, J.: Core-based trees (CBT) – an architecture for scalable inter-domain multicast routing. Comput. Commun. Rev. **23**(4), 85–95 (1993)
3. Chen, S., Günlük, O., Yener, B.: Optimal packing of group multicastings. In: Proceedings of IEEE INFOCOM'98, San Francisco, CA, pp. 980–987. IEEE, New York (1998)
4. Cormen, T.H., Leiserson, C.E., Rivest, R.L., Stein, C.: Introduction to Algorithms, 2nd edn. MIT Press/McGraw-Hill, Cambridge (2001)
5. Deering, S., Estrin, D.L., Farinacci, D., Jacobson, V., Liu, C.-G., Wei, L.: The PIM architecture for wide-area multicast routing. IEEE ACM Trans. Netw. **4**(2), 153–162 (1996)

6. Dijkstra, E.W.: A note on two problems in connexion with graphs. Numer. Math. **1**(1), 269–271 (1959)
7. Du, D.-Z., Lu, B., Ngo, H., Pardalos, P.M.: Steiner tree problems. In: Floudas, C.A., Pardalos, P.M. (eds.) Encyclopedia of Optimization, vol. 5, pp. 227–290. Kluwer Academic Publishers, Dordrecht (2001)
8. Gupta, S.K.S., Srimani, P.K.: Adaptive core selection and migration method for multicast routing in mobile ad hoc networks. IEEE Trans. Parallel Distrib. Syst. **14**(1), (2003)
9. Hong, S., Lee, H., Park, B.H.: An efficient multicast routing algorithm for delay-sensitive applications with dynamic membership. In: Proceedings of IEEE INFOCOM'98, pp. 1433–1440 (1998)
10. Kompella, V.P., Pasquale, J.C., Polyzos, G.C.: Optimal multicast routing with quality of service constraints. J. Netw. Syst. Manag. **4**(2), 107–131 (1996)
11. Kou, L., Markowsky, G., Berman, L.: A fast algorithm for Steiner trees. Acta Informatica **15**(2), 141–145 (1981)
12. Oliveira, C.A.S., Pardalos, P.M.: A survey of combinatorial optimization problems in multicast routing. Comput. Oper. Res. (2004)
13. Oliveira, C.A.S., Pardalos, P.M.: Mathematical Aspects of Network Routing Optimization. Springer, New York (2010)
14. Oliveira, C.A.S., Pardalos, P.M.: A hybrid metaheuristic for routing on multicast networks. In: Batsyn, M.V., Kalyagin, V.A., Pardalos, P.M. (eds.) Models, Algorithms and Technologies for Network Analysis, pp. 97–110. Springer, Berlin (2014)
15. Oliveira, C.A.S., Prokopyev, O.A., Pardalos, P.M., Resende, M.G.C.: Streaming cache placement problems: Complexity and algorithms. Int. J. Comput. Sci. Eng. **3**(3), 173–183 (2007)
16. Prytz, M.: On optimization in design of telecommunications networks with multicast and unicast traffic. Ph.D. thesis, Dept. of Mathematics, Royal Institute of Technology, Stockholm (2002)
17. Salama, H., Reeves, D., Viniotis, Y.: A distributed algorithm for delay-constrained unicast routing. In: Proceedings of IEEE INFOCOM'97, Kobe (1997)
18. Sriram, R., Manimaran, G., Siva Ram Murthy, C.: A rearrangeable algorithm for the construction of delay-constrained dynamic multicast trees. IEEE/ACM Trans. Netw. **7**(4), 514–529 (1999)
19. Vadera, A.: SCAMP: SCAlable multicast protocol for communication in large groups. Ph.D. thesis, India Institute of Technology, Kanpur (1999)
20. Wang, C.-F., Liang, C.-T., Jan, R.-H.: Heuristic algorithms for packing of multiple-group multicasting. Comput. Oper. Res. **29**(7), 905–924 (2002)
21. Zhuge, H., Zhang, J.: Topological centrality and its e-science applications. J. Am. Soc. Inf. Sci. Technol. **61**(9), 1824–1841 (2010)

Minimizing the Fuel Consumption of a Multiobjective Vehicle Routing Problem Using the Parallel Multi-Start NSGA II Algorithm

Iraklis-Dimitrios Psychas, Magdalene Marinaki, Yannis Marinakis, and Athanasios Migdalas

Abstract In this paper, a new multiobjective formulation of the Vehicle Routing Problem, the Multiobjective Fuel Consumption Vehicle Routing Problem (MFCVRP), using two different objective functions is presented. The first objective function corresponds to the optimization of the total travel time and the second objective function is the minimization of the fuel consumption of the vehicle taking into account the travel distance, the load of the vehicle, and other route parameters. We solve two cases of the Multiobjective Fuel Consumption Vehicle Routing Problem. In the first case the problem is symmetric and in the second case the problem is asymmetric. The problem is solved with the Parallel Multi-Start NSGA II that uses more than one initial population of individuals and a Variable Neighborhood Search algorithm for the improvement of each produced solution. The instances that are used for the solution of the problem are modified instances based on the classic Euclidean Traveling Salesman Problem benchmark instances taken from the TSP library.

Keywords Multiobjective Fuel Consumption Vehicle Routing Problem • Parallel Multi-Start NSGA II • VNS • GRASP

I.-D. Psychas • M. Marinaki • Y. Marinakis (✉)
School of Production Engineering and Management, Technical University of Crete, 73100 Chania, Greece
e-mail: ipsychas102@gmail.com; magda@dssl.tuc.gr; marinakis@ergasya.tuc.gr

A. Migdalas
Department of Civil Engineering, Aristotle University of Thessaloniki, 54124 Thessaloniki, Greece

Industrial Logistics, Luleå University of Technology, 97187 Luleå, Sweden
e-mail: samig@gen.auth.gr; athmig@ltu.se

V.A. Kalyagin et al. (eds.), *Models, Algorithms and Technologies for Network Analysis*, Springer Proceedings in Mathematics & Statistics 156,
DOI 10.1007/978-3-319-29608-1_5

1 Introduction

In real world applications, very few optimization problems have only one objective function. Usually for the real world problems more than one objective have to be optimized in order to lead to a set of non-dominated solutions which is called Pareto front [6]. The **Vehicle Routing Problem (VRP)** is a generalization of the Traveling Salesman Problem [14, 28] and is an NP-hard optimization problem [18]. For more information on the VRP, please see [27, 44, 45]. In real world applications the optimization only of one criterion, for example, the distance for the Capacitated VRP may not be enough to prove that the quality of the routes is good enough and that these routes can lead to a decrease of the cost of the routing plan. Thus, in recent years the publications on **Multiobjective Vehicle Routing Problems (moVRPs)** are increased [20]. For a more recent review please see [25].

In recent years, the optimization of energy reduction and of fuel consumption in the Vehicle Routing Problems has been studied. An easy way to calculate the fuel consumption of a vehicle is to calculate the tonne-kilometers of the vehicle taking into account the covering distance and the load of the vehicle [34]. The tonne-kilometers (tonne-km or tkm) is the load of a vehicle measured in tonnes multiplying with the distance that the vehicle is covering measured in kilometers. If we multiply this number (tonne-km) with the average CO_2 emission factor (that is measured in gCO_2/tonne-km) [35], the result is the CO_2 emissions of the vehicle measured in grams (gCO_2). A model for the Energy Minimizing VRP was proposed by Kara et al. where the total cost of a route was calculated by multiplying the traveled distance with the total weight of a vehicle (tare plus the load) [21]. Considering the Leonardi et al.'s research [29] the "Efficiency of the vehicle use" can be calculated by a ratio tonne-kilometers/mass-kilometers. To calculate mass-kilometers, the weight of the empty vehicle is added to the load of the freight, resulting in the total weight of a vehicle. Also, in order to calculate the "CO_2 Efficiency" they assume that there are some other real route parameters such as the vehicle class, the driver's behavior, and other environmental and route parameters that are multiplied with the "Efficiency of the vehicle use" in order to give the CO_2 consumption of a vehicle. Another parameter that can be taken into account, especially for time windows routing problems, is the parameter of speed [1, 2, 13, 24, 41]. Also, in [11] a Green VRP model is presented. This model can be used if the fuel tank capacity as well as the fuel consumption of a vehicle (gallons per mile) is known. In [43] the "CO_2 emissions" can be calculated if the vehicle's load, the average distance, and the average carbon dioxide emission factor are known. The Fuel Consumption Rate (FCR) for a VRP (FCVRP) was proposed by Xiao et al. [46] and the objective of the proposed model was to minimize the fuel consumption. In their research, both the distance traveled and the load are considered for the calculation of the fuel costs (FCR). Zhang et al. [47] multiply the FCR with the CO_2 emission rate (CER) and the distance between the nodes in order to calculate the amount of CO_2 emissions of a vehicle. In [17] a bi-objective Green Vehicle Routing Problem is proposed for the minimization of the total traveled distance and the CO_2 emissions using the

NSGA II algorithm. The CO_2 emissions are calculated by multiplying the travel distance with an emission factor. Also, in this research it is referred that in real world there are more than one factor that could affect the fuel economy of a vehicle such as the vehicle's load, the vehicle's speed, the weather conditions (head-winds, back-winds, or the use of air conditions in hot weather), and the traffic congestion (stop-and-go movement or traveling at steady-state consumption). However, none of those factors were included in their multiobjective model. Another factor that could affect the CO_2 emissions is the slope of the road [5]. In this research, it is referred that the safe roads for the vehicles have slope grades between 0 and 10 %. In [10] a bi-objective Pollution Routing problem's model is proposed where the first objective function is minimizing the CO_2 emissions considering the speed, the load, and other parameters of a vehicle and the other objective function is the minimization of the driving time. In [32, 42] two energy Pickup and Delivery VRP models that are taking into account a large number of parameters are analyzed. Other CO_2 emissions minimizing models are presented in [22, 23, 30]. Another suggestion for the minimization of the emissions of a vehicle is the reduction of the traveling time by making shortest routes and by traveling with the best speed for the environment [40]. A larger survey until 2014 for the Energy and Green Vehicle Routing Problems can be found in [26, 33].

In this paper, the proposed multiobjective problem is a variant of the Multiobjective Energy Reduction VRP [38] where the time duration, the fuel consumption for delivery routes, and the fuel consumption for pickup routes were minimized using the Parallel Multi-start NSGA II algorithm.

In real life in order to travel from a point A to a point B different routes may be used. We may use a highway road in order to reach B more quickly, covering more distance with more speed or to reach B slower using smaller road networks, covering less distance with lower speed. Considering that the increase of the traveled distance (which may sometimes decrease the traveled time) with constant load will, also, increase the fuel consumption of a vehicle [46], we can conclude that the **traveled time** and the **fuel consumption** of a vehicle could be two "under minimization" competitive parameters of a multiobjective problem.

In this paper, the **Multiobjective Fuel Consumption Vehicle Routing Problem** is proposed. In this problem, two different objective functions are used and minimized simultaneously. In the first objective function, the total travel time is minimized. For the second objective function there are two cases, in the first case (symmetric case) the **Fuel Consumption** (*FC*) is minimized taking into account the traveled distance and the load of the vehicle [46] and considering that there are no other route parameters or that there are perfect route conditions (for example, there is no wind or there are routes without uphills and downhills). In the second case (asymmetric case), the **Route based Fuel Consumption** (*RFC*) is minimized taking into account some other asymmetric parameters of real life (weather conditions or uphill and downhill routes) in addition to the load and the traveled distance [17, 29]. We solved a number of instances with two objective functions. This formulation is an improvement of the formulation that we presented in [38], where the Multiobjective Energy Reduction Vehicle Routing Problem was formulated and solved. In the previous formulation [38], three different objective functions were used, where the first two are the same as in the proposed formulation while the third one was a completely different objective function as it was focused in a route where only pickups were allowed. Instead of this objective function, in the present formulation, an objective function that describes more realistic conditions of the environment is added and the problem includes only deliveries in the routes. Another difference of the two formulations is that in the Multiobjective Energy Reduction Vehicle Routing Problem proposed by Psychas et al. [38] a symmetric multiobjective energy VRP was solved while in this paper with the addition of the real conditions of the road an asymmetric multiobjective energy VRP is solved.

For this research the **Parallel Multi-Start NSGA II (PMS-NSGA II)** algorithm for the solution of the proposed problem is used and is compared with another variant of NSGA II. **PMS-NSGA II** is an improved version of **NSGA II (Non-dominated Sorting Genetic Algorithm)** [8, 9] and was originally proposed in [38]. A number of variants of the NSGA II algorithm have been used for solving Multiobjective Vehicle Routing Problems, e.g., for solving VRP with route balancing [19], for solving multiobjective VRP problems with time windows [16], and for solving a Green Vehicle Routing Problem [17]. Other Multiobjective Genetic Algorithms for the solution of Multiobjective Vehicle Routing Problems have been used in [3, 37].

The structure of the paper is as follows. In Sect. 2, the optimization model of the MFCVRP is described. In Sect. 3, an analytical description of the PMS-NSGA II algorithm is presented. In Sect. 4, the evaluation measures used in the comparisons are presented. In Sect. 5, the computational results are presented and, finally, concluding remarks and the future research are given in the last section.

2 Multiobjective Fuel Consumption Vehicle Routing Problem

In this section, the formulation of the **Multiobjective Fuel Consumption Vehicle Routing Problem (MFCVRP)** is given with two different objective functions. The first objective function $OF1$ is the minimization of the time needed for a vehicle to travel between two customers or a customer and the depot **and the second objective function $OF2$ is the minimization of the RFC when the decision maker plans delivery routes where all the customers have only demands and there are, also, route and environmental parameters in each road**. For the creation of the second objective function, in order to calculate the RFC from a node i to a node j taking into account the covering distance and the load of the vehicle, we were based on the FCR of Xiao et al. [46] and, thus, the RFC is calculated from the following equation:

$$RFC_{ij} = c_{ij}x_{ij}(1 + \frac{y_{ij}}{Q})r_{ij} = FC_{ij}r_{ij} \tag{1}$$

where c_{ij} is the distance from node i to node j, Q is the maximum capacity of the vehicle, x_{ij} denotes that the vehicle visits customer j immediately after customer i

with load y_{ij}, and r_{ij} are real route parameters (head-wind and back-wind, uphills and downhills, and traffic congestion) that are proposed in [17, 29].

The parameter r_{ij} is always a positive number and it may be $r_{ij} \neq r_{ji}$. If the value of r_{ij} is between 0 and 1, we consider that the route from i to j is a downhill route or the wind is back-wind. If r_{ij} is larger than 1, we consider that the route from i to j is an uphill route or a head-wind. The product $c_{ij}r_{ij}$ leads to an asymmetric formulation of the whole problem due to the fact that it may be $r_{ij} \neq r_{ji}$. If the $r_{ij} = 1\forall(i,j)$ that belongs to the route, then the problem is a symmetric problem. In that case, the $RFC_{ij} = FC_{ij}$.

In order to calculate the r_{ij} parameter in real life problems the following method is used: Based on Cicero et al. [5] the roads where the vehicles can travel safe have slope **in Grades** between 0 and 10 %. Also, the **Beaufort Scale** that is used for the measurement of the Wind Speed consists of an integer number between 0 and 12 [7]. In the proposed model, the **Grade Index** (G_{ij}) and the **Beaufort Index** (B_{ij}) are calculated in two different ways considering the road (if it is uphill or downhill) and if the wind is head-wind or back-wind taking into account Table 1. Taking into account the "Scania" truck's (engine capacity: 9 L, emission compliance: Euro 5, and 230 Hp at 1900 rpm (rounds per minute)) rpm clock the economic driving rpm value is between 1000 rpm and 1500 rpm (green rpm area) (http://www.topspeed.com/trucks/truck-reviews/scania/2010-scania-p-series-ar126354.html). If the driver drives with smooth shifting, the rpm index must not exceed the 1500 rpm. On the other hand if the driver drives with aggressive shifting, the rpm index will exceed the 1500 rpm. The upper limit of the rpm clock is 3000 rpm. The more the rpm index exceeds the 1500 rpm the more fuel consumption is succeeded. Also, the less the rpm index exceeds the 1500 rpm the less fuel consumption is succeeded.

Considering the G_{ij}, the B_{ij}, and the rpm_{ij}, the r_{ij} parameter is calculated as follows:

$$r_{ij} = \frac{G_{ij} + B_{ij}}{2} * \frac{rpm_{ij}}{1500} \quad (2)$$

For example, if from a point A to a point B the FC is equal to ten units, the Grade is 5 % uphill, the wind is 4 Beaufort head-wind, and the rpm is equal to 1500, then the RFC is calculated as follows:

$$RFC = FC * r = 10 * (\frac{(1 + \frac{0.05}{0.1}) + (1 + \frac{4}{12})}{2}) * 1 = 14.1\, units \quad (3)$$

Table 1 Grade Index and Beaufort Index

Grade Index (G_{ij})		Beaufort Index (B_{ij})	
Uphill	Downhill	Head-wind	Back-wind
1+(Grade/10%)	1-(Grade/10%)	1+(Beaufort/12)	1-(Beaufort/12)

On the other hand if the FC is equal to ten units, the Grade is 5 % downhill, the wind is 4 Beaufort back-wind, and the rpm is equal to 1500, then the RFC is calculated as follows:

$$RFC = FC * r = 10 * (\frac{(1 - \frac{0.05}{0.1}) + (1 - \frac{4}{12})}{2}) * 1 = 5.8\,units \tag{4}$$

For the **Multiobjective Fuel Consumption Vehicle Routing Problem (MFCVRP)**, the first objective function is used for the minimization of the total travel time. Thus, if $t_{ij}^{i_1}$ is the time needed to visit customer j immediately after customer i using vehicle i_1 and $s_j^{i_1}$ is the service time of customer j using vehicle i_1, then the first objective function is [38]:

$$\min OF1 = \sum_{i=1}^{n}\sum_{j=1}^{n}\sum_{i_1=1}^{m}(t_{ij}^{i_1} + s_j^{i_1})x_{ij}^{i_1} \tag{5}$$

where n is the number of nodes, m is the number of homogeneous vehicles, and the depot is denoted by $i = j = 1$.

The first case of the second objective function ($OF2a$) is used for the minimization of the **Fuel Consumption** (FC) taking into account the traveled distance and the load of the vehicle when the vehicle travels between two customers or a customer and the depot in the case that the vehicle performs only deliveries in its route and we consider that the $r_{ij} = 1\forall(i,j)$ that belongs to the route. The vehicle should begin with full load and after a visitation of a customer the load is reduced based on the demand of the customer. If we consider that the most loaded is the vehicle the more fuel it consumes, we take the following objective function [38]:

$$\min OF2a = \sum_{j=1}^{n}\sum_{i_1=1}^{m} c_{1j}x_{1j}^{i_1}(1 + \frac{y_{1j}^{i_1}}{Q}) + \sum_{i=2}^{n}\sum_{j=1}^{n}\sum_{i_1=1}^{m} c_{ij}x_{ij}^{i_1}(1 + \frac{y_{i-1,i}^{i_1} - D_i}{Q}) \tag{6}$$

with the maximum capacity of the vehicle denoted by Q, the i customer has demand equal to D_i (with $D_1 = 0$), $x_{ij}^{i_1}$ denotes that the vehicle i_1 visits customer j immediately after customer i with load $y_{ij}^{i_1}$, $y_{1j}^{i_1} = \sum_{i=1}^{n} D_i$ for all vehicles as the vehicle begins with load equal to the summation of the demands of all customers assigned in its route, and c_{ij} is the distance from node i to node j.

The second case of the second objective function ($OF2b$) is used for the minimization of the **Route based Fuel Consumption** (RFC) taking into account the route parameters in addition to the traveled distance and the load of the vehicle when the vehicle travels between two customers or a customer and the depot in the case that the vehicle performs only deliveries in its route. The vehicle should begin with full load and after a visitation of a customer the load is reduced based on the demand of the customer. If we consider that the most loaded is the vehicle the more fuel it consumes, we take the following objective function:

$$\min OF2b = \sum_{j=1}^{n} \sum_{i_1=1}^{m} c_{1j} x_{1j}^{i_1} (1 + \frac{y_{1j}^{i_1}}{Q}) r_{1j} + \sum_{i=2}^{n} \sum_{j=1}^{n} \sum_{i_1=1}^{m} c_{ij} x_{ij}^{i_1} (1 + \frac{y_{i-1,i}^{i_1} - D_i}{Q}) r_{ij} \quad (7)$$

The constraints of the problems are the following:

$$\sum_{j=1}^{n} \sum_{i_1=1}^{m} x_{ij}^{i_1} = 1, i = 1, \cdots, n \quad (8)$$

$$\sum_{i=1}^{n} \sum_{i_1=1}^{m} x_{ij}^{i_1} = 1, j = 1, \cdots, n \quad (9)$$

$$\sum_{j=1}^{n} x_{ij}^{i_1} - \sum_{j=1}^{n} x_{ji}^{i_1} = 0, i = 1, \cdots, n, i_1 = 1, \cdots, m \quad (10)$$

$$\sum_{j=1, j \neq i}^{n} y_{ji}^{i_1} - \sum_{j=1, j \neq i}^{n} y_{ij}^{i_1} = D_i, i = 1, \cdots, n, i_1 = 1, \cdots, m \quad (11)$$

$$Q x_{ij}^{i_1} \geq y_{ij}^{i_1}, i, j = 1, \cdots, n, i_1 = 1, \cdots, m \quad (12)$$

$$x_{ij}^{i_1} = \begin{cases} 1, & \text{if } (i,j) \text{ belongs to the route} \\ 0, & \text{otherwise} \end{cases} \quad (13)$$

Constraints (8) and (9) represent that each customer must be visited only by one vehicle; constraints (10) ensure that each vehicle that arrives at a node must leave from that node. Constraints (11) indicate that the reduced load (cargo) of the vehicle after it visits a node is equal to the demand of that node. Constraints (12) are used to limit the maximum load carried by the vehicle and to force $y_{ij}^{i_1}$ to be equal to zero when $x_{ij}^{i_1} = 0$.

3 Parallel Multi-Start NSGA II Algorithm

In this paper, a **Parallel Multi-Start NSGA II (PMS-NSGA II)** algorithm is used for the solution of the problem. The **Parallel Multi-Start NSGA II (PMS-NSGA II)** algorithm is a suitable modified version of NSGA II (Non-dominated Sorting Genetic Algorithm II) [9] for the solution of routing type problems. The algorithm was first presented in [38] and its main characteristic is the use of a Multi-Start method based on Greedy Randomized Adaptive Search Procedure (GRASP) [12] for the creation of the initial population. A second characteristic is that the algorithm uses more than one population that are evolved in parallel and a number of Pareto fronts used (equal to the number of populations). The algorithm, also, incorporates an external archive with the Pareto front of the whole population based on the crowding distance and the rank of each of the populations. Finally, in order to increase the exploitation abilities of each member of the populations a Variable Neighborhood Search (VNS) [15] algorithm is used in each one of them separately.

For all the algorithms, the solutions are represented with the path representation of the tour. For example, if we have a solution with five nodes, a possible path representation would be the "1 2 3 4 5". If these routes do not start with node 1, we find it and put it at the beginning of the route. This representation of solutions is used in the main algorithm and in the local search phases of the algorithm. Only for the calculation of the values of each objective function of the multiobjective problem, for each solution, at first, we separate the nodes of each solution into vehicle routes considering the time and the capacity of the vehicle constraints, for example, "1 2 3 1 4 5 1" and, then, the value of each objective function for each solution is calculated.

Initially, it is assumed that X different initial populations with W individuals for each population exist. Each population is separated in as many sub-populations of w individuals as the number of the objective functions is. If K is the number of objective functions, then $w = W/K$.

- The first solutions of the first 40 % of the populations are produced as follows. The first solution of the first sub-population of w individuals is a solution that is produced by solving a single objective problem with the VNS algorithm [38] using the first objective function, the first solution of the second set of w is a solution that is produced by solving a single objective problem with the VNS algorithm using the second objective function, etc. For these first solutions, the value of the other objective function is calculated without affecting the procedure in this phase of the algorithm. As we would like to start with a good solution, we increase the number of the two main parameters of the VNS (vns_{max} and $local_{max}$ [38]).
- The first solutions of the next 20 % of the populations are produced as follows. The first solution of the first sub-population of w individuals is a solution that is produced by solving a single objective problem by using the Nearest Neighborhood method [28] using the first objective function, the first solution of the second set of w is a solution that is produced by solving a single objective problem by using the Nearest Neighborhood method using the second objective function, etc.
- The first solutions of the last 40 % of the populations are produced as follows. The first solution of the first sub-population of w individuals is a solution that is produced by solving a single objective problem by using a variant of GRASP method [12] using the first objective function, the first solution of the second set of w is a solution that is produced by solving a single objective problem by using a variant of GRASP method using the second objective function, etc. In this GRASP algorithm instead of using the Restricted Candidate List (RCL), we use the following procedure: Node 1 is always used as a starting node. Then, a random number (*Rand*) equal to 0 or 1 is generated. If $Rand = 0$, then the nearest node to node i is visited. If $Rand = 1$, the second nearest node to node i is visited.

Table 2 How to produce the initial solutions for a two objective functions problem

		Number of individuals	Method used
Individuals in the first 40 % of the populations			
W	w for OF1	1	VNS
		2 to w/3	Swap method
		(w/3)+1 to 2w/3	2-opt method
		(2w/3)+1 to w	Random solutions
	w for OF2	1	VNS
		2 to w/3	Swap method
		(w/3)+1 to 2w/3	2-opt method
		(2w/3)+1 to w	Random solutions
Individuals in the next 20 % of the populations			
W	w for OF1	1	Nearest Neighborhood
		2 to w/3	Swap method
		(w/3)+1 to 2w/3	2-opt method
		(2w/3)+1 to w	Random solutions
	w for OF2	1	Nearest Neighborhood
		2 to w/3	Swap method
		(w/3)+1 to 2w/3	2-opt method
		(2w/3)+1 to w	Random solutions
Individuals in the last 40 % of the populations			
W	w for OF1	1	GRASP
		2 to w/3	Swap method
		(w/3)+1 to 2w/3	2-opt method
		(2w/3)+1 to w	Random solutions
	w for OF2	1	GRASP
		2 to w/3	Swap method
		(w/3)+1 to 2w/3	2-opt method
		(2w/3)+1 to w	Random solutions

For the calculation of the rest members of the populations we use the procedure that follows next. For each population and for each objective function (each set of w) the Swap method [28] is applied to the first solution and is used for the calculation of the second to $w/3$ individuals while the 2-opt method [31] is applied to the first solution and is used in order to produce the $(w/3) + 1$ to $2w/3$ individuals. Finally, the last individuals are produced at random. In Table 2, it is given an analytical description of how the initial population of a 2-objective functions problem is produced.

After the calculation of the initial population the classic steps of the NSGA II are followed. The *rank* and the *crowding distance* [8, 9] of each member of the populations are calculated. Then, in order to find the two parents' solutions that we need we make the following procedure. For each parent solution two candidate solutions are selected randomly from the population and the parent solution is the one with the best *rank* or the best *crowding distance* in the case that the *rank* is the same. Then, a crossover procedure between two parents is performed in order to produce two offspring [38]. The equations for the two offspring are:

$$offspring_l(t) = (1 - g) * parent_m(t) + g * parent_n(t) \tag{14}$$

$$offspring_f(t) = g * parent_m(t) + (1 - g) * parent_n(t) \tag{15}$$

where g is a random number in (0,1), m, n are the indices denoting the two parents ($m, n = 1, \cdots, W$) and l, f are the indices denoting the two offspring ($l, f = 1, \cdots, W$). We repeat the two previous steps until W solutions (offspring) are created. The next step is to improve the solutions using the VNS [15] algorithm as it was proposed in [38] for the solution of the Multiobjective Energy Reduction Vehicle Routing Problem, where the vns_{max} was set equal to 20 and the $local_{max}$ was set equal to 10 in order not to increase the computational time of the algorithm. In the next step, the parents ($parent_i$) and offspring ($offspring_i$) vectors are combined in a new one ($offspring'_i$) and, then, the members of the $offspring'_i$ are sorted using the *rank* and the *crowding distance* as in the previous step. From these solutions, a number of individuals for each population equal to the initial population survive in the next iteration. At the end of each iteration, the solutions with rank equal to 1 from all populations are combined into one single population and a new Best Pareto Front, using the *rank* and the *crowding distance*, is calculated. A pseudocode of the algorithm is the following:

Do while the maximum number of initial Populations has not been reached:
    Initialization
    Selection of the number of individuals
    Generation of the initial population
    Evaluation of the population for each objective function
    Selection of the mutation operator
    Initialization of the Population's Pareto Front
    Main Phase
    Do while the maximum number of generations has not been reached:
        Calculate the *rank* and the *crowding distance*
        For every two parents
            Produce two offspring using crossover operator
            Evaluation of the offspring for each objective function
        Endfor
        Application of VNS on each offspring
        Calculate the *rank* and the *crowding distance* of all parents and offspring
        Sort parents and offspring according to *rank* and *crowding distance*
        Select W individuals
        Evaluation of the individuals for each objective function
        Update of the Pareto front
    Enddo

Return Population's Pareto Front
Enddo
Return Best Pareto Front

In order to test the efficiency of the proposed algorithm, another version of the NSGA II is used to compare it with the proposed algorithm. The difference of this variant with the PMS-NSGA II is that in this variant one population of initial solutions is created. The Nearest Neighborhood algorithm is used for the creation of the first member of the initial population of w individuals and is improved using the proposed variant of VNS [38]. The Swap method [28] is used for the calculation of the second to $w/3$ individuals while the 2-opt method [31] is used in order to produce the $w/3 + 1$ to $2w/3$ individuals. All the other members of the population are created at random. The algorithm continues as the proposed algorithm does.

4 Evaluation Measures

The evaluation measures that are used for the comparison of the Pareto fronts of the two methods are the same measures that are used in [38] except the Spacing [39] that is replaced from Δ measure.

In this paper, as the optimum Pareto front is not known, four different measures are used:

- The range to which the front spreads out is described by the following equation [48]:

$$M_k = \sqrt{\sum_{i=1}^{K} max\{\| p' - q' \|\}} \tag{16}$$

where K is the number of objectives and p' and q' are the values of the objective functions of two solutions that belong to the Pareto front.
- The number of solutions of the Pareto front (L).
- The Δ measure includes information about both spread and distribution of each solutions [36]. For the calculation of the Δ measure the following equation is used:

$$\Delta = \frac{ds_f + ds_l + \sum_{i=1}^{|S|-1} |dist_i - \overline{dist}|}{ds_f + ds_l + (|S| - 1)\overline{dist}} \tag{17}$$

where S is the number of the intermediate solutions between the extreme solutions, ds_f and ds_l are the Euclidean distances between the extreme solutions [36] and the intermediate solutions [36] of the obtained non-dominated set [9],

$dist_i$ is the distance from an intermediate solution i to the next intermediate solution, $i = 1, 2, \ldots, (S-1)$, and $\overline{dist}$ the average value of all $dist_i$ distances.

- The Coverage [48] is calculated for a pair (A_1, B_1) of approximation sets of Pareto solutions of two different algorithms. In this metric the fraction of solutions in B_1 that are weakly dominated by one or more solutions in A_1 is calculated. The Coverage measure (C measure) is calculated by the following equation:

$$C(A_1, B_1) = \frac{|\{b \in B_1; \exists a \in A_1 : a \leq b\}|}{|B_1|}. \quad (18)$$

5 Computational Results

The algorithm was implemented in Visual C++ and it was tested in a set of instances. As there was not available a set of benchmark instances in the literature for the problem solved in this paper, we have to generate a number of instances that are also have been used in [38]. As we didn't want to test the algorithm in random set of benchmark instances, we created a number of sets based on instances for the solution of the Traveling Salesman Problem and the VRP. More precisely, from the TSPLIB five instances (kroA100, kroB100, kroC100, kroD100, and kroE100) with 100 nodes were selected. However, these instances, as they are used for the solution of the Traveling Salesman Problem, include only data for the coordinates of the nodes. All the other data needed for the Multiobjective Fuel Consumption VRP (capacity, time limits, and demands) were taken from the third instance (par3) of the classic Christofides benchmark instances [4] used for the solution of the Capacitated Vehicle Routing Problem (CVRP).

In each new test instance that was created, a combination of one or more instances from the five kro#100 instances (where # corresponds to A or B or C or D or E) with the par3 instance is performed. In this new generated instance the coordinates are taken from the corresponding kro#100 data set and the corresponding demand of each of the nodes was taken from the par3 instance. Also, the maximum tour length, the service time, and the capacity of each vehicle were taken from the par3 instance. We created 2-objective functions problems by combining these five instances. For example, in order to create a 2-objective functions asymmetric problem using the OF1 and the OF2b, we used kroA100par3, kroB100par3, kroC100par3, and kroD100par3 as the data used for all objective functions, respectively. More precisely, the first objective function is used as described in Sect. 2 and the necessary data are taken from kroA100par3 and the second objective function (OF2b) is used as described in Sect. 2 and the necessary data are taken from kroB100par3, kroC100par3, and kroD100par3 as it is described in the following. In order to create the route parameters of the network, we have to create an asymmetric table with positive numbers. In order to create those numbers we divided the route parameters' table into two parts. In the first part all elements that are downside of the main diagonal are calculated using the Euclidean distance between the nodes corresponding to the kroC100par3 set and in the second part all elements that are upside of the main diagonal are calculated using the Euclidean distance between the nodes corresponding to the kroD100par3 set. Then, each element of the route parameters' table is divided with the corresponding element of a table created by calculating the Euclidean distance between the nodes corresponding to the kroB100par3 set. Finally, the Euclidean distance between the nodes corresponding to the kroB100par3 set corresponds to the distance matrix.

In case of solving a 2-objective functions symmetric problem with first objective function the OF1 and second objective function the OF2a (OF1–OF2a) the combination "A–B" in all the following tables and figures (in all tables and figures A corresponds to kroA100par3, B corresponds to kroB100par3, and so on) means that for the new generated instance the Euclidean distance between the nodes using the kroA100par3 corresponds to the time duration between the nodes and the Euclidean distance between the nodes using the kroB100par3 corresponds to the distance between the nodes. In case of solving a 2-objective functions asymmetric problem with first objective function the OF1 and second objective function the OF2b (OF1–OF2b) the combination "A–B–CD" means that the Euclidean distance between the nodes using the kroA100par3 corresponds to the time duration between the nodes, the Euclidean distance between the nodes using the kroB100par3 corresponds to the distance between the nodes, and the Euclidean distance between the nodes for the instances kroC100par3, kroD100par3, and kroB100par3 is used for the calculation of the route parameters table as it was explained previously.

A number of different alternative values for the parameters of the algorithm were tested and the ones selected are those that gave the best computational results concerning both the quality of the solution and the computational time needed to achieve this solution and, also, taking into account the fact that we would like to test the algorithms with the same function evaluations. Thus, the selected parameters for all algorithms are given in the following:

Single population NSGA II

- Number of individuals: 1000.
- Number of generations: 500.
- Number of initial populations: 1.

Parallel Multi-Start NSGA II

- Number of individuals for each initial population: 100.
- Number of generations: 500.
- Number of initial populations: 10.

After the selection of the final parameters, the two versions of the Non-dominated Sorting Genetic Algorithm II (NSGA II) were tested for ten combinations for two and three objective functions, respectively. In the following tables and figures, the comparisons performed based on the four evaluation measures presented previously and the Pareto fronts are given. More precisely, we use the number of solutions (L)

Table 3 Results of the four measures for the two algorithms and for ten instances for OF1–OF2a

Objective Function 1–Objective Function 2a								
	Single population NSGA II				Parallel Multi-Start NSGA II			
	L	M_k	Δ	C	L	M_k	Δ	C
A–B	49	524.73	0.71	**0.67**	**55**	**538.68**	**0.55**	0.18
A–C	60	**547.70**	**0.63**	0.30	**66**	531.08	0.64	**0.47**
A–D	51	**519.59**	**0.53**	0.27	**56**	518.08	0.64	**0.51**
A–E	**47**	**521.57**	0.60	0.22	41	518.20	**0.57**	**0.66**
B–C	58	**520.09**	**0.57**	0.10	58	516.10	0.60	**0.78**
B–D	61	512.11	0.68	0.37	**62**	**512.60**	**0.50**	**0.57**
B–E	61	527.52	0.62	**0.44**	**62**	**541.17**	**0.59**	0.36
C–D	51	514.01	0.67	0.33	51	**520.74**	**0.64**	**0.37**
C–E	**60**	527.87	**0.55**	0.12	49	**537.60**	0.77	**0.75**
D–E	64	528.72	**0.57**	0.52	**66**	**539.37**	0.59	**0.42**

Table 4 Results of the four measures for the two algorithms and for ten instances for OF1–OF2b

Objective Function 1–Objective Function 2b								
	Single population NSGA II				Parallel Multi-Start NSGA II			
	L	M_k	Δ	C	L	M_k	Δ	C
A–B–CD	59	538.37	**0.62**	0.33	**70**	**545.13**	0.66	**0.36**
A–C–BD	49	533.50	**0.62**	**0.48**	**66**	**543.23**	0.65	0.35
A–D–BE	**58**	**538.93**	0.61	0.26	54	527.69	**0.59**	**0.53**
A–E–BD	**58**	**538.83**	**0.63**	0.26	54	533.71	0.70	**0.60**
B–C–AD	**63**	517.91	**0.59**	0.26	46	**527.89**	0.60	**0.40**
B–D–AC	**55**	**532.20**	**0.64**	0.26	53	527.31	0.68	**0.69**
B–E–AD	57	534.61	**0.55**	**0.51**	**63**	**537.84**	0.60	0.35
C–D–AE	**48**	538.25	**0.57**	0.44	45	**539.32**	0.62	**0.48**
C–E–AB	46	515.14	0.68	**0.51**	**51**	**530.63**	**0.65**	0.33
D–E–BC	49	**515.16**	**0.61**	0.33	**61**	503.39	0.63	**0.49**

in the non-dominated set, the maximum extend in each dimension (M_k), the spread of solutions (Δ), and the Coverage (C *measure*) for evaluation measures.

In Table 3, the results of the four measures for the two algorithms and for ten combinations of instances are presented when a 2-objective functions problem is solved using the objective functions OF1 and OF2a, while in Table 4 the corresponding results when a 2-objective functions problem is solved using the objective functions OF1 and OF2b are given. In Tables 3 and 4 with bold letters we denote the best values found for every metric in each instance. In the four figures (Figs. 1, 2, 3, and 4) two representative Pareto fronts for each one of Tables 3 and 4 are given, respectively. More precisely, in Figs. 1 and 2 two Pareto fronts from the

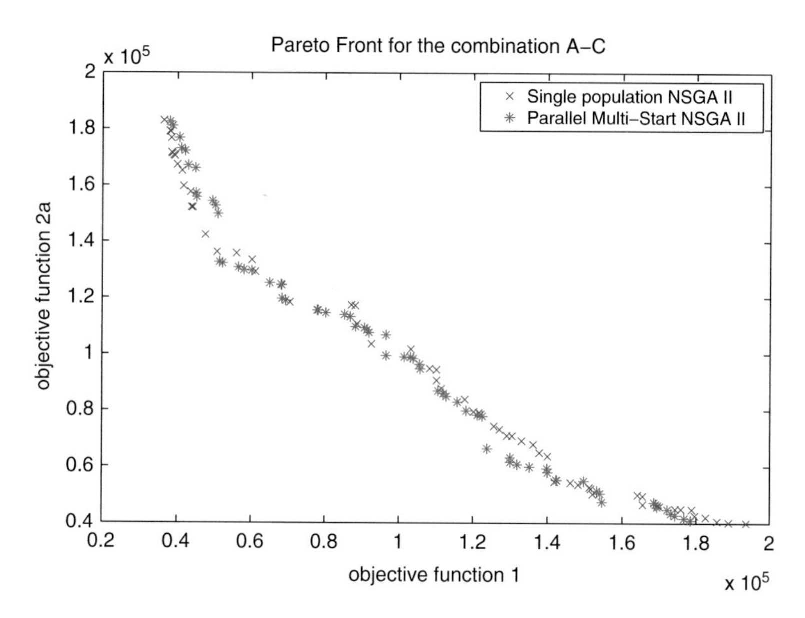

Fig. 1 Pareto front for the instance kroA100par3–kroC100par3 (A–C) for OF1–OF2a

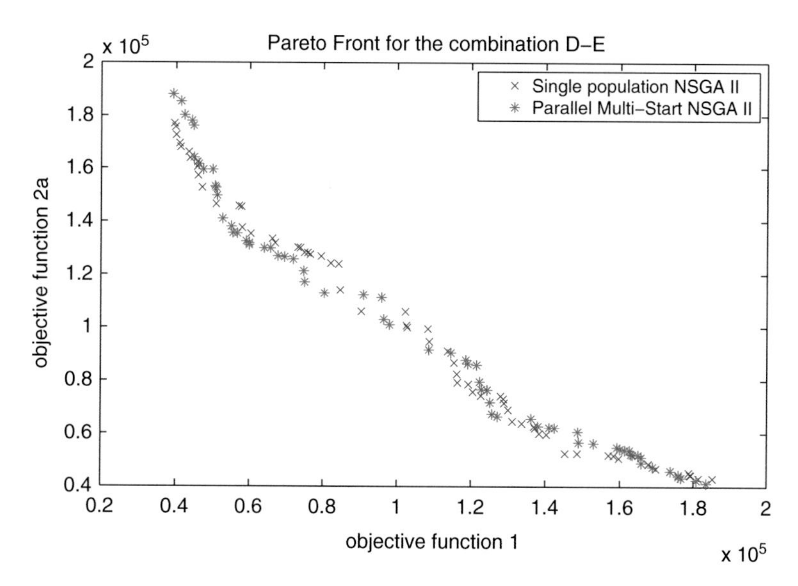

Fig. 2 Pareto front for the instance kroD100par3–kroE100par3 (D–E) for OF1–OF2a

instances of Table 3 are given while in Figs. 3 and 4 the two Pareto fronts are from the instances corresponding to Table 4.

In general, it is preferred to find as many as possible non-dominated solutions, the expansion of the Pareto front to be as large as possible which shows that better

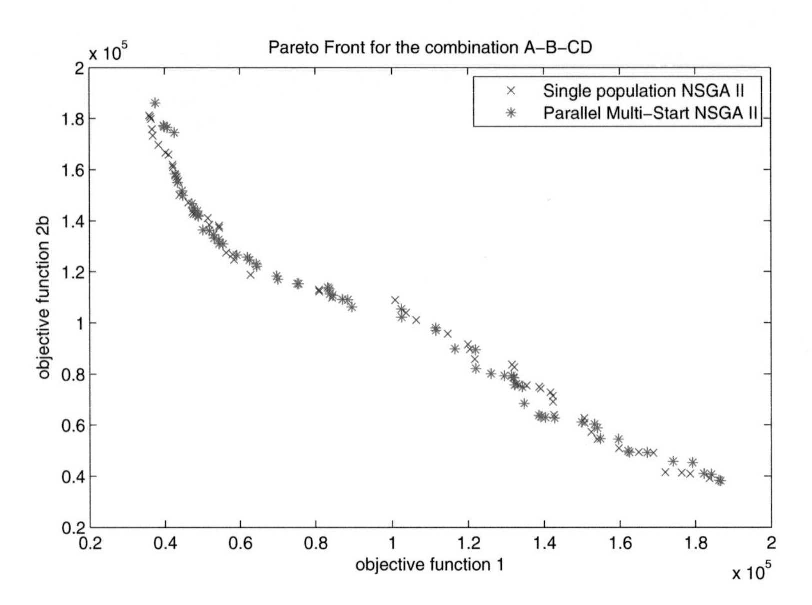

Fig. 3 Pareto front for the instance kroA100par3–kroB100par3–kroC100par3–kroD100par3 (A–B–CD) for OF1–OF2b

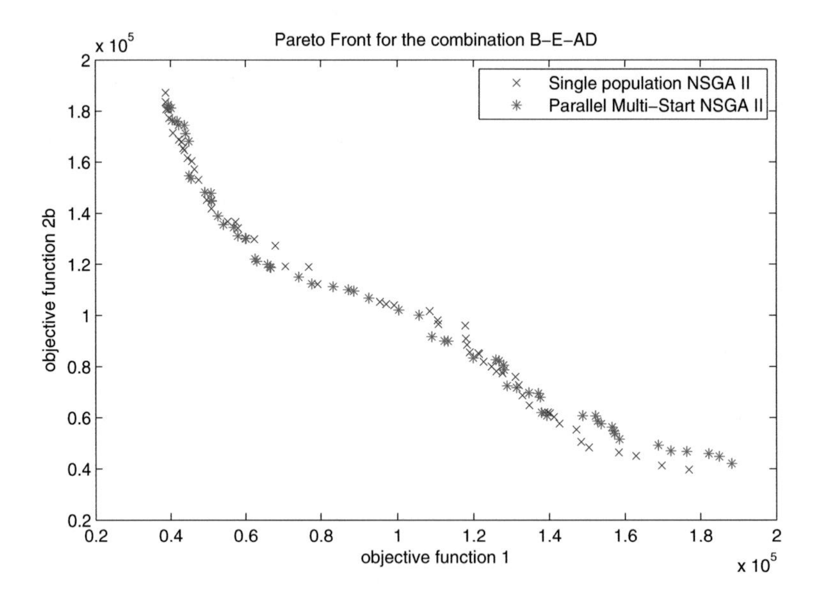

Fig. 4 Pareto front for the instance kroB100par3–kroE100par3–kroA100par3–kroD100par3 (B–E–AD) for OF1–OF2b

solutions have been found in every dimension, the spacing of solutions to be as smaller as possible which means that the non-dominated solutions are close between them, and the Coverage measure to be as large as possible. For the calculation of the Coverage measure $C(A_1, B_1)$, A_1 corresponds to the Pareto solutions of the Single population NSGA II and B_1 corresponds to the Pareto solutions of the Parallel Multi-Start NSGA II.

In Table 3, the results of the two methods in a two objective problem using OF1–OF2a objective functions are presented. Regarding the instances, the proposed algorithm (**Parallel Multi-Start NSGA II**) performs better than the other in all the measures except the Δ measure that performs equal with the other method. According to the number of Pareto solutions L, the proposed algorithm performs better than the other at six instances. Taking into account the M_k measure and the C measure the proposed algorithm performs better at six and eight instances, respectively.

In Table 4, the results of the two methods in a two objective functions problem using OF1–OF2b objective functions are presented. In the ten instances, the proposed algorithm (**Parallel Multi-Start NSGA II**) performs better than the other for two measures. For the number of Pareto solutions measure, L, the two algorithms have similar performance. Taking into account the M_k measure and the C measure the proposed algorithm performs better at six and seven instances, respectively. Considering the Δ measure, the Single population NSGA II algorithm performs better than the proposed at eight instances.

In total, according to the results of all the previous tables the proposed method, **Parallel Multi-Start NSGA II**, performs better at the three tested measures. More specifically, the proposed method performs better at the 55 % of the instances for the L and for the Δ measures, it performs better at 35 % of the instances. Also, it performs better at the 60 and 75 % of the combinations for the M_k and C measures, respectively.

6 Conclusions and Future Research

In this paper, the Multiobjective Fuel Consumption Vehicle Routing Problem is formulated and solved using a variant of NSGA II algorithm, the Parallel Multi-Start NSGA II. In this work, our main target was to minimize the fuel consumption that a vehicle consumes during the routes in addition with the minimization of the total travel time. In the formulation the real route conditions were added and, thus, the problem was formulated as an asymmetric VRP. A variant of the NSGA II algorithm was proposed for the solution of the problem and it was compared with a simpler version of NSGA II. The main difference of the two algorithms is the Parallel Multi-Start strategy that was proposed in order to have an efficient initial population for the NSGA II algorithm and, thus, to converge faster in better Pareto fronts. A number of instances were created for the comparisons of the two algorithms and the results produced by the proposed algorithm proved the effectiveness of the algorithm. Our future research will be focused on the application of the whole procedure in more difficult Multiobjective Routing Problems and the incorporation of the Parallel Multi-Start in other Evolutionary Multiobjective Optimization algorithms in order to see its effectiveness in different from NSGA II algorithms.

References

1. Bandeira, J.M., Fontes, T., Pereira, S.R., Fernandes, P., Khattak, A., Coelho, M.C.: Assessing the importance of vehicle type for the implementation of eco-routing systems. Transp. Res. Procedia **3**, 800–809 (2014)
2. Bektas, T., Laporte, G.: The pollution-routing problem. Transp. Res. B **45**, 1232–1250 (2011)
3. Chand, P., Mohanty, J.R.: Multi objective genetic approach for solving vehicle routing problem. Int. J. Comput. Theory Eng. **5**(6), 846–849 (2013)
4. Christofides, N., Mingozzi, A., Toth, P.: The vehicle routing problem. In: Christofides, N., Mingozzi, A., Toth, P., Sandi, C. (eds.) Combinatorial Optimization. Wiley, Chichester (1979)
5. Cicero-Fernandez, P., Long, J.R., Winer, A.M.: Effects of grades and other loads on-road emissions of hydrocarbons and carbon monoxide. J. Air Waste Manag. Assoc. **47**, 898–904 (1997)
6. Coello Coello, C.A., Van Veldhuizen, D.A., Lamont, G.B.: Evolutionary Algorithms for Solving Multi-Objective Problems. Springer, Berlin (2007)
7. Cullen, S.: Trees and wind: wind scales and speeds. J. Arboriculture **28**(5), 237–242 (2002)
8. Deb, K., Agrawal, S., Pratab, A., Meyarivan, T.: A fast elitist non-dominated sorting genetic algorithm for multi-objective optimization: NSGA-II. In: Schoenauer, M., Deb, K., Rudolph, G., Yao, X., Lutton, E., Merelo, J.J., Schwefel, H.-P. (eds.) Proceedings of the Parallel Problem Solving from Nature VI Conference. Lecture Notes in Computer Science, vol. 1917, pp. 849–858. Springer, Berlin (2000)
9. Deb, K., Pratap, A., Agarwal, S., Meyarivan, T.: A fast and elitist multiobjective genetic algorithm: NSGA-II. IEEE Trans. Evolut. Comput. **6**(2), 182–197 (2002)
10. Demir, E., Bektas, T., Laport, G.: The bi-objective pollution-routing problem. Eur. J. Oper. Res. **232**, 464–478 (2014)
11. Erdogan, S., Miller-Hooks, E.: A green vehicle routing problem. Transp. Res. E **48**, 100–114 (2012)
12. Feo, T.A., Resende, M.G.C.: Greedy randomized adaptive search procedure. J. Glob. Optim. **6**, 109–133 (1995)
13. Figliozzi, M.: Vehicle routing problem for emissions minimization. Transp. Res. Rec. J. Transp. Res. Board **2**, 1–7 (2011)
14. Gutin, G., Punnen, A.P.: The Traveling Salesman Problem and Its Variations. Kluwer Academic, Dordrecht (2002)
15. Hansen, P., Mladenović, N.: Variable neighborhood search: principles and applications. Eur. J. Oper. Res. **130**, 449–467 (2001)
16. Huayu, X., Wenhui, F., Tian, W., Lijun, Y.: An Or-opt NSGA-II algorithm for multi-objective vehicle routing problem with time windows. In: 4th IEEE Conference on Automation Science and Engineering, pp. 309–314 (2008)
17. Jemai, J., Zekri, M., Mellouli, K.: An NSGA-II algorithm for the green vehicle routing problem. In: Evolutionary Computation in Combinatorial Optimization. Lecture Notes in Computer Science, vol. 7245, pp. 37–48. Springer, Berlin (2012)
18. Johnson, D.S., Papadimitriou, C.H.: Computational complexity. In: Lawer, E.L., Lenstra, J.K., Rinnoy Kan, A.H.D., Shmoys, D.B. (eds.) The Traveling Salesman Problem: A Guided Tour of Combinatorial Optimization, pp. 37–85. Wiley, New York (1985)
19. Jozefowiez, N., Semet, F., Talbi, E.G.: Enhancements of NSGA II and its application to the vehicle routing problem with route balancing. In: Artificial Evolution. Lecture Notes in Computer Science, vol. 3871, pp. 131–142. Springer, Berlin (2006)

20. Jozefowiez, N., Semet, F., Talbi, E.G.: Multi-objective vehicle routing problems. Eur. J. Oper. Res. **189**, 293–309 (2008)
21. Kara, I., Kara, B.Y., Yetis, M.K.: Energy minimizing vehicle routing problem. In: COCOA 2007, pp. 62–71. Springer, Heidelberg (2007)
22. Koc, C., Bektas, T., Jabali, O., Laporte, G.: The fleet size and mix pollution-routing problem. Transp. Res. B **70**, 239–254 (2014)
23. Kontovas, C.A.: The green ship routing and scheduling problem (GSRSP): a conceptual approach. Transp. Res. D **31**, 61–69 (2014)
24. Kuo, Y.: Using simulated annealing to minimize fuel consumption for the time-dependent vehicle routing problem. Comput. Ind. Eng. **59**(1), 157–165 (2010)
25. Labadie, N., Prodhon, C.: A Survey on multi-criteria analysis in logistics: focus on vehicle routing problems. In: Applications of Multi-Criteria and Game Theory Approaches Springer Series in Advanced Manufacturing, pp. 3–29. Springer, London (2014)
26. Lahyani, R., Khemakhem, M., Semet, F.: Rich vehicle routing problems: from a taxonomy to a definition. Eur. J. Oper. Res. **241**, 1–14 (2015)
27. Laporte, G.: The vehicle routing problem: an overview of exact and approximate algorithms. Eur. J. Oper. Res. **59**, 345–358 (1992)
28. Lawer, E.L., Lenstra, J.K., Rinnoy Kan, A.H.G.R., Shmoys, D.B.: The Traveling Salesman Problem: A Guided Tour of Combinatorial Optimization. Wiley, New York (1985)
29. Leonardi, J., Baumgartner, M.: CO_2 efficiency in road freight transportation: status quo, measures and potential. Transp. Res. D **9**, 451–464 (2004)
30. Li, H., Lv, T., Li, Y.: The tractor and semitrailer routing problem with many-to-many demand considering carbon dioxide emissions. Transp. Res. D **34**, 68–82 (2015)
31. Lin, S.: Computer solutions of the traveling salesman problem. Bell Syst. Tech. J. **44**, 2245–2269 (1965)
32. Lin, C., Choy, K.L., Ho, G.T.S, Ng, T.W.: A genetic algorithm-based optimization model for supporting green transportation operations. Expert Syst. Appl. **41**, 3284–3296 (2014)
33. Lin, C., Choy, K.L., Ho, G.T.S, Chung, S.H., Lam, H.Y.: Survey of green vehicle routing problem: past and future trends. Expert Syst. Appl. **41**, 1118–1138 (2014)
34. McKinnon, A.: A logistical perspective on the fuel efficiency of road freight transport. In: OECD, ECMT and IEA: Workshop Proceedings, Paris (1999)
35. McKinnon, A.: Green logistics: the carbon agenda. Electron. Sci. J. Logist. **6**(3), 1–9 (2010)
36. Okabe, T., Jin, Y., Sendhoff, B.: A Critical survey of performance indices for multi-objective optimisation. Evolut. Comput. **2**, 878–885 (2003)
37. Ombuki, B., Ross, B.J., Hanshar, F.: Multi-objective genetic algorithms for vehicle routing problem with time windows. Appl. Intell. **24**, 17–30 (2006)
38. Psychas, I.D., Marinaki, M., Marinakis, Y.: A parallel multi-start NSGA II algorithm for multiobjective energy reduction vehicle routing problem. In: Gaspar-Cunha, A. et al. (eds.) Eighth International Conference on Evolutionary Multicriterion Optimization, EMO 2015, Part I. Lecture Notes in Computer Science, vol. 9018, pp. 336–350. Springer, Cham (2015)
39. Sarker, R., Coello Coello, C.A.: Assessment methodologies for multiobjective evolutionary algorithms. In: Evolutionary Optimization. International Series in Operations Research and Management Science, vol. 48, pp. 177–195. Springer, New York (2002)
40. Sbihi, A., Eglese, R.W.: Combinatorial optimization and green logistics. 4OR **5**(2), 99–116 (2007)
41. Suzuki, Y.: A new truck-routing approach for reducing fuel consumption and pollutants emission. Transp. Res. D **16**, 73–77 (2011)
42. Tajik, N., Tavakkoli-Moghaddam, R., Vahdani, B., Meysam Mousavi, S.: A robust optimization approach for pollution routing problem with pickup and delivery under uncertainty. J. Manuf. Syst. **33**, 277–286 (2014)
43. Tiwari, A., Chang, P.C.: A block recombination approach to solve green vehicle routing problem. Int. J. Prod. Econ. 164, 379–387 (2015)
44. Toth, P., Vigo, D.: The Vehicle Routing Problem. Monographs on Discrete Mathematics and Applications. SIAM, Philadelphia (2002)

45. Toth, P., Vigo, D.: Vehicle Routing: Problems, Methods and Applications, 2nd edn. MOS-Siam Series on Optimization. SIAM, Philadelphia (2014)
46. Xiao, Y., Zhao, Q., Kaku, I., Xu, Y.: Development of a fuel consumption optimization model for the capacitated vehicle routing problem. Comput. Oper. Res. **39**(7), 1419–1431 (2012)
47. Zhang, S., Lee, C.K.M., Choy, K.L., Ho, W., Ip, W.H.: Design and development of a hybrid artificial bee colony algorithm for the environmental vehicle routing problem. Transp. Res. D **31**, 85–99 (2014)
48. Zitzler, E., Deb, K., Thiele, L.: Comparison of multiobjective evolutionary algorithms: empirical results. Evolut. Comput. **8**(2), 173–195 (2000)

Manifold Location Routing Problem with Applications in Network Theory

Emre Tokgoz and Theodore B. Trafalis

Abstract A new location routing problem (LRP) named manifold location routing problem (MLRP) and the corresponding solution technique were introduced to the scientific and engineering communities for a single facility (1-MLRP) in Tokgöz et al. (Computational Management Science. Springer, Berlin, 2014). MLRP is an LRP when the surface is assumed to be a Riemannian manifold surface (RMS). An example of an RMS is the surface of the Earth, where the roads on this surface are geodesics. The shortest path and the distances between two locations on RMS can be determined by using the shortest path geodesic distances. In this work, the formulation of the 1-MLRP, the underlying theory used for introducing 1-MLRP, and a heuristic algorithm to solve 1-MLRP are explained. In addition, the generalization of 1-MLRP introduced in Tokgöz et al. (Computational Management Science. Springer, Berlin, 2014) from a single-facility to the two-facility case is implemented by introducing the 2-facility MLRP (i.e., 2-MLRP). The implementation of a method called linked chain method employed in Tokgöz et al. (Computational Management Science. Springer, Berlin, 2014) to determine the vehicle routes from the potential locations of the facilities is also explained. Riemannian geometry and the meaning of geodesics on Riemannian manifolds are also discussed briefly. In addition the implementation of 2-MLRP in networks is explained in the last section.

Keywords Manifold location routing problem • Location routing problem • Riemannian manifold • Facility location problem • Vehicle routing problem • Geodesics • Network

---

E. Tokgoz (✉)
Department of Industrial Engineering, Quinnipiac University, Hamden, CT, USA
e-mail: Emre.Tokgoz@quinnipiac.edu

T.B. Trafalis
Department of Industrial and System Engineering, University of Oklahoma, Norman, OK, USA
e-mail: ttrafalis@ou.edu

V.A. Kalyagin et al. (eds.), *Models, Algorithms and Technologies for Network Analysis*, Springer Proceedings in Mathematics & Statistics 156,
DOI 10.1007/978-3-319-29608-1_6

1 Introduction

In supply chain networks, determining locations of facilities and the item distribution to customers from these facilities have an important role in determining cost effective operations [31]. Determination of the facility locations is a well-known problem in the literature, named facility location problem (FLP) [8]. The distribution of items via vehicles is known as the vehicle routing problem (VRP) [15]. Solving VR and FL problems simultaneously can yield a robust facility location and reduced-cost distribution within the customer–supplier network [24]. The combination of FLP and VRP formulates a new problem, the so-called location routing problem (LRP) [19]. The literature on the LRP expands to cover determination of facility locations in both discrete (see, for example, [3, 5, 8, 20, 28, 32]) and continuous (see, for example, [16, 17, 26, 27, 29]) spaces with capacitated (see, for example, [4, 21]) and incapacitated (see, for example, [7]) facility assumptions. Customers are assumed to order full-truckload in location analysis that is not always practical in reality. In a case when a customer sits on more than one route within the network, it is natural to obtain a sub-optimal solution [23]. Applications of LRP in supply chain research include but not limited to grocery store location and distribution [3, 28], and general goods distribution [20].

One of the aims of the continuous LRP formulated for a single facility in the planar setting is to minimize the total Euclidean distance between the facility and customers that is known as the Weber problem [30]. In the case of multiple facilities, the corresponding problem is known to be the multi-facility Weber problem [7]. Weber problem (WP) is an NP-hard problem in continuous space [29]. One way of solving WP is by employing the iterative procedure of Weiszfeld [33]. An obstacle of continuous WP solution is the feasibility of the location determined for the facility. For instance, the solution of a WP can suggest to locate a grocery location on a lake within the network domain. The solutions obtained for WP can be either used as the bounds for the feasible solution or for approximating the discrete location solution (see, for example, [11]).

A limited number of heuristic solutions are obtained for the planar LRP using neural networks. In [26, 27], planar single-facility LRP is studied by using two similar approaches assuming inter-connected neuron ring connection between the customers and the facility. An iterative heuristic solution for the multi-depot planar LRP is developed in [25] assuming capacitated vehicles and bounded tour lengths for the vehicles. A hierarchical heuristic-based method is employed in [16] that employs continuous input from the routing results while systematically improving the facility location using the heuristic method.

A problem named manifold location routing problem (MLRP) and the corresponding solution technique are introduced to the scientific and engineering communities for a single facility (1-MLRP) in [31]. MLRP is an LRP on Riemannian manifold surface (RMS). MLRP is a generalization of a set of existing real life LRP problems: The surface of Earth is an RMS where the roads on this surface are geodesics. The shortest path and the distances between two locations on RMS can be determined by using the shortest path geodesic distances. Noting that the known techniques to solve the LRP in the literature are only obtained for the Euclidean space by using planar and spherical surfaces and distances (see, for example, [2, 10]), MLRP improves the existing theory on LRP and therefore the corresponding results in applications [31]. In the case where the surface is the Euclidean plane, the curvature of the RMS is 0 and the MLRP corresponds to LRP on the Euclidean space. When the manifold surface is a sphere's surface, the curvature of the RMS is 1 and this MLRP corresponds to LRP on a sphere. A heuristic algorithm is introduced to solve the proposed LRP on RMS with computational results displayed for a particular scenario in [31]. The numerical results obtained for the MLRP introduced in [31] are incomparable with the ones known in the literature for the traditional LRP because of the change in the surface and distance assumptions. In this work, the formulation of 1-MLRP, the underlying theory used for introducing 1-MLRP, and the heuristic algorithm introduced in [31] to solve 1-MLRP will be explained.

In addition, the generalization of 1-MLRP introduced in [31] from a single-facility to two-facility case will be implemented to introduce the 2-facility MLRP (i.e., 2-MLRP). The implementation of a method called linked chain method (LCM) employed by Tokgöz et al. [31] to determine the vehicle routes from the potential locations of the facilities will be also explained. Riemannian geometry and the meaning of geodesics on Riemannian manifolds will be covered briefly. 2-MLRP formulation with the corresponding algorithmic solution will be implemented. The implementation of 2-MLRP in the network theory will be explained in the last section.

2 Riemannian Manifolds and Geodesic Distances

The geometry of topological manifolds was studied by Bernhard Riemann in the nineteenth century [22]. In Riemannian geometry, the metric properties vary from point to point (see, for example, [9]). A Riemannian manifold (RM) is a differentiable manifold M in which each tangent space is equipped with an inner product, h, a Riemannian metric, which varies smoothly from point to point. Well-known examples of Riemannian manifolds include the n-dimensional sphere S^n, and n-dimensional Euclidean space $\mathbb{R}^n$, for all $n \geq 1$. The curvature of a RMS indicates the flatness of the local region of the manifold surface. The curvature of $\mathbb{R}^n$ is zero in the entire space due to the flatness. The curvature of S^n is 1 on the entire manifold surface. If the curvature of a surface changes from flat to spherical surface, then the curvature changes from 0 to 1 in the local neighborhood. Every Riemannian manifold can be considered as a local Euclidean space. This consideration is possible by using a one-to-one and onto map $\psi : U \subset M \to \mathbb{R}^n$ called homeomorphism. A homeomorphism is a mapping of each open neighborhood U in the topology of the manifold M to an open neighborhood $\psi(U)$ in the Euclidean space (see, for example, [9, 14].)

A connected Riemannian manifold carries the structure of a metric space whose distance function is the arc-length of a minimizing geodesic. Let M be a connected Riemannian manifold and $\gamma : [a, b] \rightarrow M$ be a parameterized differentiable curve in M with the velocity vector γ'. A path minimizing geodesic is the shortest path curve between two points on RMS. For instance, a geodesic on the surface of a sphere between the two poles is a great circle. The length of a geodesic γ on an RMS between customers c_i and c_j $(i \neq j)$ can be calculated by using the geodesic distance formula

$$L(\gamma_{ij}) = \int_M \|\nabla \gamma_{ij}\| \, d\mu$$

where ∇ is the gradient operator and μ is the standard measure on M. In this formula every geodesic distance depends on the local curvature of M. For more details on RMS and geodesics see [9].

In this work, the shortest path geodesic distances on RMS are calculated to determine the shortest path distances between the customers and the facilities. The geodesics correspond to the curves determined between the facilities and customers on the RMS. The total cost of visiting all the customers is

$$L(\gamma) = \sum_{i,j} \int_{M_{ij}} \|\gamma'_{ij}\|_{M_{ij}} \quad \text{for all } i \neq j \tag{1}$$

where M_{ij} represents the local region of M between customers x_i and x_j. The open sets $B_{ij} \subset \mathbb{R}^2$ correspond to $M_{ij} \subset M$ for all i and j noting that M is locally equivalent to Euclidean space. The facilities are located on a three- dimensional RM therefore the tangent space for each local neighborhood on the manifold is $\mathbb{R}^2$. Geodesics on each B_{ij} are obtained by projecting the local portion of the geodesic from RMS to B_{ij} in $\mathbb{R}^2$. Therefore the total distance traveled to visit all the customers is

$$L(\gamma) = \sum_{i,j} \int_{B_{ij}} \|h'_{ij}\|_{\mathbb{R}^2} \tag{2}$$

where h_{ij} is the geodesic in $\mathbb{R}^2$ corresponding to the geodesic γ_{ij} on M and B_{ij} is the region in $\mathbb{R}^2$ corresponding to M_{ij} on M. One possible norm to be employed is the standard norm in $\mathbb{R}^2$:

$$L(\gamma) = \sum_{i,j} \int_{t=a_i}^{a_j} \sqrt{1 + \left(\frac{dh_{ij}}{dt}\right)^2} \, dt \tag{3}$$

In the case when the geodesic intersects with a domain constraint on the *RMS*, the distance is calculated by considering the path on the boundary of the domain constraint.

The distances between customers and the single facility to be allocated will be measured on compact connected Riemannian manifold surfaces (CCRMS); a generalization of planar surfaces. The shortest path geodesic distances are the shortest distances between the customers and a possible location of the facility to be allocated. Manifold Weber problem (MWP), restatement of the Weber problem for manifold settings, is introduced in [31] to determine the shortest path geodesics between nodes. The distance calculations used for MWP is a generalization of the planar distances employed for the WP [30]. The distance calculation method used for MLRP is a generalization of the planar distances employed by other researchers to solve the existing facility allocation problems. For instance, consider a road path that starts changing from flat surface to a mountain surface; this continuous change in the surface results in changing the curvatures from zero (corresponding to the flat surface) to a positive curvature at each local neighborhood throughout the surface towards the mountain. In this case, the geodesics are the roads that connect certain intersections in the local regions. The shortest path geodesic between two locations on a manifold's surface is determined by determining all the geodesics between them.

Throughout this work the manifold domain M is assumed to be a compact connected RMS with non-negative curvature. This is a realistic assumption noting that there can be lakes, mountains, rivers, etc. that we call "obstacles" in the domain causing infeasible facility allocation. We assume only vehicles to be used for transportation and there exists at least one path (e.g., road) between the customers and facilities. Let V_i be a family of finitely many open sets representing the obstacles in M where facilities cannot be allocated geographically in the interior of these sets. V_i are assumed to be open noting that the facilities can be allocated on the boundaries of V_i. Therefore the customer and facility allocation domain on the RMS is

$$D = M - \bigcup_{i \in I} V_i\,, \qquad I \text{ is a finite index set}$$

Figure 1 is an example of a domain with obstacles represented by squares painted in blue.

Figure 2 displays the projected locations of the customers from RMS to $2D$ surface by using orthogonal projection which is a homeomorphism. Figure 3 displays the customer locations displayed in Fig. 2 in $3D$ without the RMS where red and blue diamonds represent the classified customers. The customer classification is due to assuming two facilities to serve two different sets of customers.

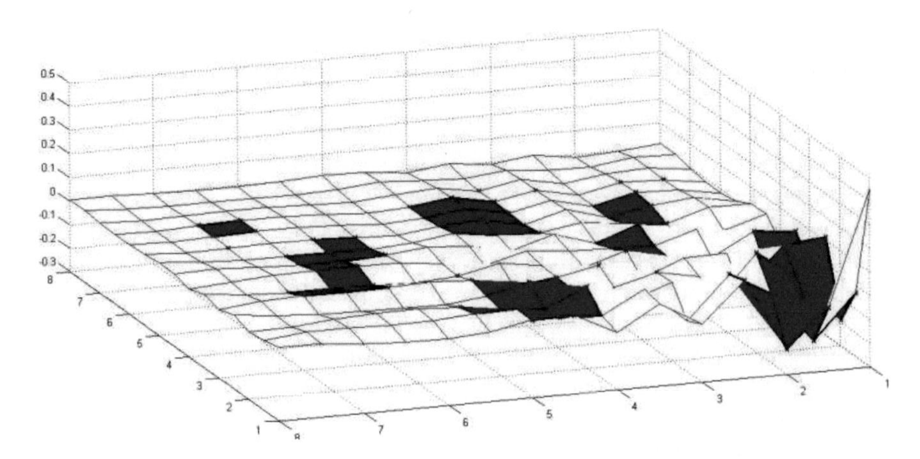

Fig. 1

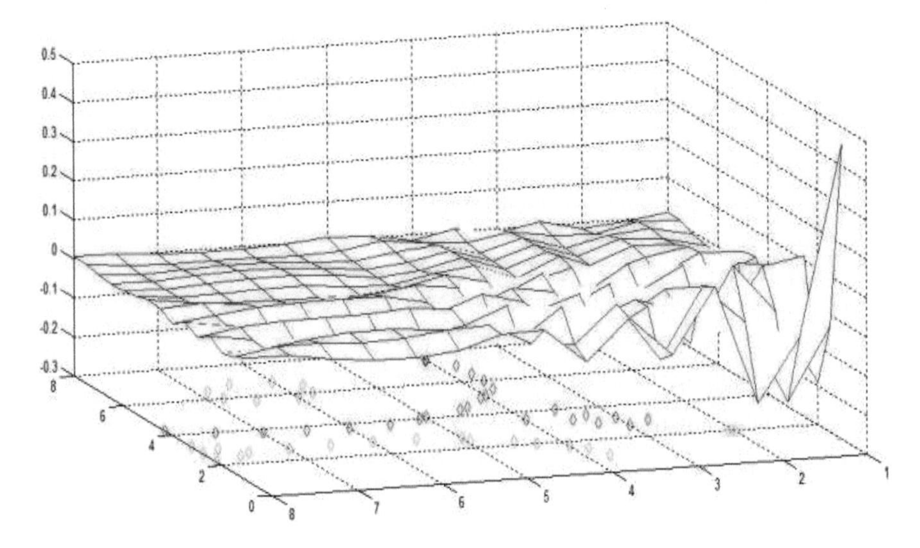

Fig. 2

3 MLRP Algorithmic Solution: Single-Facility Case

Tokgöz et al. [31] assumed

- Customers and the facility are located on a CCRMS;
- Distances between the customers and possible locations of the facility are calculated by using geodesic distances;
- Customers have known demands and locations;
- Vehicles are capacitated and homogeneous;
- Facility to be allocated is incapacitated;

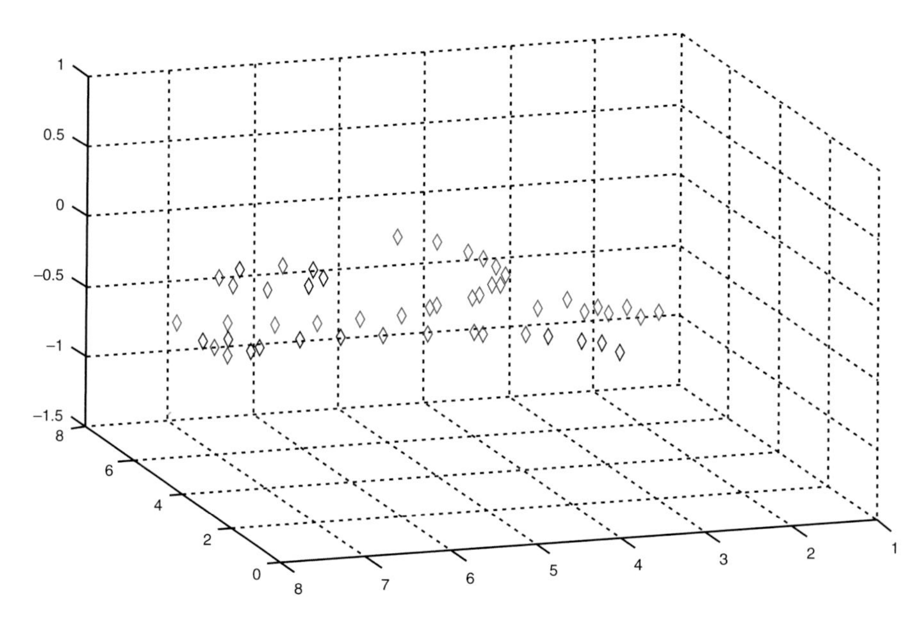

Fig. 3

- All the vehicles have the same capacity;
- The number of vehicles to be operated does not exceed the upper bound of the number of vehicles;
- The number of vehicles to be used will be derived as a by-product of the solution;
- There is no fixed cost for vehicles; and
- Each vehicle route starts and finishes at the facility to be allocated.

The following notations are used

- M : CCRMS corresponding to the local region on Earth's surface;
- C : Set of customers c_i, $i \in I = \{1, 2, \ldots, m\}$;
- φ : Homeomorphism defined for projection from m to $\mathbb{R}^2$;
- a_k : Customer locations on M with the coordinates $c_k = (\varphi(x_k), \varphi(y_k))$ on $\mathbb{R}^2$;
- a_0 : Facility location on M with the Euclidean coordinate $c_0 = (\varphi(x_0), \varphi(y_0))$ on $\mathbb{R}^2$;
- E : Combined set of customers and the facility with the facility indexed to be 0;
- γ_{ij} : Parametric geodesic on M connecting customers' a_i and a_j;
- V : Set of vehicles $v = 1, 2, \ldots, n$ with $|V| \leq n$;
- v^i_{max} : The maximum capacity of vehicle v_i; and
- D : The demand set with $d_i \in D$ corresponding to the demand of customer $c_i \in C, i \in I$;

$$\delta_{ijk} = \begin{cases} 1 \text{ if } i \text{ and } j \text{ are connected via route } k \\ 0 \qquad\qquad\qquad \textit{otherwise} \end{cases}$$

to introduce the single objective function of the 1-MLRP

$$\min z = \sum_{i,j \in E} \sum_{v \in V} \int_M \left\| \gamma'_{ij}(t) \right\| dt.\delta_{ijv} \tag{4}$$

that is designed to minimize the total transportation cost on the manifold setting M subject to the constraints

$$\sum_{j \in E} \left(\delta_{ijv} - \delta_{jiv} \right) = 0, \ \forall i \in E; \, v \in V$$

controlling the traffic flow for visiting each customer by the same vehicle;

$$\sum_{i \in I} \sum_{j \in C} \delta_{ijv} d_i \leq v^i_{\max}, \ \forall v \in V \tag{5}$$

the maximum capacity of the vehicles is violated during the transportation;

$$\sum_{i \in U} \sum_{v \in V} \sum_{j \in E-U} \delta_{ijv} \geq 1, \ \forall U \subset C \tag{6}$$

declines the existences of subtours ensuring that there exists at least one vehicle leaving any subset of customers;

$$\sum_{i \in E} \sum_{v \in V} \delta_{ijv} = 1, \ \forall j \in C \tag{7}$$

indicating that each customer belongs to one and only one tour;

$$\sum_{j \in E} \delta_{0jv} \leq 1, \ \forall v \in V \tag{8}$$

ensuring each vehicle leaving the facility either once or never used for transportation noting that not all the vehicles are necessarily used for transportation;

$$\delta_{ijv} \in \{0, 1\}, \ \forall i, j \in E; \ v \in V$$

are the decision variables conditions. This proposed 1-MLRP is a mixed integer non-linear programming problem (MINLP) with 3 continuous variables and $n(m+1)^2$ discrete variables. Two of the continuous variables are the locations of the customers in the Euclidean space and the third continuous variable is the parameter t used for the geodesics on the manifold M. A heuristic algorithm is developed to solve 1-MLRP in [31] under certain assumptions. The proposed algorithm required projecting customer locations and the geodesics between customers from M to

$\mathbb{R}^2$ as well as projecting the facility locations from the manifold surface $\mathbb{R}^2$ to M by using a homeomorphism. The geodesic distances between customers are calculated on the manifold surface by using the metric defined on M. These geodesic distances include all possible roads between customers within the compact connected domain on the surface of M. These lengths correspond to the edge lengths of the road network formed by the customers in the Euclidean space. Therefore the metric choice on M determines the metric to be used in the Euclidean space. A homeomorphism can be employed to project the customer locations and the geodesic distances from M to $\mathbb{R}^2$.

The second step of the solution algorithm of 1-MLRP is designed to solve the LRP in the Euclidean space by employing a heuristic approach [31]. Weiszfeld's (1937) formula [33] is used in [16] for determining the initial location of the facility. Ellipsoids are used for heuristic calculations that also formed the topology of the Euclidean space in [16]. In [31] open balls are assumed to form the topology of $\mathbb{R}^n$ for heuristic calculations. The ease of using open balls is calculating the radius of the circles rather than calculating two different radii for an ellipsoid. The open-ball heuristic method used in [31] is called the LCM: A cost effective facility allocation after chaining z open balls with the center of the ith open ball being the best location of the center of the step $(i-1)^{st}$ open ball, $i = 0, 1, \ldots, z$. The center of each B_i is determined by solving the routing problem within the disk obtained by the interior of B_i for all i. The radius of the circle B_i is determined by calculating the distance between the ith and $(i-1)^{st}$ open balls. The radius of each consecutive circle is modified dynamically. The stopping criterion for adding circles to the LCM is when a sufficiently small distance between the $(z-1)^{st}$ and zth circles is obtained.

In [31], third and last step of the algorithm is defined to project the results obtained in the second step from $\mathbb{R}^2$ to M for determining a feasible location of the facility on M. The cost effective facility location determined on $\mathbb{R}^2$ may or may not be a feasible location on M, therefore a local neighborhood search is necessary on M for the best allocation of the facility in the case when the solution is infeasible on M.

3.1 *Algorithmic Solution for 1-MLRP*

In this section, the steps of the algorithm to solve 1-MLRP, computational complexity of the algorithm, and the heuristic methodology to solve the 1-MLRP introduced in [31] will be explained. This methodology employed to solve the 1-MLRP is particularly important to understand the heuristic algorithm solution introduced for 2-MLRP.

3.2 Computational Complexity

LRP in the Euclidean space is an NP-hard problem noting that both routing and location problems are NP-hard on planar surfaces [29]. Therefore 1-MLRP is also an NP-hard problem noting that the special case of the MLRP is the Euclidean space when the curvature of the RMS is zero. The two main computational challenges in the algorithmic solution are calculating the lengths of the pathways between the customers on the RMS and solving the routing problem in the Euclidean space at every circle generated by the LCM. On one hand the algorithm proposed has the advantage of calculating a radius per circle compared to the algorithm of Manzour-al-Ajdad et al. [16] with computations including two radii for each ellipsoid, on the other hand the computational algorithm proposed for 1-MLRP in [31] includes the challenge of calculating the objective function on the manifold surface M rather than $\mathbb{R}^2$. The time complexity of projection from $\mathbb{R}^2$ to M and determining the best location of the facility on the manifold surface can be a constant, therefore can be negligible in computational complexity calculations in the case when the feasible solution is within a close region on the RMS. It can also be the complexity of the proposed algorithm in the case when the feasible location of the facility is too far from the solution on the *CCRMS* after using the solution algorithm for 1-MLRP.

3.3 Algorithmic Solution Steps to Solve 1-MLRP

All possible pathway lengths between customers can be calculated and the corresponding geodesic functions can be pre-determined by using geographic information systems (GIS). The geodesics determined between customers on the RMS are projected from M to $\mathbb{R}^2$ and these lengths are assigned to be the edge lengths between the customers on the customer network formed in $\mathbb{R}^2$. It is important to note that the metric used on the manifold defines the metric used on the RMS. This is due to the fact that the norm used in [31] for the distance calculations is the norm used on the projected Euclidean surface.

3.4 Facility Locations on R^2

The initial location of the facility to be allocated is determined by using Weiszfeld formula [33] for circular objects in [31]. The corresponding point generation used within each circle of the LCM is given by an algorithm. According to this algorithm, the points chosen in each circle are particularly chosen in different directions for homogenous point distribution. Homogenous distribution of points is necessary since the best location of the facility can be in different directions within each circle. The choice of homeomorphism φ used for the projection from M to $\mathbb{R}^2$ can yield to a single circle with the center of the circle being the location of the facility. In the infeasible case when the customer and facility locations are the same, this non-optimal solution will be recovered by reallocating the facility at a distance from the customer.

The next step of the algorithm is solving the routing problem from the initial location of the facility to determine the possible routes to the customers. These local areas are circles in $\mathbb{R}^2$ corresponding to the local areas on the surface of M since M is a locally Euclidean space.

After determining the best location of the facility within the first circle by solving the routing problems for t_1 points (locations), the next step of the algorithm is application of the LCM by determining the circles $k = 2, 3, \ldots, z$. Consecutive circles are generated by using the radius determined for each circle. At every kth step of the LCM, there are t_k numbers within the disk region formed by the interior of the circle k, $k = 2, \ldots, z$, to solve the corresponding routing problem. Following the vehicle and facility assumptions for the MLRP, the main objective of the routing problem in [31] was to find the minimal length route for each one of the t_k points generated at the kth circle. The route duplication between customers i and j is prevented by using the technique of Altinel and Oncan [1] with computational complexity $O(m \log(m))$. Circle generation used for the LCM continues until a sufficiently small circle with radius (if exists) is obtained. This condition is a result of determining a sufficiently small distance between two locations. Determination of the best possible center of the last circle with this radial condition is the final step of the LCM on the Euclidean surface. The numbers of points chosen within each consecutive circle used for the LCM are forced to decrease by using an algorithm yielding to the convergence of the circle radii for a randomly chosen sufficiently small ϵ [31].

It is important to note that the final location of the facility determined is close to the best feasible solution that does not necessarily reflect the best feasible location of the facility. The projection of the determined solutions from $\mathbb{R}^2$ to M is the last step of the solution to be explained next.

3.5 Projection from $\mathbb{R}^2$ to M

The last step of the algorithm developed in [31] is to map the shortest path distribution routes and the facility location back to the RMS by using the inverse map φ^{-1} of the homeomorphism. Customer location data is pre-existing on M, therefore it is not necessary to map the customer data from the Euclidean surface back to the RM surface. A discrete feasible location close to the infeasible facility location solution is determined on M in the case of an infeasible facility location on M after projection from $\mathbb{R}^2$ to M.

4 Manifold Location Routing Problem: 2-Facility Case

Two-facility case of the MLRP (2-MLRP) can be particularly important for networks in which production and distribution require two facilities to be located. In general terms, one of the two facilities in the network can be employed as the receiver from the suppliers/providers while the other facility is assumed to serve to the customers/clients. In this case we assume the domain of the MLRP to be

$$D = M - \bigcup_{i \in I_1} V_i$$

4.1 *Statement of 2-MLRP*

In this section we introduce the assumptions, notations, and parameters to formulate a 2-MLRP:

Assumptions

1. Customers have known demands and locations;
2. Two types of customers x and y with f_1 serving customers of type x, and f_2 serving customers of type y;
3. Capacitated homogeneous vehicles;
4. Facilities are incapacitated in the items that they are required to distribute to their customers and capacitated in the other items that they are not required to distribute;
5. All vehicles have the same capacity;
6. The number of vehicles to be operated does not exceed the upper bound of the number of vehicles;
7. There is no fixed cost for vehicles;
8. Each vehicle route starts and finishes at the facility that they initiated its route; and
9. Facilities f_1 and f_2 are required to be allocated as close as they can be to each other.

The goal of 2-MLRP is to find the best location of the two facilities with the minimal transportation cost by determining the vehicle routes on RMS.

Notations

- C_1 : Set of customers x
- C_2 : Set of customers y
- $C = C_1 \cup C_2$
- F : Set of facilities $f_i, \ i = 1, 2$
- $\aleph_a$: Demand of customer a
- υ : Maximum capacity of each vehicle
- V_i : Set of vehicles $\upsilon_i = 1, \ldots, n_i$ with ρ_i being the maximum number of vehicles for facility f_i

Parameters

- $x_i = (l_i^1, r_i^1)$ and $y_j = \left(l_j^2, r_j^2\right)$: Customer coordinates
- $f_k = (p_k, q_k)$: Coordinates of the facility f_k, $k = 1, 2$
- $\gamma_{ij}(t)$: The parametric geodesics connecting x_i, $x_j, f_1 \in C_1 \cup \{f_1\}$, $1 \leq i, j \leq n_1$
- $\alpha_{ij}(t)$: The parametric geodesics connecting y_i , $y_j, f_2 \in C_2 \cup \{f_2\}$, $1 \leq i, j \leq n_2$
- $\beta_{ij}(t)$: The parametric geodesics connecting the two facilities f_1 and f_2

$$\mu_{ijk} = \begin{cases} 1 \text{ if node } i \text{ follows node } j \text{ on route } k \\ 0 \text{ otherwise} \end{cases}$$

Using the notation above, the 2-MLRP problem to be solved on D is the following 3-objective mixed integer non-linear programming problem (MINLP):

$$\min \sum_{i,j \in C_1 \cup \{f_1\}} \sum_{k \in V_1} \int_D \left\| \gamma'_{ij}(t) \right\| dt . \mu_{ijk} \tag{9}$$

$$\min \sum_{i,j \in C_2 \cup \{f_2\}} \sum_{k \in V_2} \int_D \left\| \alpha'_{ij}(t) \right\| dt . \mu_{ijk} \tag{10}$$

$$\min \int_D \left\| \beta'_{ij}(t) \right\| dt \tag{11}$$

subject to

$$\sum_{i \in C \cup F} \sum_{k \in V_s} \mu_{ijk} = 1 \quad \forall j \in C \cup F \text{ and } s = 1, 2 \tag{12}$$

$$\sum_{j \in C} \aleph_j \sum_{i \in C \cup F} \mu_{ijk} \leq \upsilon \quad \forall k \in V_s \text{ and } s = 1, 2 \tag{13}$$

$$\sum_{i \in C \cup F} \mu_{ijk} - \sum_{i \in C \cup F} \mu_{jik} = 0 \quad \forall j, k \tag{14}$$

$$\sum_{j \in C \cup \{f_s\}} \mu_{sjk} \leq 1 \quad \forall k \text{ and } s = 1, 2 \tag{15}$$

$$\sum_{k \in C} \sum_{i \in S} \sum_{j \in (C \cup F - S)} \mu_{ijk} \geq 1 \quad \forall S \subseteq C \text{ and } \{f_t\} \subset F \tag{16}$$

$$\mu_{ijk} \in \{0, 1\} \text{ for } \forall i, j \in C \cup \{f_t\}, k \in V_t \tag{17}$$

$$(m_i, z_i) \;\&\; \left(l_j, r_j\right) \in D_1 \tag{18}$$

The objective functions (9) and (10) are designed for minimizing the total transportation costs for facilities f_1 and f_2, respectively. Objective function (11) determines the minimum distance between the two facilities to be allocated. Constraints (12) indicate that every customer belongs to exactly one tour for each facility. Constraints (13) are designed to prohibit agent capacity violation for each facility. Constraints (14) represent the flow conservation ensuring that every node is entered and left by the same agent for each customer network. Constraints (15) represent that any given agent should leave the corresponding facility depot at most once. Constraints (16) show that for any subset of customers in each network there is at least one agent leaving the subset guaranteeing that subtours cannot exist. Constraints (17) and (18) are related to the nature of the decision variables.

5 Solution Methodology for 2-MLRP

Sequential, iterative, and hierarchical methods have been employed to solve LRP in $\mathbb{R}^2$ (see, for example, [16, 25, 33]). The hierarchical method has the advantage of determining a solution to the LRP by finding a solution to the vehicle routing and FLPs simultaneously in comparison to the sequential and iterative methods. Ellipses are used for heuristics yielding to computationally effective results compared to the methods known in the literature [16]. Hierarchical solution approach follows determining the location of the facility within a region and then solving the VRP. The sequential and iterative methods are compared in [23] and concluded that the iterative method can yield more accurate solutions. In this section the main steps of the heuristic algorithm used for solving the formulated 2-MLRP will be explained.

5.1 Main Steps of the Heuristic Algorithm

The first step of the heuristic algorithm we employed to solve the 2-MLRP is projection of the customer locations and geodesic distances from M to $\mathbb{R}^2$ by using a homeomorphism. At this step, the geodesic distances between the customers are calculated on the manifold surface by using the metric defined on M. These geodesic distances include all possible roads between customers on the *CCRMS*. These lengths correspond to the edge lengths of the road network formed by the customers in the Euclidean space, therefore the metric choice of M determines the metric to be used for the Euclidean space. The homeomorphism used in the first step is also used to project the customer locations and the geodesic distances from M to $\mathbb{R}^2$ after solving the MWP for each customer sets C_1 and C_2.

The third step of the algorithm is to initiate the locations of the two facilities f_1 and f_2 by using a formula similar to the one used in [31] for 1-MLRP.

The fourth step of the algorithm is designed to solve the LRP in the Euclidean space by employing a heuristic approach. We employ open balls B_i as the topology for the heuristic calculations and use the LCM as explained in Sect. 3.4.

Fifth and last step of the algorithm is projecting the results obtained in the fourth step of the algorithm from $\mathbb{R}^2$ to M and determining feasible locations for the facilities on M. The cost effective facility location determined on $\mathbb{R}^2$ may or may not be feasible on M, therefore a local neighborhood search is necessary on M for the best allocation of the facility in the case when the solution is infeasible on M. The main steps of the algorithm are summarized in the following Section and the details of it will be explained in Sect. 5.

5.2 *Heuristic Algorithm for the Proposed 2-MLRP*

The following algorithm with a heuristic solution approach will be used to solve the 2-MLRP stated in the previous section.

2-MLRP Heuristic Algorithm

First Step **Projections from $M$to $\mathbb{R}^2$**

(1.*a*) Use a homeomorphism φ to project the customer locations from the surface of M to $\mathbb{R}^2$

(1.*b*) Map the obstacles from M to $\mathbb{R}^2$ using φ introduced in (1.*a*)

Second Step **Geodesic Projection**

(2.*a*) Solve the MWP by calculating the geodesic distances between customers of C_1 and C_2 separately on M to determine the shortest path between the customers

(2.*b*) Use the homeomorphism φ introduced in (1.*a*) for mapping the determined geodesics between the customers from M to $\mathbb{R}^2$

(2.*c*) Assign the distances determined in (2.*b*) for the customer sets C_1 and C_2 as the distances between the corresponding customer locations on $\mathbb{R}^2$

Third Step **Facility Locations' Initialization**

(3.*a*) Initiate the location of the facility f_1 to serve customer of the set C_1 by using a formula similar to the one introduced in [31] for 1-MLRP. Calculate the radius of the initial circle. Initiate a counter $k_1 = 1$ for the corresponding number of circles used.

(3.*b*) Initiate the location of the facility f_2 to serve customers of the set C_2 by using a formula similar to the one introduced in [31] for 1-MLRP. Calculate the radius of the initial circle used. Initiate a counter $k_2 = 1$ for the corresponding number of circles used.

Fourth Step **The Linked Chain Method (LCM)—Heuristic Solution**
Solve (4.*a*)–(4.*c*) simultaneously

(4.*a*) Choose t_{k_1} random numbers within the disk region formed by the interior of the circle k_1, $k_1 = 1, 2, \ldots, z_1$ when the distances between f_1 and x are considered;
(4.*b*) Choose t_{k_2} random numbers within the disk region formed by the interior of the circle k_2, $k_2 = 1, 2, \ldots, z_2$ when the distances between f_2 and y are considered;
(4.*c*) Choose t_{k_3} random numbers within the disk region formed by the interior of the circle k_3, $k_3 = 1, 2, \ldots, z_3$ when the distances between f_1 and f_2 are considered;
(4.*d*) Solve the routing problem for each circle k_i ($i = 1, 2, 3$) for determining the next center of the circle (if exists and different from the previous center of the circle) with the best possible routing between all the customers $x \in C_1$ and f_1, all the customers $y \in C_2$ and f_2, and the facilities f_1 and f_2, independently;
(4.*e*) Employ the LCM to determine the best center for circle k_i after determining the center of circle $k_i - 1$ for each $i = 1, 2, 3$;
(4.*f*) For each $i = 1, 2, 3$, design the new circle k_i in the LCM by assigning its center as the location determined from the VRP solution with the radius of the new circle k_i to be the distance between the centers of the circles k_i and $k_i - 1$;
(4.*g*) Continue linking circles to the LCM until the distance between the circles z_i and $z_i - 1$ is sufficiently small for each $i = 1, 2, 3$; and
(4.*h*) Determine the best possible location of the final circle z_i as the last step of LCM for each $i = 1, 2, 3$.

Fifth Step **Projection from $\mathbb{R}^2$ to M**

(5.*a*) Use the inverse map of the homeomorphism determined in (1.*a*) to project the allocated facilities and the corresponding minimum cost routes from $\mathbb{R}^2$ to M;
(5.*b*) Determine feasible locations for the facilities on M if the projected locations from $\mathbb{R}^2$ to M are not feasible by employing a local search within the local region; and
(5.*c*) IF one or more facility locations are infeasible, DO a neighborhood search to determine a discrete feasible location on M. Recalculate the distances between f_1 and f_2.

5.3 *Computational Complexity*

It is shown in [31] that 1-MLRP is an NP-hard problem noting that both routing and location problems are NP-hard. Therefore, 2-MLRP is also an NP-hard problem noting that it is clearly a more complicated problem than 1-MLRP. There are three main computational challenges in the proposed algorithm for solving the 2-MLRP:

First Challenge Calculating the lengths of the pathways (integrals given in (9)–(11)) between the customers on the RMS;

Second Challenge Solving the routing problem in the Euclidean space at every circle of the LCM; and

Third Challenge Determining the locations of f_1 and f_2 simultaneously based on their dependence to their respective customers in addition to their dependence on each other.

On one hand the proposed 2-MLRP heuristic algorithm has the advantage of calculating a radius per circle compared to the heuristic algorithm employed in [16] for the Euclidean case with computations including two radii for each ellipsoid, on the other hand the computational algorithm we proposed includes the challenge of calculating the objective functions (given by (9)–(11)) on the manifold surface M rather than $\mathbb{R}^2$. The heuristic algorithm we designed to solve the proposed MLRP has computational complexity

$$O\left(T_1 + \sum_{i=1}^{3}\left(t_i m_i \log(m_i) \sum_{j=0}^{z_i-1} p^j\right) + T_2 + T_3\right)$$

where

- T_1 : The time it takes to compute the lengths of the geodesics between customers on the surface of M;
- m_i: The number of customers in set C_i
- t_i : The number of points chosen within the initial circle for each $i = 1, 2, 3$;
- $t_i p^j$: The number of points chosen within the jth circle for each $i = 1, 2, 3$;
- The expression

$$\sum_{i=1}^{3}\left(t_i m_i \log(m_i) \sum_{j=0}^{z_i-1} p^j\right)$$

  is the maximum time allocated to compute the routing paths between the facilities f_1 and f_2 $(i = 3)$, the routing paths between f_1 and the customers x $(i = 2)$, and the routing paths between f_2 and the customers y $(i = 1)$. The summation over i is taken due to the fact that the computation of all paths between customers and the facilities depends on each other;
- T_2 : The time it takes to determine a feasible location of the facility on the manifold's surface after determining the corresponding locations in $\mathbb{R}^2$; and
- T_3 : Additional computational effort is needed in case the locations of the facilities determined are infeasible on M.

The routing sub-problem's computational complexity is $O(m_i \log(m_i))$ for each $i = 1, 2, 3$; the time complexity obtained in [6]. The following figure displays the linked chains between a single facility and its customers in $\mathbb{R}^2$ that follows the information from Figs. 1, 2 and 3.

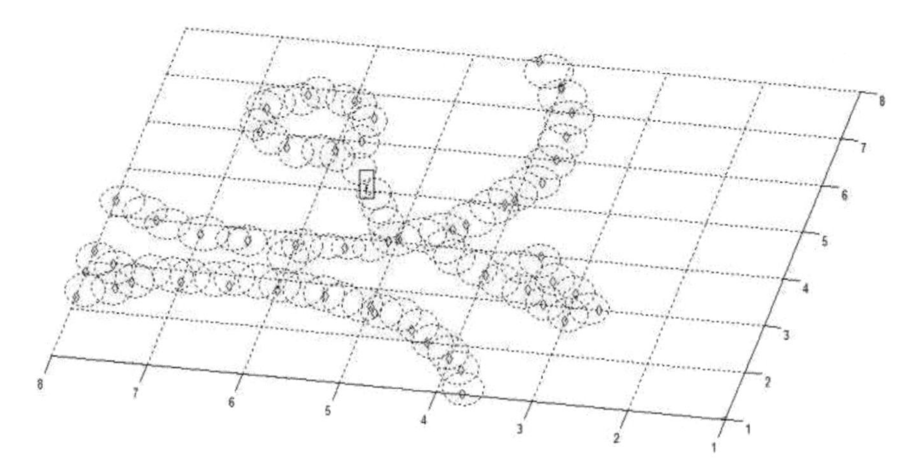

Fig. 4

5.4 *Algorithmic Solution Details of 2-MLRP*

In this section we explain the details of the 2-MLRP heuristic algorithm given in the previous section. Instead of using *D* domain we use *M* (Fig. 4).

5.4.1 Projections from M to R^2

The first and second steps of the heuristic algorithm require the use of a homeomorphism $\varphi : M \to \mathbb{R}^2$ to project the customer locations and all possible routes between the two sets of customers C_1 and C_2 from M to $\mathbb{R}^2$. This projection from M to $\mathbb{R}^2$ is necessary due to the difficulty of calculations on Riemannian manifolds.

For 2-MLRP, similar to the projection mapping defined in [31], a homeomorphism that can be employed to map the locations of the customers from the Riemannian manifold surface M to the Euclidean surface $\mathbb{R}^2$ is

$$\varphi : M \to \mathbb{R}^2$$

$$x_i \mapsto \left(l_i^1 \cos\left(\theta_1\right), r_i^1 \sin\left(\theta_1\right)\right) \tag{19}$$

$$y_j \mapsto \left(l_j^2 \cos\left(\theta_2\right), r_j^2 \sin\left(\theta_2\right)\right) \tag{20}$$

where $x_i = \left(l_i^1, r_i^1\right)$ and $y_j = \left(l_j^2, r_j^2\right)$ represent the customer locations, and θ_s $(s = 1, 2)$ are the angles used for homeomorphic projection [9]. Changing the homeomorphisms defined in Eqs. (19) and (20) would effect the appearance of the customer locations in the Euclidean space but would not affect the actual locations of customers on M. Our choice of homeomorphism φ defined in (19) and (20) yields the projection of customer locations to circular regions. The radii of these circles formed for both set of customers C_1 and C_2 can be easily obtained by using the well-known distance formula. The distance between the customers on M then becomes the weights of the edges between the customers in the customer network.

5.4.2 Initial Facility Locations and LCM: Heuristic Solution

In this work, the initial locations of the facilities to be allocated are determined by using the formula similar to the one used in [31] for circular objects. Therefore, the formula we employ for the initial allocation of the facilities is

$$l_0^k = \frac{\sum_{j=1}^{n_k} \frac{l_j^k \cos(\theta_k)}{L\left(p_0^{j-1}, \varphi(x_j)\right)}}{\sum_{j=1}^{n_k} \frac{1}{L\left(p_0^{j-1}, \varphi(x_j)\right)}} \quad \text{and} \quad r_0^k = \frac{\sum_{j=1}^{n_k} \frac{r_j^k \cos(\theta_k)}{L\left(p_0^{j-1}, \varphi(x_j)\right)}}{\sum_{j=1}^{n_k} \frac{1}{L\left(p_0^{j-1}, \varphi(x_j)\right)}}$$

for each $k = 1, 2$ where $L\left(p_0^{j-1}, \varphi\left(x_j\right)\right)$ represents the geodesic distances between the point p_0^{j-1} and the customer $\varphi\left(x_j\right)$. The following algorithm used in [31] for point generation within each circle is also used for the 2-MLRP case:

Algorithm of Generating t_k Number of Points at the kth Circle

For all $k \geq 1$ let $t_{k-1} > t_k$ and define

$$t_k = \lceil 2\pi R_k \lambda_k \rceil$$

where λ_k is a randomly chosen number from the interval $(0.5, 1)$ and R_k is the radius of the kth circle. The sth point is allocated within the kth circle by using the formula

$$(l_{ks}, r_{ks}) = \left(l_k \cos\left(\frac{2\pi R_k}{t_k} s\right), r_k \sin\left(\frac{2\pi R_k}{t_k} s\right)\right)$$

for every s satisfying $0 \leq s \leq t_k - 1$. The chosen points within each circle follow different directions within the circles for homogeneous point distribution. Homogenous distribution of points is necessary since the best location of the facility can be in different directions within each circle. The choice of homeomorphism used for the projection from M to $\mathbb{R}^2$ can yield a single circle with the center of the circle being the location of the facility. This process is implemented on every circle generated for determining the routing paths between the facilities f_1 and f_2, between f_1 and all customers $x \in C_1$, and the routing paths between f_2 and all customers $y \in C_2$.

In the case when the customer and facility locations are the same, this non-optimal solution will be recovered by reallocating the facility to the closest feasible location after a neighborhood search.

The solution to the routing problem from the initial location of each facility to their respective customers determines the possible routes to these customers. These local areas are circles in $\mathbb{R}^2$ corresponding to the local areas on the surface of M since M is a locally Euclidean space. Recall that we chose t_i to be the number of points for the first circle generated to solve the routing problem when $i = 1, 2, 3$ represent the routes between customers of C_1, customers of C_2, and the facilities. After determining the best location of each facility within the first circle by solving the corresponding routing problems for t_i points (locations) for each $i = 1, 2, 3$, the next step of the algorithm is application of the LCM by determining the circles $k = 2, 3, \ldots, z_i$. Consecutive circles are generated by using the radius determined for each circle. At every kth step of the LCM, we choose t_k number of points within the disk region formed by the interior of the circle k ($k = 2, \ldots, z$) to solve the corresponding routing problem. The following formula introduced in [1] with computational complexity $O(m \log(m))$ is used for fulfilling the main objective of the routing problem: determining the minimal length route for each one of the t_k points generated at the kth by preventing from route duplication between customers i and j:

$$d_{ij} = \begin{cases} (1 + w_1) L(\varphi_{i0}) + (1 - w_1) L(\varphi_{0j}) - w_2 L(\varphi_{ij}) + w_3 \upsilon & \text{if } \varphi_{i0} > \varphi_{0j} \\ (1 - w_1) L(\varphi_{i0}) + (1 + w_1) L(\varphi_{0j}) - w_2 L(\varphi_{ij}) + w_3 \upsilon & \text{otherwise} \end{cases}$$

In this formula $L(\varphi_{i0})$ is the distance on the RMS between the ith customer and the possible location of the facility indexed to be "0"; υ is the total demands of customers i and j divided by the average of the total demand; and (w_1, w_2, w_3) are the weights assigned to the equation components.

Circle generation used for the LCM continues until a sufficiently small circle with radius (if exists) is obtained. This condition is a result of the calculation $|R_z - R_{z-1}| < \epsilon$ for a sufficiently chosen ϵ where R_k is the radius of the circle k. Determination of the best possible center of the last circle with this radial condition is the final step of the LCM on the Euclidean surface. It is important to note that the final location of the facility determined is close to the best feasible solution but does not necessarily reflect the best feasible location of the facility. The last step of the algorithm to be explained next is the projection of the determined solutions from $\mathbb{R}^2$ to M.

5.4.3 Mapping from $\mathbb{R}^2$ to M

The projection of the customers from *CCRMS* to the Euclidean space is accomplished by using the homeomorphism φ. The geodesics determined between customers on the RMS are also projected by using φ from M to $\mathbb{R}^2$ and these lengths are assigned to be the edge lengths between the customers on the customer network formed in $\mathbb{R}^2$. It is important to note that the metric used on the manifold defines the metric used on the RMS. This is due to the fact that the norm we use for the distance calculations is the norm used on the projected Euclidean surface.

The customer locations on M are fixed therefore mapping these locations back to M would be a redundant process. The locations of the two facilities determined on $\mathbb{R}^2$ need to be mapped back to the manifold surface by using the inverse map of $\varphi : \varphi^{-1} : M \rightarrow \mathbb{R}^2$. The path minimizing route between f_1 and f_2 needs to be also determined by using the inverse mapping since the locations of these two facilities are determined initially on $\mathbb{R}^2$, not on M.

6 Applications in Network Theory

In this section, the theoretical solution described in the previous sections to solve the 2-MLRP will be explained for a large data network with the corresponding graphs.

Determining facility locations for a customer network is an interest of researchers. For instance, in [12], assuming that the p-center problem with an additional assumption that the facility at a node fails to respond to demands from the node, dynamic programming is implemented for the location determination on a path network in the Euclidean space. The findings of this work are particularly useful for large emergency and health care services in big cities. For other applications of data network and implementation of location analysis see [13, 18].

Figure 5 is an example of a manifold (M) surface on which the customers are assumed to be located.

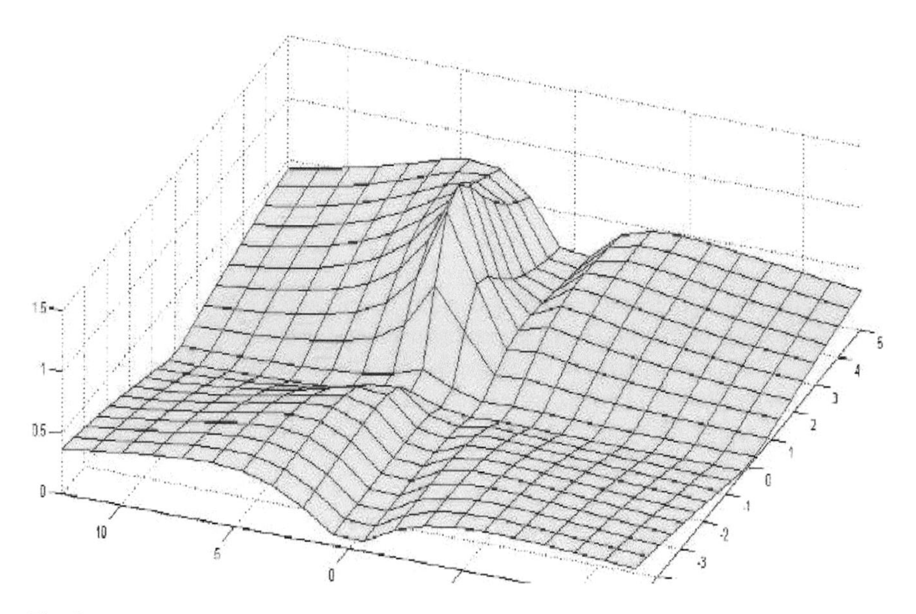

Fig. 5

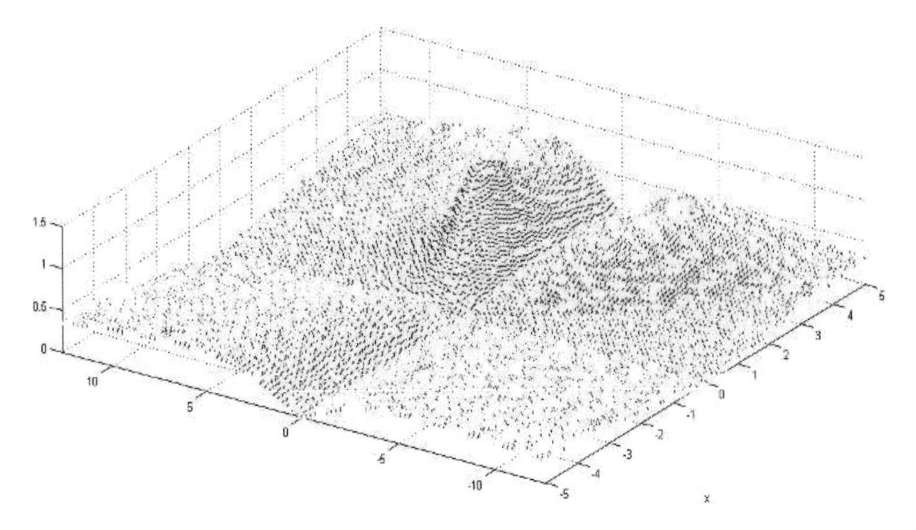

Fig. 6

Figure 6 is visualization of the large customer network on the manifold surface where customers are represented with dots. This graph represents the locations of the customers before they are split into two set of customers C_1 and C_2.

In Fig. 7, large circles consisting of blue and red dots represent the projected customers from the manifold surface to $\mathbb{R}^2$ by using the homeomorphism φ explained in Sect. 5.4.1. The interior regions of the three solid lines represent the locations where facilities cannot be allocated (i.e., obstacles). Therefore the large data network represented in the figure above is re-represented and mapped on to the circles in the figure below as a result of the mapping $\varphi : M \to \mathbb{R}^2$.

The two boxes labeled $f1$ and $f2$ in Fig. 8 below represent the initial locations of the two facilities we want to allocate that are determined by using the algorithm introduced in Sect. 5.4.2. The two circles linking $f1$ and $f2$ represent the chain of circles as a result of the LCM application.

The circles wrapping the great circles in Fig. 9 represent the circles generated by LCM for a set of customers. LCM is implemented on the entire customer set.

Figure 10 illustrates the case when the facility allocation is finalized and one of the facilities is allocated in the same place with a customer. In this case, the facility with the infeasible location is relocated to a close-by reasonable location after a neighborhood search in domain D.

7 Summary

In this work, MLRP for one- and two-facility cases is explained with the corresponding algorithmic solutions. 1-MLRP is introduced in [31] as a mixed integer non-linear programming problem that's determined to be an NP-hard problem.

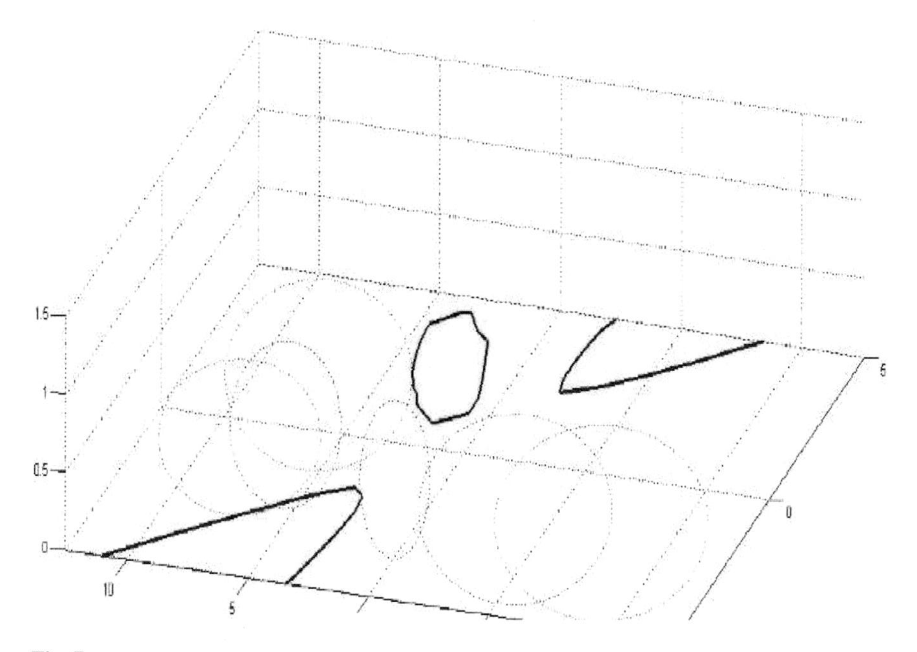

Fig. 7

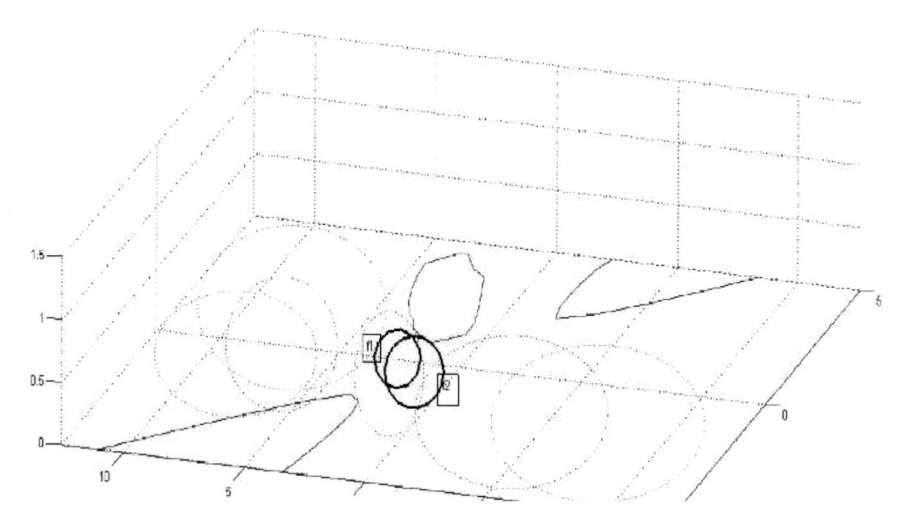

Fig. 8

Similarly, 2-MLRP is an NP-hard problem formulated as a mixed integer non-linear programming problem with 3 objective functions. As a part of the heuristic algorithm solution, Weiszfeld's formula [33] is used for determining the initial location of the facility in the 1-MLRP case. In the case of 2-MLRP, a similar formula to the one used in [31] is used for determining the initial locations of

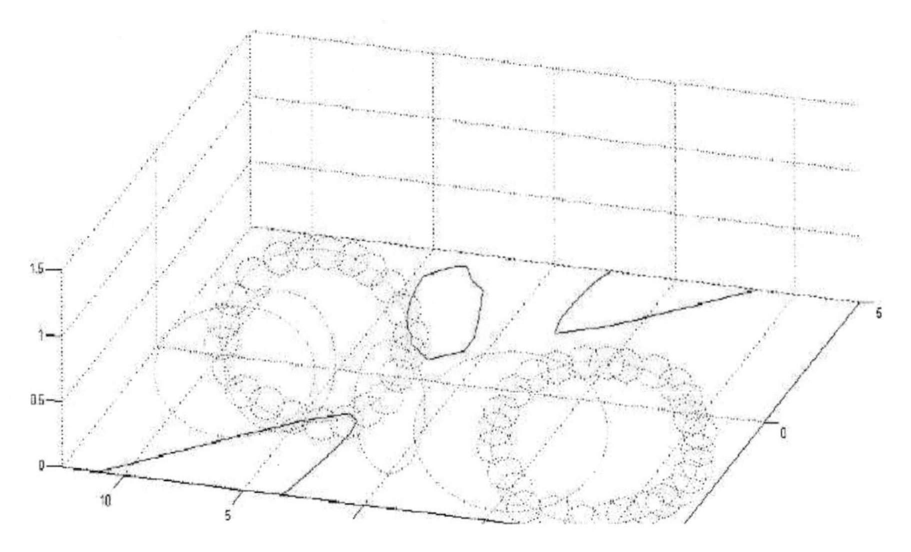

Fig. 9

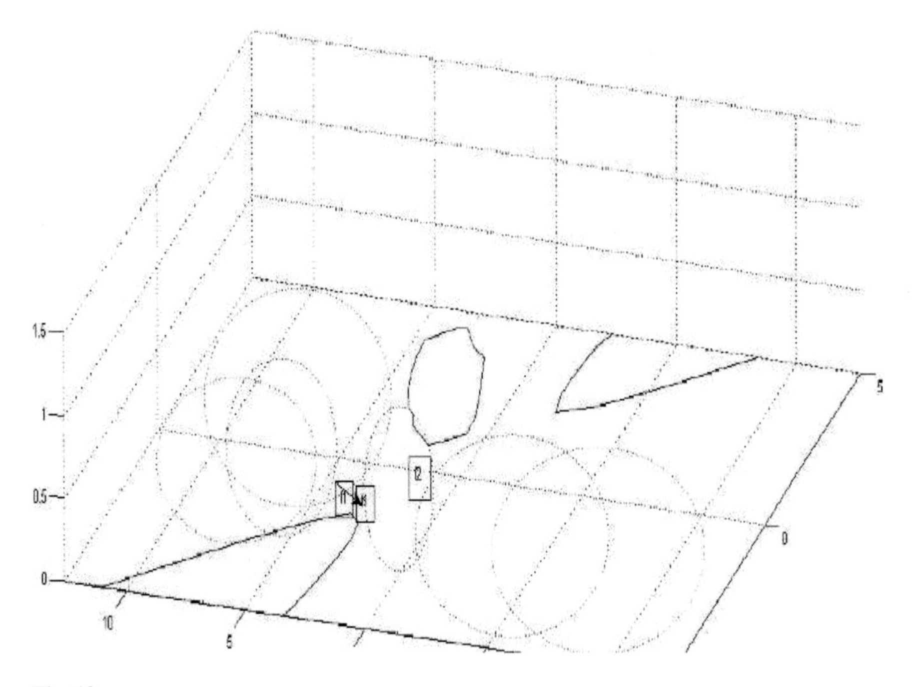

Fig. 10

the facilities. The homeomorphism used in [31] for 1-MLRP is similarly used for 2-MLRP when mapping from the manifold surface to the Euclidean space is implemented. The point generation algorithm of Altinel et al. [1] and LRP solutions are employed to solve the 2-MLRP similar to 1-MLRP. We explained the visual network representation for both 1- and 2-MLRP cases throughout this work. The solutions to both 1- and 2-MLRP cases are expected to yield more accurate results than their special cases on planar and spherical surface cases. This is due to the fact that a manifold surface is a generalization of planar and spherical surfaces. In addition, the geodesic distance calculation on Earth's surface (noting that Earth's surface is a manifold) is a more realistic approach than calculating the distances on planar and spherical surfaces.

Acknowledgements The work of Theodore Trafalis was conducted at National Research University Higher School of Economics and supported by RSF (Russian Science Foundation) grant 14-41-00039.

References

1. Altinell, K., Oncan, T.: A new enhancement of the Clarke and Wright savings heuristic for the capacitated vehicle routing problem. J. Oper. Res. Soc. **56**, 954–61 (2005)
2. Aly, A.A., Kay, D.C., Litwhiler, D.W. Jr.: Location dominance on spherical surfaces. Oper. Res. **27**, 972–981 (1979)
3. Ambrosino, D., Sciomachen, A., Grazia Scutella, M.: A heuristic based on multi-exchange techniques for a regional fleet assignment location-routing problem. Comput. Oper. Res. **36**, 442–60 (2009)
4. Barreto, S., Ferreira, C., Paixao, J., Sousa Santos, B.: Using clustering analysis in a capacitated location-routing problem. Eur. J. Oper. Res. **179**, 968–77 (2007)
5. Bruns, A., Klose, A., Stahly, P.: Restructuring of Swiss parcel delivery services. Oper. Res. Spectr. **22**, 285–302 (2000)
6. Clarke, G., Wright, J.W.: Scheduling of vehicle from central depot to a number of delivery points. Oper. Res. **12**, 568–581 (1964)
7. Cooper, L.: Heuristic methods for location-allocation problems. SIAM Rev. **6**, 37–53 (1964)
8. Daskin, M.S.: What you should know about location modeling. Nav. Res. Logist. **55**, 283–294 (2008)
9. doCarmo, M.P.: Differential Geometry of Curves and Surfaces. Prentice Hall Inc, Englewood Cliffs (1976)
10. Drezner, Z., Wesolowsky, G.O.: Facility location on a sphere. J. Oper. Res. Soc. **29**, 997–1004 (1978)
11. Gamal, M.D.H., Salhi, S.: Constructive heuristics for the uncapacitated continuous location-allocation problem. J. Oper. Res. **52**, 821–829 (2001)
12. Huang, R., Kim, S., Menezes, M.B.C.: Facility location for large-scale emergencies. Ann. Oper. Res. **181**(1), 271–286 (2010)
13. Johnson, M.P., Gorr, W.L., Roehrig, S.: Location of service facilities for the elderly. Ann. Oper. Res. **136**(1), 329–349 (2005)
14. Jost, J.: Riemannian Geometry and Geometric Analysis, 6th edn. Springer, Berlin (2011)
15. Laporte, G.: What you should know about the vehicle routing problem. Nav. Res. Logist. **54**, 811–819 (2007)

16. Manzour-al-Ajdad, S.M.H., Torabi, S.A., Salhi, A.: A hierarchical algorithm for the planar single-facility location routing problem. Comput. Oper. Res. **39**, 461–470 (2012)
17. Manzour-al-Ajdad, S.M.H., Torabi, S.A., Eshghi, K.: Single-source capacitated multi-facility Weber problem: an iterative two phase heuristic algorithm. Comput. Oper. Res. **39**, 1465–1476 (2012)
18. Marianov, V.: Location of multiple-server congestible facilities for maximizing expected demand, when services are non-essential. Ann. Oper. Res. **123**(1-4), 125–141 (2003)
19. Nagy, G., Salhi, S.: Location-routing: Issues, models, and methods. Eur. J. Oper. Res. **177**, 649–672 (2007)
20. Perl, J., Daskin, M.S.: A warehouse location-routing problem. Transp. Res. B **19**, 381–396 (1985)
21. Prins, C., Prodhon, C., Soriano, P., Ruiz, A., Wolfler-Calvo, R.: Solving the capacitated location-routing problem by a cooperative Lagrangean relaxation-granular tabu search heuristic. Transp. Sci. **41**, 470–483 (2007)
22. Riemann, B.: Grundlagen für eine allgemeine Theorie der Functionen einer veränderlichen complexen Grösse. Inauguraldissertation, Göttingen (1851)
23. Salhi, S., Rand, G.K.: The effect of ignoring routes when locating depots. J. Oper. Res. **39**, 150–156 (1989)
24. Salhi, S., Nagy, G.: Consistency and robustness in location routing. Stud. Locat. Anal. **13**, 3–19 (1999)
25. Salhi, S., Nagy, G.: Local improvement in planar facility location using vehicle routing. Ann. Oper. Res. **167**, 287–296 (2009)
26. Schwardt, M., Dethloff, J.: Solving a continuous location routing problem by use of a self-organizing map. Int. J. Phys. Distrib. Logist. Manag. **35**, 390–408 (2005)
27. Schwardt, M., Fischer, K.: Combined location-routing problems—a neural network approach. Ann. Oper. Res. **167**, 253–269 (2009)
28. Semet, F., Taillard, E.: Solving real-life vehicle routing problems efficiently using tabu search. Ann. Oper. Res. **41**, 469–488 (1993)
29. Sherali, H.D., Noradi, F.L.: NP-hard, capacitated, balanced p-median problems on a chain graph with a continuum of link demands. Math. Oper. Res. **13**, 32–49 (1988)
30. Tellier, L.-N.: The Weber problem: solution and interpretation. Geo Anal. **4**(3), 215–233 (1972)
31. Tokgöz, E., Alwazzi, S. & Theodore, T.B. A heuristic algorithm to solve the single-facility location routing problem on Riemannian surfaces, Computational Management Science, Springer 12, 297–415 (2015)
32. Wasner, M., Zapfel, G.: An integrated multi-depot hub location vehicle routing model for network planning of parcel service. Int. J. Prod. Econ. **90**, 403–419 (2004)
33. Weiszfeld, E.: Sur le point pour lequel la somme des distances de n points donnes est minimum. Tohoku Math. J. **43**, 355–386 (1937)

A Branch and Bound Algorithm for the Cell Formation Problem

Irina Utkina and Mikhail Batsyn

Abstract The cell formation problem (CFP) is an NP-hard optimization problem considered for cell manufacturing systems. Because of its high computational complexity several heuristics have been developed for solving this problem. In this paper we present a branch and bound algorithm which provides exact solutions of the CFP. This algorithm finds optimal solutions for 13 problems of the 35 popular benchmark instances from the literature.

Keywords Cell formation problem • Branch and bound algorithm • NP-hard problems • Combinatorial optimization • Upper bound • Exact solution

1 Introduction

The first paper on the cell formation problem (CFP) was by Flanders [10] in 1925. In Russia the Group Technology was introduced by Mitrofanov [17]. The main problem in the Group Technology (GT) is to find an optimal partitioning of machines and parts into cells, in order to maximize the number of parts which are processed inside cells and to minimize the number of parts which are processed outside cells. This problem is called the CFP. Burbidge [4] developed product flow analysis (PFA) approach to this problem and introduced the GT and the CFP in his book [4].

Ballakur and Steudel [2] have shown that the CFP is an NP-hard problem for different objective functions. That is why in order to solve this problem there have been developed several heuristic approaches for this problem and almost no exact ones.

In this paper we present an exact algorithm for the CFP based on a branch and bound method.

I. Utkina (✉) • M. Batsyn
Laboratory of Algorithms and Technologies for Networks Analysis, National Research University Higher School of Economics, 136, Rodionova Str., Nizhny Novgorod, Russia
e-mail: iutkina@hse.ru; mbatsyn@hse.ru

V.A. Kalyagin et al. (eds.), *Models, Algorithms and Technologies for Network Analysis*, Springer Proceedings in Mathematics & Statistics 156,
DOI 10.1007/978-3-319-29608-1_7

2 Formulation

The CFP is a bi-clustering problem in which we simultaneously cluster machines and parts into cells. The objective of the CFP is to find an optimal partitioning of machines and parts into groups (production cells, or shops) in order to minimize the inter-cell movement of parts from one cell to another and to maximize the number of intra-cell processing operations. The input data for this problem is matrix A which contains zeroes and ones. The size of this matrix is $m \times p$ which means that it has m machines and p parts. The element a_{ij} of input matrix is equal to one if machine i processes part j. The objective is to minimize the number of zeroes inside cells and the number of ones outside cells. Because it is not possible to minimize these two parameters at the same time there have been suggested several objective functions which combine these two goals.

3 Objective Functions

The following objective functions are well-known in the literature:

1. The number of exceptions (the number of ones outside cells) and voids (the number of zeroes inside cells)

$$E + V = n_1^{out} + n_0^{in}$$

2. Grouping efficiency by Chandrasekharan and Rajagopalan [9]

$$\eta = q * \eta_1 + (1 - q) * \eta_2,$$

where

$$\eta_1 = \frac{n_1^{in}}{n^{in}}$$

$$\eta_2 = \frac{n_0^{out}}{n^{out}}$$

n_1—the number of ones in the input matrix

n_0—the number of zeroes in the input matrix

n^{in}—the number of elements inside cells

n^{out}—the number of elements outside cells

n_1^{in}—the number of ones inside cells

n_1^{out}—the number of ones outside cells

n_0^{in}—the number of zeroes inside cells

n_0^{out}—the number of zeroes outside cells

This means that η_1 is the ratio of ones inside cells to the total number of elements inside cells and η_2 is the ratio of zeroes outside cells to the total number of elements outside cells. Parameter q is a coefficient, which reflects the weights of intra-cell and inter-cell processing operations ($0 \leq q \leq 1$). Usually it is taken 0.5 which means that it is equally important to maximize the number of ones inside cells and maximize the number of zeroes outside cells.

3. Grouping efficacy suggested by Kumar and Chandrasekharan [14]

$$f = \frac{n_1^{in}}{n_1 + n_0^{in}}$$

4. Group capability index (GCI) suggested by Hsu [11]

$$GCI = 1 - \frac{n_1^{out}}{n_1} = \frac{n_1^{in}}{n_1}$$

In this work we use the grouping efficacy as the objective function because of its good properties (see Kumar and Chandrasekharan [14]).

4 Definitions

$$A(m \times p) = [a_{ij}] = \begin{cases} 1 & \textit{if machine } \mathrm{i} \textit{ processes part } \mathrm{j} \\ 0 & \textit{otherwise} \end{cases} \text{—an input matrix}$$

$M(1 \times m)$—a vector which contains the assignment of machines to cells

$P(1 \times p)$—a vector which contains the assignment of parts to cells

f^*—the optimal value of function f

$$f = \frac{n_1^{in}}{n_1 + n_0^{in}}$$

As an example M = [1231] and P=[11321] mean that machines 1, 4 and parts 1, 2, 5 are assigned to cell 1, machine 2 and part 4 are assigned to cell 2, and machine 3 and part 3 are assigned to cell 3.

5 Branch and Bound Algorithm

5.1 Branching

Because of the bi-clustering structure of the CFP branching goes by two parameters, the algorithm has branching on machines and parts sequentially changing each other: machines–parts–machines–...
Branching on machines makes changes in vector M. It starts from assigning the first machine to cell 1. After that, when the algorithm branches on machines, it finds the first machine which is not assigned to any cell and tries to assign it to the existing cells with numbers from 1 to k or creates a new cell $(k + 1)$ for this machine. An example of branching is shown in Fig. 1.

Branching on parts makes changes in the vector P. It starts with all zeroes inside P which means that no parts are assigned to any cell. When the algorithm branches on parts it finds the first part which is not assigned to any cell and tries to assign it to the existing cells from 1 to k or to a new cell $(k + 1)$ if there are some unassigned machines which can be also added to this new cell. We assume that the number of parts is greater than the number of machines. Examples are shown in Figs. 2 and 3.

The algorithm branches on parts and machines successively. It starts with $M = [100\ldots0]$ and $P = [00\ldots0]$. Next it changes vector P as illustrated in Figs. 2 and 3 then—vector M as illustrated in Fig. 1 and so on. This way the algorithm builds the search tree. The leaves of the search tree contain complete solutions and nodes contain partial solutions. The complete search tree does not depend on data in the input matrix. It depends only on its size. The complete search tree for the input matrix with two machines and three parts is shown in Fig. 4. Two first numbers in square brackets represent vector M, three others—vector P. If vector M or P contains zeroes, this means that these machines or parts are not yet assigned to any cell. If vector P contains a number with a star, this means that this part can lie in a cell with a number equal or greater than this number. The complete tree contains all feasible solutions.

5.2 Bounds

In order to reduce the search tree size the algorithm calculates an upper bound (UB) at each partial solution. After that, it is compared with the current best value of function f found by the algorithm by this time. If it is less or equal, then the algorithm prunes this branch and goes to the next alternative at this level. The UB is calculated as follows.

1. Calculate the values of variables n_1, n_1^{in}, and n_0^{in} in the current partial solution.
2. For every machine (row) which is not yet assigned to available cells the algorithm tries to put it into available cells or a new cell. The algorithm takes into account

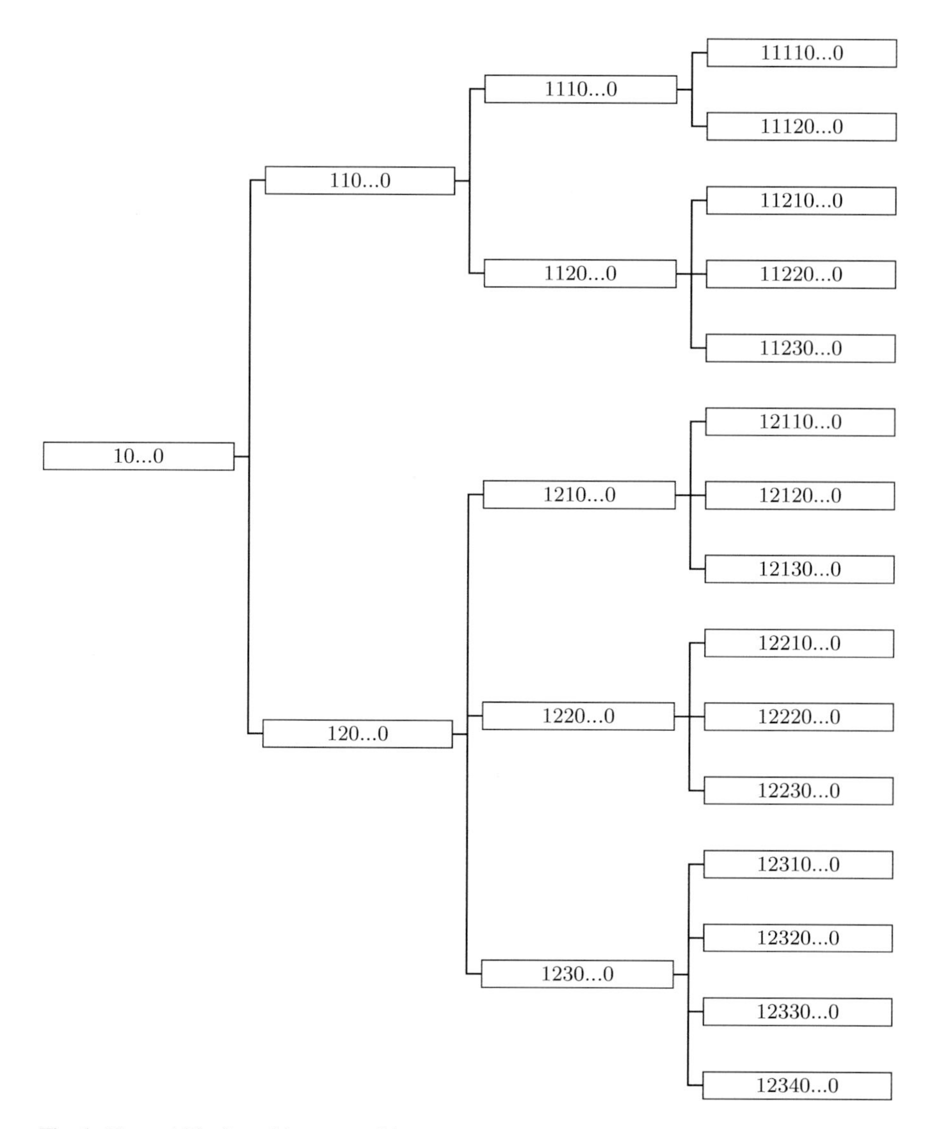

Fig. 1 Vector M for branching on machines

all ones and zeroes in this row which can get into existing cells and all ones in this row which are not yet assigned to any cell.

3. Choose the best alternative for all machines.
4. For every part (column) which is not yet assigned to available cells the algorithm tries to put it into available cells or no one. The algorithm takes into account only items lying in the rows which are already assigned.

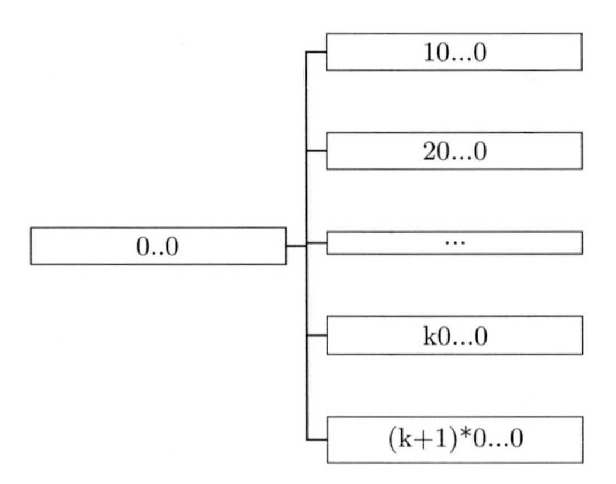

Fig. 2 Vector *P* for branching on parts when there are unassigned machines

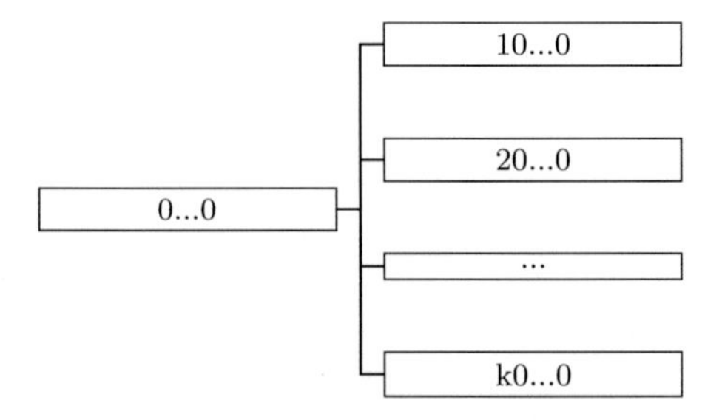

Fig. 3 Vector *P* for branching on parts when all machines are assigned to cells

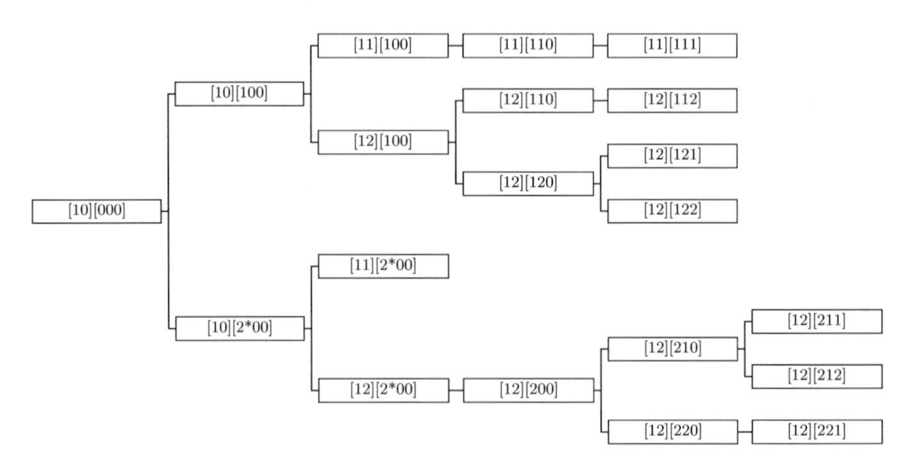

Fig. 4 A complete search tree

5. Choose the best alternative for all parts.
6. Calculate UB as function f with all parameters which have been calculated at previous steps.

An example is shown below.

M \ P	1	1	1	1	2	0	0
1	**1**	**1**	**1**	**1**	0	0	0
1	**1**	**1**	**0**	**1**	0	0	0
2	0	0	0	0	**1**	1	0
0	0	0	0	0	1	1	1
0	0	0	0	0	1	0	1

1. We have $n_1 = 14$, $n_1^{in} = 8$, and $n_0^{in} = 1$.
2. Calculate the number of new zeroes and ones which get inside cells after the corresponding machine assignment:

	cell 1	cell 2	a new cell
machine 4	$+4_0, +2_1$	$+\mathbf{0}_\mathbf{0}, +\mathbf{3}_\mathbf{1}$	$+0_0, +2_1$
machine 5	$+4_0, +1_1$	$+\mathbf{0}_\mathbf{0}, +\mathbf{2}_\mathbf{1}$	$+0_0, +1_1$

3. The following assignments are the best. Machine 4 is assigned to cell 2. Machine 5 is assigned to cell 2.
4. Calculate the number of new zeroes and ones which get inside cells after the corresponding part assignment:

	cell 1	cell 2	no one
part 6	$+2_0, +0_1$	$+\mathbf{0}_\mathbf{0}, +\mathbf{1}_\mathbf{1}$	$+0_0, +0_1$
part 7	$+2_0, +0_1$	$+1_0, +0_1$	$+\mathbf{0}_\mathbf{0}, +\mathbf{0}_\mathbf{1}$

5. The following assignments are the best. Part 6 is assigned to cell 2. Part 7 is not assigned to any cell.
6. $UB = \frac{n_1^{in} + \Delta n_1^{in}}{n_1 + n_0^{in} + \Delta n_0^{in}}$, where Δn_1^{in} and Δn_0^{in} are the total number of ones and zeroes which get inside cells after the chosen machine and part assignment.
$UB = \frac{8+3+2+1+0}{14+1+0+0+0+0} = \frac{14}{15} = 0.9333$

6 Results

The suggested branch and bound algorithm has been able to solve 13 of 35 popular benchmark instances from the literature. The results are presented in Table 1. All computations were run on Intel Core i7 with 16 GB RAM. We start the algorithm

Table 1 Results

#	Name	Size	$f*$	Number of branches without initial solution	Time, s, without initial solution	Initial solution by James [12]	Number of branches with initial solution	Time to find initial solution, s	Time, s, with initial solution	Bychkov [5] time, s
1	King [13]	5×7	0.8235	116	0.00	0.8235	21	0.44	0.00	0.63
2	Waghodekar and Sahu [22]	5×7	0.6957	103	0.00	0.6957	55	0.43	0.00	2.29
3	Seifoddini [20]	5×18	0.7959	231	0.00	0.7959	51	0.46	0.00	5.69
4	Kusiak [15]	6×8	0.7692	158	0.00	0.7692	32	0.44	0.00	1.86
5	Kusiak and Chow [16]	7×11	0.6087	5441	0.03	0.6087	2538	0.54	0.02	9.14
6	Boctor [3]	7×11	0.7083	525	0.00	0.7083	242	0.52	0.00	5.15
7	Seifoddini and Wolfe [19]	8×12	0.6944	388	0.00	0.6944	275	0.54	0.00	13.37
8	Chandrasekharan and Rajagopalan [7]	8×20	0.8525	916	0.00	0.8525	44	0.56	0.00	18.33
9	Chandrasekharan and Rajagopalan [8]	8×20	0.5872	12,340	0.11	0.5872	12,206	0.57	0.09	208.36
10	Mosier and Taube [18]	10×10	0.7500	2267	0.02	0.7500	651	0.53	0.00	6.25
11	Chan and Milner [6]	10×15	0.9200	865	0.00	0.9200	35	0.54	0.00	2.93
12	Askin and Subramanian [1]	14×24	0.7206	512,339	12.49	0.7206	100,743	0.78	2.63	259.19
13	Stanfel [21]	14×24	0.7183	923,594	22.38	0.7183	193,276	0.79	4.89	179.21

with and without an initial solution. For an initial solution we apply the heuristic suggested by James [12]. We use this heuristic because it always provides the optimal solutions for the considered CFP instances in very small time.

The results show that the developed algorithm is more efficient than the CPLEX model of Bychkov [5].

Acknowledgements This research is partly supported by LATNA Laboratory, NRU HSE, RF government grant 11.G34.31.0057.

References

1. Askin, R.G., Subramanian, S.P.: A cost-based heuristic for group technology configuration. Int. J. Prod. Res. **25**(1), 101–113 (1987)
2. Ballakur, A., Steudel, H.J.: A within cell utilization based heuristic for designing cellular manufacturing systems. Int. J. Prod. Res. **25**, 639–655 (1987)
3. Boctor, F.F.: A linear formulation of the machine-part cell formation problem. Int. J. Prod. Res. **29**(2), 343–356 (1991)
4. Burbidge, J.L.: The new approach to production. Prod. Eng. **40**, 3–19 (1961)
5. Bychkov, I., Batsyn, M., Pardalos, P.: Exact model for the cell formation problem. Optim. Lett. **8**, 2203–2210 (2014)
6. Chan, H.M., Milner, D.A.: Direct clustering algorithm for group formation in cellular manufacture. J. Manuf. Syst. **1**(1), 64–76 (1982)
7. Chandrasekharan, M.P., Rajagopalan, R.: MODROC: an extension of rank order clustering for group technology. Int. J. Prod. Res. **24**(5), 1221–1233 (1986)
8. Chandrasekharan, M.P., Rajagopalan, R.: An ideal seed non-hierarchical clustering algorithm for cellular manufacturing. Int. J. Prod. Res. **24**(2), 451–464 (1986)
9. Chandrasekharan, M.P., Rajagopalan, R.: Groupability: analysis of the properties of binary data matrices for group technology. Int. J. Prod. Res. **27**(6), 1035–1052 (1989)
10. Flanders, R.E.: Design manufacture and production control of a standard machine. Trans. ASME **46**, 691–738 (1925)
11. Hsu, C.P.: Similarity coefficient approaches to machine-component cell formation in cellular manufacturing: a comparative study. Ph.D. Thesis, Department of Industrial and Manufacturing Engineering, University of Wisconsin Milwaukee (1990)
12. James, T.L., Brown, E.C., Keeling, K.B.: A hybrid grouping genetic algorithm for the cell formation problem. Comput. Oper. Res. **34**(7), 2059–2079 (2007)
13. King, J.R.: Machine-component grouping in production flow analysis: an approach using a rank order clustering algorithm. Int. J. Prod. Res. **18**(2), 213–232 (1925)
14. Kumar, K.R., Chandrasekharan, M.P.: Grouping efficacy: a quantitative criterion for goodness of block diagonal forms of binary matrices in group technology. Int. J. Prod. Res. **28**(2), 233–243 (1990)
15. Kusiak, A.: The generalized group technology concept. Int. J. Prod. Res. **25**(4), 561–569 (1987)
16. Kusiak, A., Chow, W.S.: Efficient solving of the group technology problem. J. Manuf. Syst. **6**(2), 117–124 (1987)
17. Mitrofanov, S.P.: Nauchnye osnovy gruppovoy tekhnologii. Lenizdat, Leningrad, 435 p. (in Russian) (1959)
18. Mosier, C.T., Taube, L.: The facets of group technology and their impact on implementation. OMEGA **13**(6), 381–391 (1985)

19. Seifoddini, H.: A note on the similarity coefficient method and the problem of improper machine assignment in group technology applications. Int. J. Prod. Res. **27**(7), 1161–1165 (1989)
20. Seifoddini, H., Wolfe, P.M.: Application of the similarity coefficient method in group technology. IIE Trans. **18**(3), 271–277 (1986)
21. Stanfel, L.: Machine clustering for economic production. Eng. Costs Prod. Econ. **9**, 73–81 (1985)
22. Waghodekar, P.H., Sahu, S.: Machine-component cell formation in group technology MACE. Int. J. Prod. Res. **22**, 937–948 (1984)

Part II
Network Data Mining

Hybrid Community Detection in Social Networks

Hongwei Du, Weili Wu, Lei Cui, and Ding-Zhu Du

Abstract Community detection is an important subject in the study of social networks. In this article, we point out several ideas to design hybrid methods for community detection.

Keywords Social networks • Community detection • Modularity • Hybrid algorithms

1 Introduction

Social relationships have been used widely to develop online social networks (OSNs) such as FaceBook, Twitter, ResearchGate, and LinkIn. They allow individuals to present themselves, articulate their social networks, and establish or maintain personal data and connections with others. Many OSNs have gone through a rapid progression from their birth. For example, Facebook states to have more than 400 million of active users, whereas Twitter counts a rate of 300,000 new users per day. Collaboration and interaction among users in a social network pave the road for a pervasive experience, since online communities and applications are available on a large set of computing devices, ranging from personal computers to smart phones.

Often, OSNs have community structure. People in the same community may have a lot of interests in common and they may influence each other strongly. This makes the community structure play an important role in the study of social networks. For example, if the community structure is known, then the influence maximization problem can be approximated with good performance [3, 4, 14, 22], while in general networks, the approximation performance ratio cannot be better

H. Du
Harbin Institute of Technology Shenzhen Graduate School, Shenzhen, China
e-mail: hongwei@ieee.org

W. Wu • L. Cui • D.-Z. Du (✉)
Department of Computer Science, University of Texas at Dallas, Richardson, TX 75080, USA
e-mail: weiliwu@utdallas.edu; cxl131130@utdallas.edu; dzdu@utdallas.edu

V.A. Kalyagin et al. (eds.), *Models, Algorithms and Technologies for Network Analysis*, Springer Proceedings in Mathematics & Statistics 156,
DOI 10.1007/978-3-319-29608-1_8

than $O \log n$). Knowing community structure is also very helpful in studying the rumor blocking [10]. While the rumor spreads too fast inside a community, blocking is to cut its spread to outside the community.

When the community structure is unknown, the community detection becomes a very important problem. There exist many algorithms in the literature for study of the community detection. Most of them are heuristics, i.e., algorithms without theoretical performance analysis. Therefore, research issues and challenges exist in giving them theoretical analysis and designing new algorithms with guaranteed performance. In this article, we would like to present some ideas and indicate some possibilities.

Given a network, the community detection aims at clustering vertexes according to some quality function or condition. Since the definition of community is ambiguous, the quality function or condition has many choices. Each of them is designed based on certain consideration. For example, we may consider the following properties:

(A1) Connections in the same community are more than connections between different communities.
(A2) The influence inside a community is stronger than the influence to outside.

If an algorithm is designed based on (A1), then it is called the connection-based detection algorithm. If based on (A2), then it is called the influence-based detection algorithm.

A very well-known quality function is called *modularity* defined by Newman [18]. An algorithm is called a modularity-based algorithm if it is designed based on the modularity function.

It has been known that communities obtained from an algorithm based on a quality function may not be qualified based on another quality function. For example, a connection-based algorithm may produce a community partition with low modularity value. A modularity-based algorithm may produce community partition not satisfying connection property. This article is motivated from this fact. We propose hybrid algorithms which may be based on two or more quality functions so that they may produce community partitions with high quality, meeting the need of applications.

2 Connection-Based Detection

There are four different understandings for connection property (A1).

(C1) Each community has more connections inside than connections to outside.
(C2) Each community has more connections inside than connections to any other community.
(C3) Each node in a community has more connections inside than connections to outside.
(C4) Each node in a community has more connections inside than connections to any other community.

Consider an OSN with a graph $G = (V, E)$ as its mathematical model. A partition of V, $(V_1, V_2, .., V_h)$ is called a *community partition* under condition (Ci) if condition (Ci) holds. Let $A(G) = (a_{ij}$ be the adjacency matrix of G, i.e.,

$$a_{ij} = \begin{cases} 1 & \text{if edge } (v_i, v_j) \text{ exists} \\ 0 & \text{otherwise.} \end{cases}$$

For any two node subsets U and W, define

$$L(U, W) = \sum_{i \in U, j \in W} a_{ij}.$$

Then we can write these four conditions into the following formulations:

$$\begin{aligned}
&(\text{C1}) \quad (\forall s)[L(V_s, V_s) \geq L(V_s, \bar{V}_S)].\\
&(\text{C2}) \quad (\forall s)(\forall t \neq s)[L(V_s, V_s) \geq L(V_s, V_t)].\\
&(\text{C3}) \quad (\forall s)(\forall v \in V_s)[L(v, V_s) \geq L(v, \bar{V}_S)].\\
&(\text{C4}) \quad (\forall s)(\forall v \in V_s)(\forall t \neq s)[L(v, V_s) \geq L(v, V_t)].
\end{aligned}$$

Clearly,

$$(\text{C4}) \Rightarrow (\text{C3}) \Rightarrow (\text{C1}), (\text{C4}) \Rightarrow (\text{C2}) \Rightarrow (\text{C1}).$$

Therefore, a community partition satisfying (C1) is also called in *the most weak sense* [12] while a community partition satisfying (C2) is called in *weak sense* [21].

It is easy to see that if a community partition satisfies condition (C1) (or (C3)), then putting some communities together to form a big community results in a community partition still satisfying condition (C1) (or (C3)). Motivated from this property, based on the above conditions, many researchers design algorithms for community detection by maximizing the number of communities although this property does not hold for conditions (C2) and (C4).

$$\begin{aligned}
(\text{D}i) \ & \text{maximize } h \\
& \text{subject to } (\text{C}i).
\end{aligned}$$

Zhang et al. [24] showed that (D1) is NP-hard and formulated (D1) into a 0–1 linear programming. Lu et al. [16] showed that all (D2), (D3), and (D4) are NP-hard and gave two heuristics for all four maximizations.

Based on different understandings, one has designed different algorithms in the literature [16, 23].

3 Modularity-Based Detection

Newman [18] proposed to find community partition $(V_1, V_2, \ldots, V_k)$ of node set V by maximizing modularity function

$$Q(V_1, V_2, \ldots, V_k) = \sum_{s=1}^{k} \left[\frac{L(V_s, V_s)}{L(V, V)} - \left(\frac{L(V_s, V_s) + L(V_s, \bar{V}_s}{L(V, V)} \right)^2 \right].$$

Brandes et al. [2] that this is an NP-hard problem. Actually, Q is a nonlinear function. Its maximization is a typical nonlinear combinatorial optimization problem. A popular property of this class of problems is that it is NP-hard even on trees [8]. Dasgupta and Desai [6] showed several interesting results, including the following two:

Theorem 1. *For regular* $(n-4)$ *graphs, there exists a number* $\varepsilon > 0$ *such that there is no polynomial-time* $(1+\varepsilon)$*-approximation for the modularity maximization unless NP=P where n is the number of nodes.*

Theorem 2. *For d-regular graphs with* $d < \frac{n}{2 \ln n}$, *there exists a polynomial-time* $O(\log d)$*-approximation.*

Many algorithms have been designed for community detection using the modularity maximization [1, 5–8, 11, 19, 20]. While most of them are heuristics, Dinh and Thai [7] gave polynomial-time constant-approximation for power-law graphs. Especially, the low-degree following algorithm designed by them is quite simple and clever.

4 Hybrid Detection

In this section, we would like to give several ideas to design hybrid community detection methods.

A naive idea is to maximize Q subject to one of (C1)–(C4). The question is how to give a good formulation for this optimization problem so that exact solution or approximation solution can be computed efficiently. In the following, let us point out the possibility of realization of this idea.

First, we introduce some formulation given in [24]. Let x_{is} be a 0–1 variable which indicates if node v_i belongs to community V_s, and z_{hs} a 0–1 variable which indicates if edge e_h belongs to community V_s. Then for edge $e_h = (v_i, v_j)$, we have

$$z_{hs} \le x_{is}, z_{hs} \le x_{js}, x_{is} + x_{js} - 1 \le z_{hs}.$$

Constraint (C1) can be represented by that for all $s = 1, 2, \ldots, k$,

$$2\sum_{h=1}^{m} z_{hs} \geq \sum_{j=1}^{n}\sum_{i=1}^{n} x_{is}a_{ij} - 2\sum_{h=1}^{m} z_{hs}.$$

This means that constraint (C1) can be represented by a set of linear inequalities. Moreover, Q can be written as a quadratic function with respect to variables x_{is} and z_{hs}. Therefore, maximizing Q subject to (C1) can be formulated as a 0–1 quadratic programming. In [9], we may find a method to transform a 0–1 quadratic programming into a semi-definite programming with which it is quite hopeful to find an approximation solution with guaranteed performance. (C2)–(C4) can also be represented by a set of linear inequalities. It may be worth mentioning that semi-definite programming has been successfully used for community detection [25–27].

Li et al. [15] proposed the modularity density function,

$$D(V_1, V_2, \ldots, V_k) = \sum_{s=1}^{k} \left[\frac{L(V_s, V_s) - L(V_s, \bar{V}_s)}{|V_s|} \right].$$

They claim that this function is superior than the modularity function Q in both theoretical and numerical comparison. It provides equivalence with the objective function of the kernel k means.

Could we optimize D subject to one of (C1)–(C4)? This would induce an optimization problem with a fractional objective function, linear constraints, and 0–1 variables, which is called a 0–1 geometric programming. It is a quite challenging problem. Indeed, it is hard to find a reference in the literature on how to find an exact solution or a good approximate solution for 0–1 geometric programming.

Lu et al. [17] gave an influence-based detection method. They proposed to do community detection by solving the following:

$$\begin{aligned} &\text{maximize} \quad I(V_1, V_2, \ldots, V_k) = \sum_{s=1}^{k} \sigma_m(V_s) \\ &\text{subject to} \quad (V_1, V - 2, \ldots, V_k) \text{ is a partition of } V, \end{aligned}$$

where m is a diffusion model such as independent cascade model, linear threshold model, etc. This method is based on consideration that each community tends to enlarge its group influence.

Could we design a hybrid method by combining a connection-based method and an influence-based method? This combination would induce a maximization problem with objective function I with linear constraints. Note that I is a monotone nondecreasing submodular function because each $\sigma_m(V_s)$ is. This would give a very interesting research direction. Again, if variables are continuous, then it is a tractable problem. However, currently, we are considering discrete variables which would generate a lot of research issues.

Our discussion as above is on undirected graphs. Actually, all connection-based detection methods, modularity-based detection methods have been extended to directed graphs [6, 7, 13]. Therefore, above ideas of ours can also be studied in directed graphs.

Following above ideas, we are working in detail in the design and analysis of several hybrid methods. Both theoretical and computer experimental results will be published in the future.

References

1. Agarwal, G., Kempe, D.: Modularity-maximizing graph communities via mathematical programming. Eur. Phys. J. B **66**, 409–418 (2008)
2. Brander, U., Delling, D., Gaertler, M., Görke, R., Hoefer, M., Nikoloski, Z., Wagner, D.: On modularity clustering. IEEE Trans. Knowl. Data Eng. **20**(2), 172–188 (2007)
3. Chen, W., Wang, C., Wang, Y.: Scalable influence maximization for prevalent viral marketing in large-scale social networks. In: Proceedings of KDD, pp. 1029–1038 (2010)
4. Chen, W., Yuan, Y., Zhang, L.: Scalable influence maximization in social networks under the linear threshold model. In: Proceedings of ICDM, pp. 88–97 (2010)
5. Clauset, A., Newman, M.E.J., Moore, C.: Finding community structure in very large networks. Phys. Rev. E **70**, 066111 (2004)
6. DasGupta, B., Desai, D.: On the complexity of Newman's community finding approach for biological and social networks. J. Comput. Syst. Sci. **79**, 50–67 (2013)
7. Dinh, T.N., Thai, M.T.: Community detection in scale-free networks: approximation algorithms for maximizing modularity. IEEE J. Sel. Areas Commun. **31**(6), 997–1006 (2013)
8. Dinh, T.N., Thai, M.T.: Towards optimal community detection: from trees to general weighted networks. Internet Math. (2014, to appear)
9. Du, D.-Z., Ko, K.-I., Hu, X.: Design and Analysis of Approximation Algorithms. Springer, New York (2012)
10. Fan, L., Lu, Z., Wu, W., Thuraisingham, B.M., Ma, H., Bi, Y.: Least cost rumor blocking in social networks. In: Proceedings of ICDCS 2013, pp. 540–549 (2013)
11. Girvan, M., Newman, M.E.J.: Community structure in social and biological networks. Proc. Natl. Acad. Sci. **99**(12), 7821–7826 (2002)
12. Hu, Y., Chen, H., Zhang, P., Li, M., Di, Z., Fan, Y.: Comparative definition of community and corresponding identifying algorithm. Phys. Rev. E **78**(2), 026121 (2008)
13. Karrer, B., Newman, M.J.: Random graph models for directed acyclic networks. Phys. Rev. E **80**, 046110 (2009)
14. Kimura, M., Saito, K.: Tractable models for information diffusion in social networks. In: Proceedings of the 10th European Conference on Principles and Practice of Knowledge Discovery in Databases, pp. 259–271 (2006)
15. Li, Z., Zhang, S., Wang, R.S., Zhang, X.S., Chen, L.: Quantitative function for community detection. Phys. Rev. E **77**(3), 036109 (2008)
16. Lu, Z., Wu, W., Chen, W., Zhong, J., Bi, Y., Gao, Z.: The maximum community partition problem in networks. Discret. Math. Alg. Appl. **5**(4) (2013)
17. Lu, Z., Zhu, Y., Li, W., Wu, W., Cheng, X.: Influence-based community partition for social networks. Comput. Soc. Netw. **1**, 1 (2014)
18. Newman, M.E.J.: Scientific collaboration networks. II. Shortest paths, weighted networks, and centrality. Phys. Rev. E **64**(1), 016132 (2001)
19. Newman, M.E.J.: Fast algorithm for detecting community structure in networks. Phys. Rev. E **69**, 066133 (2004)

20. Newman, M.E., Girvan, M.: Finding and evaluating community structure in networks. Phys. Rev. E **69**(2), 026113 (2004)
21. Radicchi, F., Castellano, C., Cecconi, F., Loreto, V., Parisi, D.: Defining and identifying communities in networks. Proc. Natl. Acad. Sci. U.S.A. **101**(9), 2658–2663 (2004)
22. Wang, Y., Cong, G., Song, G., Xie, K.: Community-based greedy algorithm for mining top-k influential nodes in mobile social networks. In: Proceedings of KDD, pp. 1039–1048 (2010)
23. Xie, J., Kelley, S., Szymanski, B.K.: Overlapping community detection in networks: the state of the art and comparative study. ACM. Comput. Surv. Article No. 43. (2013)
24. Zhang, X.-S., Li, Z.-P., Wang, R.-S., Wang, Y.: A combinatorial model and algorithm for globally searching community structure in complex networks. J. Comb. Optim. **23**(4), 425–442 (2012)
25. Zhu, Y., Wu, W., Willson, J., Ding, L., Wu, L., Li, D., Lee, W.: An approximation algorithm for client assignment in client/server systems. In: Proceedings of the 33rd IEEE International Conference on Computer Communications (INFOCOM 2014) (2014)
26. Zhu, Y., Li, D., Xu, W., Wu, W., Fan, L., Willson, J.: Mutual relationship based partitioning for social networks. IEEE Trans. Emerg. Top. Comput. (2014). doi:10:1109/TETC.2014.2380391
27. Zhu, Y., Lu, Z., Bi, Y., Wu, W., Jiang, Y., Li, D.: Influence and profit: two sides of the coin. In: Proceedings of ICDM 2013, pp. 1301–1306 (2013)

Spectral Properties of Financial Correlation Matrices

Maxim Kazakov and Valery A. Kalyagin

Abstract Random matrix theory (RMT) is applied to investigate the cross-correlation matrix of a financial time series in four different stock markets: Russian, American, German, and Chinese. The deviations of distribution of eigenvalues of market correlation matrix from RMT global regime are investigated. Specific properties of each market are observed and discussed.

Keywords Random matrices • Wishart–Laguerre ensemble • Stock market • Market network • Random Graphs

1 Introduction

The study of correlation (or covariance) matrices has a long history in finance and it is an important aspect of risk management and one of the cornerstone of Markowitz's theory of optimal portfolios [6, 10]. Besides, equal-time cross-correlation matrices play a major role in market network analysis when it comes to constructing different network structures, such as maximum spanning tree or market graph [1, 2].

When stock market consists of several hundreds of individual stocks, it becomes a high-dimensional and complex system. To study these systems some methods of statistical physics have been employed, in particular, random matrix theory (RMT) [3, 8, 11, 13]. The idea is to compare the properties of an empirical correlation matrix to the ones of purely *random* matrix. In case of covariance matrix ensemble of such random matrices is called Wishart–Laguerre ensemble [5]. Possible deviations from the random case may reveal some peculiarities of empirical correlation matrices and it may give some insight into the market structure.

M. Kazakov (✉)
National Research University Higher School of Economics, N. Novgorod, Russia
e-mail: max.a.kazakov@gmail.com

V.A. Kalyagin
Lab LATNA, National Research University Higher School of Economics, N. Novgorod, Russia
e-mail: vkalyagin@hse.ru

V.A. Kalyagin et al. (eds.), *Models, Algorithms and Technologies for Network Analysis*, Springer Proceedings in Mathematics & Statistics 156,
DOI 10.1007/978-3-319-29608-1_9

The main goal of the present paper is a comparative investigation of empirical correlation matrices for different markets. The main problem addressed is to understand whether the markets are different from RMT point of view and to point out these differences. We investigate four different markets corresponding to different levels of economic development: the US, German, Russian, and Chinese. We analyze spectral properties of empirical correlation matrices and compare them to global regimes provided by RMT. In addition, we test the stability of observed deviations and their dependence on the distribution of the data.

The paper is organized as follows. In Sect. 2 we remind the main facts from RMT and discuss the results of previous related studies. In Sect. 3 we present our methods and describe the data used in numerical experiments. In Sect. 4 we conduct a comparative analysis of correlations matrices for indicated markets. Section 5 is devoted to a stability analysis of observed phenomena. Section 6 contains concluding remarks.

2 Theoretical Background

2.1 Random Matrix Theory

We want to compare spectral properties of empirical correlation matrices of stock market with the spectral properties of random matrices. In case of covariance (or correlation) matrices this is so-called Wishart–Laguerre ensemble [5]. Consider rectangular ($N \times T$) matrix H whose elements $H_{i,t}$ are independent, identically distributed random variables. Then the product $W = (1/T) \cdot H \cdot H^*$ is a positive definite symmetric ($N \times N$) matrix that represents the normalized covariance matrix of the data. When elements $H_{i,t}$ are drawn from a Gaussian distribution, the product matrices $W = \frac{1}{T} \cdot H \cdot H^*$ constitute Wishart–Laguerre ensemble of random matrices.

For the case when $T \geq N$ (the number of samples is larger than the dimension) the spectral properties of these matrices are well studied and it is known that in limit ($N \to \infty$ and $T \to \infty$ and $Q = T/N \geq 1$ fixed) all eigenvalues are positive and density distribution of the eigenvalues is given by the Marchenko–Pastur function [9, 14]:

$$\rho_{WL}(\lambda) = \frac{Q}{2\pi} \cdot \frac{\sqrt{(\lambda_+ - \lambda)(\lambda - \lambda_-)}}{\lambda}, \quad \lambda_- < \lambda < \lambda_+, \tag{1}$$

where the lower and upper bounds of eigenvalues are calculated as follows:

$$\lambda_\pm = 1 + \frac{1}{Q} \pm 2\sqrt{\frac{1}{Q}}. \tag{2}$$

Note that above results are valid only in limit when $N \to \infty$.

2.2 *Related Works*

Recently series of studies has been conducted [3, 8, 11, 13] to analyze spectral properties of empirical correlation matrices and compare them to RMT global regime discussed in previous section. Following observations have been made:

- There is one largest eigenvalue λ_{max}, which is significantly higher than the upper bound λ_+. It is also tends to be relatively close to the product $N \cdot \overline{C}$, where $\overline{C}$ is the average of non-diagonal elements of correlation matrix C. The associated eigenvector is connected with global market index.
- There are also several eigenvalues slightly greater than λ_+. They may reflect sector behavior.
- There are a number of eigenvalues below the lower bound λ_-, which can be explained by repulsion effect which we will talk about later. It may also correspond particularly to highly correlated pair of stocks.
- Finally, most of the eigenvalues fall within a range predicted by RMT. These eigenvalues are called bulk of eigenvalue spectrum. Nonetheless, it was shown that these eigenvalues also may contain useful information [7].

These results may differ for emerging markets [4, 12]. Such as, in emerging markets the largest eigenvalue appears to be higher with respect to λ_+ and there are fewer eigenvalues above the edge. At the same time, there is a large proportion of eigenvalues below λ_- and, consequently, less number of eigenvalues in the bulk. Also, average value of non-diagonal elements of correlation matrix is higher and fluctuates more dynamically.

3 Method and Data

3.1 *Method*

We consider a set of N stocks over a period of T trading days. Let $P_i(t)$ be a closing price of stock $i(i = 1, \ldots, N)$ in the day $t(t = 1, \ldots, T)$. Then the daily log return $R_i(t)$ of stock i is defined by

$$R_i(t) = \ln \frac{P_i(t)}{P_i(t-1)}. \tag{3}$$

We normalize R_i with respect to its standard deviation σ_i as follows:

$$r_i(t) = \frac{R_i(t) - \overline{R_i}}{\sigma_i}, \tag{4}$$

where $\overline{R_i}$ denotes the average return over the period studied and standard deviation (or volatility) defined as $\sigma_i = \sqrt{\overline{R_i^2} - \overline{R_i}^2}$.

Then, the equal-time cross-correlation matrix C is expressed it terms of $r_i(t)$:

$$C_{i,j} = \sum_{t=1}^{T} r_i(t) \cdot r_j(t). \tag{5}$$

The element $C_{i,j}$ of matrix C denotes correlation coefficient between stock i and stock j. Correlation matrix C also can be expressed in matrix notation as

$$C = \frac{1}{T} \cdot R \cdot R^T, \tag{6}$$

where R is an $(N \times T)$ matrix with elements $r_i(t)$.

The N eigenvalues λ_i and their corresponding eigenvectors u_i are calculated by diagonalizing C. One has

$$C \cdot u_i = \lambda_i \cdot u_i, \quad i = 1, \ldots, N. \tag{7}$$

Note that $\sum \lambda_i$ is always equal to sum of the diagonal elements of C (the trace), which is always constant and equal to N since for all elements $C_{i,i} = 1$. Hence, if some eigenvalues increase, then some others must decrease to compensate, and vice versa. This is called *eigenvalue repulsion* [3].

3.2 Data

In order to analyze spectral properties of empirical financial correlation matrices we consider four different stock markets, representing different types of economies: the US, Russian, German, and Chinese stock market. For Russian market we consider stocks traded on The Moscow Interbank Currency Exchange (MICEX). For American market we consider equities of S&P 500 traded on The New York Stock Exchange (NYSE). For German market we consider equities of HDAX traded on The Frankfurt Stock Exchange (FWB). And for Chinese market we consider stocks traded on The Hong Kong Stock Exchange (HKEx).

We want $Q = T/N$ to be relatively equal for all markets and we eliminate stocks if they haven't been traded long enough. For Russian market we also apply cleaning procedure in order to eliminate stocks with low liquidity. One exception here is an American market. In this case we allow Q to be essentially smaller than in other markets so we can apply our method to larger data set. Dates and the number of chosen stocks of each market are summarized in Table 1.

Table 1 Characteristics of considered markets

Market	Number of stocks, N	Length of time series, T	Q = T/N	Starting date	Ending date
Russia	101	1418	14.04	10/01/2008	06/06/2014
USA	316	3008	9.52	01/03/2003	12/12/2014
Germany	90	1282	14.24	01/05/2010	12/12/2014
China	78	1016	13.03	01/03/2011	12/12/2014

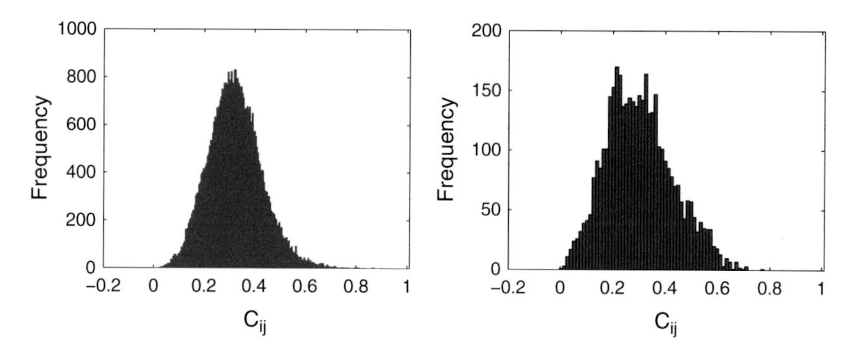

Fig. 1 Distribution of correlations. *Left*—the US market. *Right*—German market

4 Comparative Analysis of Different Markets

In this section we present the results of the analysis of empirical correlation matrices for four different stock markets. We compare the empirical distribution of eigenvalue with predictions of RMT and discuss some deviations.

4.1 Distribution of the Correlation Coefficients

First, we take a look at the statistical properties of empirical cross-correlation matrices. Figures 1 and 2 show histograms of correlation coefficients (i.e., non-diagonal elements of correlation matrix C) for all four markets. Other comparative characteristics are given in Table 2.

We notice that average value $\overline{C}$ is quite large for all cases. The interesting fact here is that it is almost the same and around 0.3 for all considered markets, except Russian. Furthermore, standard deviations are also relatively high and close to each other, this time including Russian market. For American, German, and Chinese markets almost all elements of correlation matrix are positive.

Next we test the assumption of normal distribution of cross-correlation matrix elements. Histograms on Figs. 1 and 2 don't show distribution similar to normal (or maybe just for American market). Lilliefors test rejected hypothesis of normal distribution at the 5 % significance level for all markets. We also use skewness and

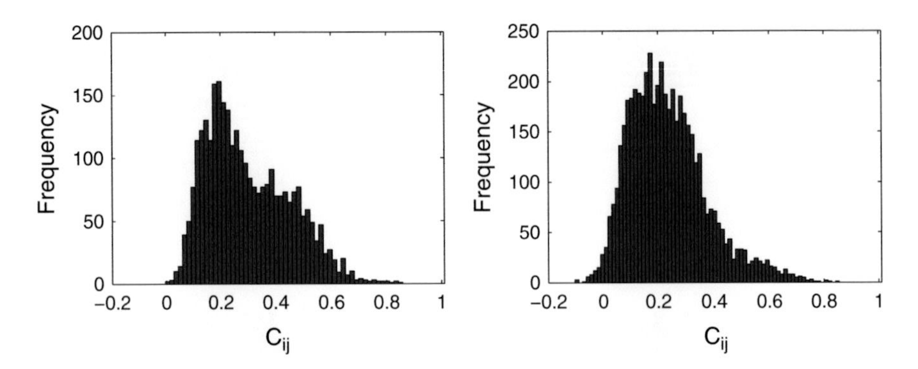

Fig. 2 Distribution of correlations. *Left*—Chinese market. *Right*—Russian market

Table 2 Statistics for cross-correlation

Market	Mean	Standard deviation	Skewness	Kurtosis
USA	0.3220	0.1060	0.3957	3.4929
Germany	0.3021	0.1282	0.3846	2.7806
China	0.2984	0.1514	0.5968	2.6833
Russia	0.2349	0.1408	0.8391	3.8482

kurtosis measures to see how much the deviations are. Skewness is a measure of the asymmetry of the data around the sample mean and kurtosis is a measure of how outlier-prone a distribution is (respectively, 0 and 3 for the normal distribution). As shown in Table 2 for all markets skewness is positive which indicates that correlations are skewed right meaning that the right tail is long with respect to the left tail. Kurtosis measure, in contrast, deviates in different directions, indicating more peaked distribution for American and Russian markets, and more flat distribution for German and Chinese. The deviations are relatively small though.

4.2 *Eigenvalue Distribution*

In this section we analyze spectral properties of empirical cross-correlation matrices and compare them to the predictions of RMT given by formulas (1) and (2). The eigenvalue spectrum is shown in Figs. 3b, 4b, 5b, and 6b with the spectrum predicted by RMT in Figs. 3a, 4a, 5a, and 6a. Table 3 presents the more detailed characteristics.

As in the previous studies, we found that there is one largest eigenvalue λ_{max} in every case which exceeds significantly the upper bound λ_{+}. We also noticed the similarity between λ_{max} and $N{\cdot}\overline{C}$ presented in Table 3 by their ratio close to the value of 1. This explains the exceptionally large value of λ_{max} for American market with respect to the others: since average value of correlation is similar for each market

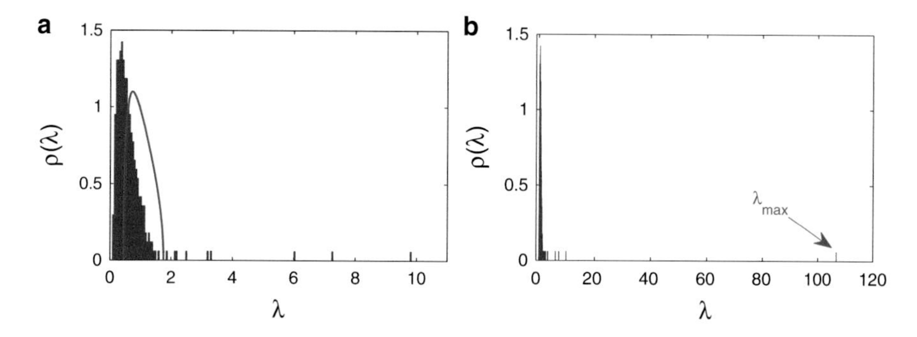

Fig. 3 American market. (**a**) Probability density of λ in comparison with RMT density (the *red solid line*) and (**b**) including the largest eigenvalue λ_{max}

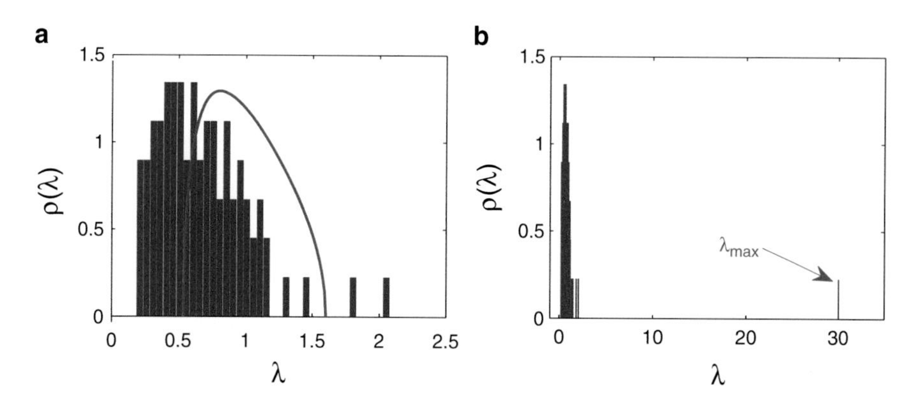

Fig. 4 German market. (**a**) Probability density of λ in comparison with RMT density (the *red solid line*) and (**b**) including the largest eigenvalue λ_{max}

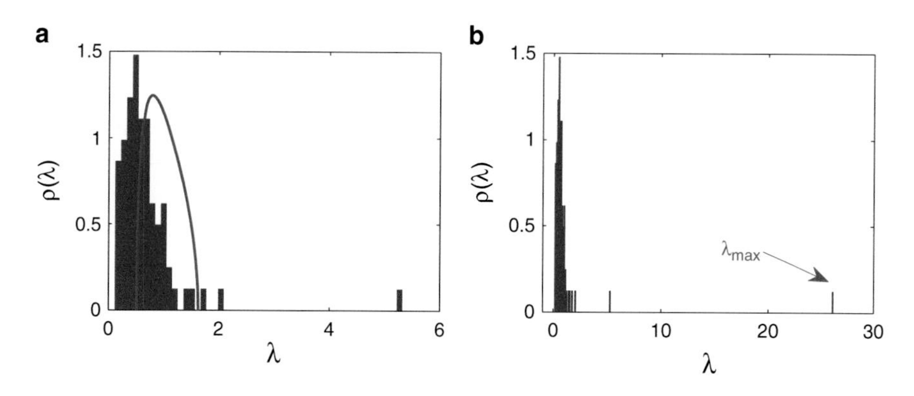

Fig. 5 Chinese market. (**a**) Probability density of λ in comparison with RMT density (the *red solid line*) and (**b**) including the largest eigenvalue λ_{max}

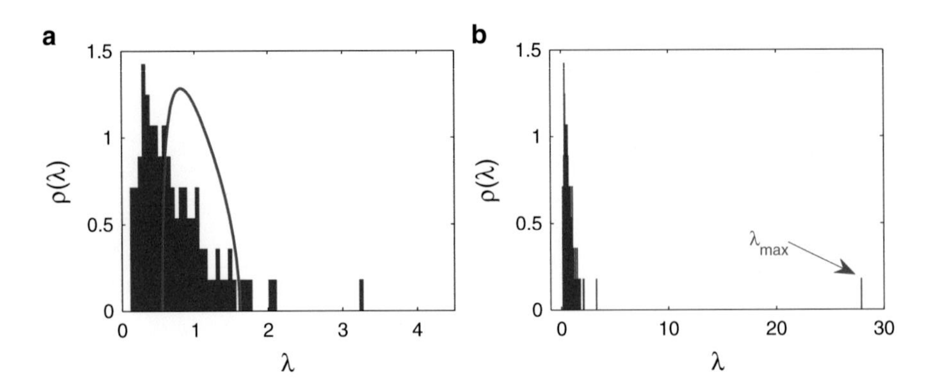

Fig. 6 Russian market. (**a**) Probability density of λ in comparison with RMT density (the *red solid line*) and (**b**) including the largest eigenvalue λ_{max}

Table 3 Eigenvalues statistics

Market	N	$\overline{C}$	λ_-	λ_+	λ_{max}	$\frac{\lambda_{max}}{\lambda_+}$	$\frac{\lambda_{max}}{N \cdot \overline{C}}$	$\lambda > \lambda_+$
USA	316	0.3220	0.4568	1.7533	106.7823	60.9039	1.0495	10
Germany	90	0.3021	0.5403	1.6001	30.0333	18.7694	1.1046	3
China	78	0.2984	0.5226	1.6309	26.1665	16.0440	1.1240	4
Russia	101	0.2349	0.5375	1.6050	27.9230	17.3975	1.1772	7

and American market is presented by data set greater by 3–4 times (with respect to the number of stocks), the value of the largest eigenvalue is also greater by 3–4 times.

The number of eigenvalues above the edge λ_+ differs for considered markets. In German and Chinese markets there are, respectively, 2 and 3 such eigenvalues, besides λ_{max}, which is small and in accordance with previous studies [11]. In American and Russian markets this number is relatively high (9 and 6, respectively) and for American market it is greater than what was observed before [8].

Furthermore, we noticed that about half of the eigenvalues falls into the range $[\lambda_-, \lambda_+]$ predicted by RMT. A little less number of eigenvalues fall below the edge λ_-. Most of this may be explained by eigenvalue repulsion effect we talked about in Sect. 3. These observations also support some previous results [11].

5 Stability Analysis

In this section we present the results of analysis of stability of observed phenomena. We want to see whether the observed deviations from RMT predictions are specific for a certain market or not. In order to do this we use bootstrap method. We

Table 4 Characteristics

Market	$\overline{C}$	λ_{max}	$\frac{\lambda_{max}}{N \cdot \overline{C}}$	$\lambda > \lambda_{+}$
USA	0.3220	106.7823	1.0495	10
Germany	0.3021	30.0333	1.1046	3
China	0.2984	26.1665	1.1240	4
Russia	0.2349	27.9230	1.1722	7

also test dependence of the deviations on type of distribution of the data using multivariate normal distribution and multivariate Student distribution. We test following characteristics:

- The mean value of correlation coefficient $\overline{C}$
- The value of the largest eigenvalue λ_{max}
- The ratio $\lambda_{max}/(N \cdot \overline{C})$, where N denotes the number of stocks
- The number of eigenvalues above the upper bound λ_{+} predicted by RMT

These characteristics are summarized for all four considered markets in Table 4.

5.1 Bootstrapping

To test the stability of observed characteristics and, consequently, their deviations from predictions of RMT we apply the bootstrap method. First, we resample the data with replacement, saving the size of the resample $(N \times T)$ the same as it was in the original data set. Note that sample here is a vector R_t corresponding to a trade day t characterized by daily returns of N stocks. Next we apply our method, defined by formulas (4)–(7), to compute characteristics of interest. We repeat this routine $10,000$ times for each considered market.

Figures 7, 8, 9, and 10 present histograms of analyzed characteristics which provide an estimate of the shape of the distribution. We found that almost all of them are stable for each market indicating that considered deviations from RMT predictions are specific for empirical correlation matrices. One exception here is the number of eigenvalues above the upper bound λ_{+}. For German and Chinese markets the value is quite robust (Fig. 10b, c), but for American and Russian cases results show that the observed values are not reliable (Fig. 10a, d). The surprising result is that the test revealed a greater number of those eigenvalues (about 12 for the USA and 9 for Russia in average).

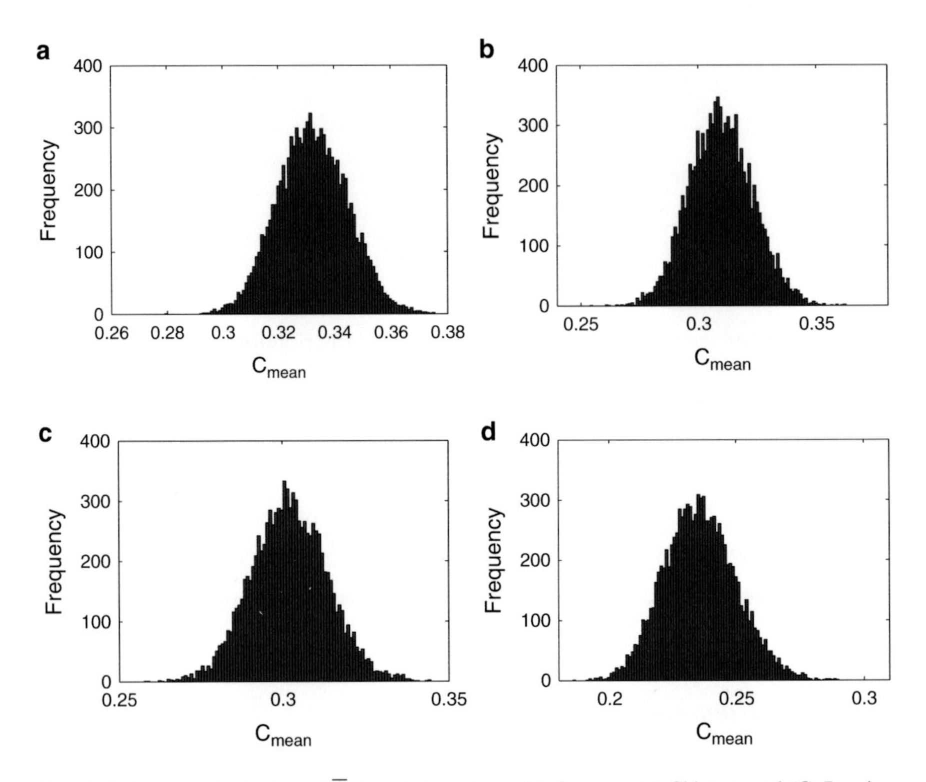

Fig. 7 Estimated distribution of $\overline{C}$ for (**a**) American, (**b**) German, (**c**) Chinese, and (**d**) Russian markets

5.2 Multivariate Normal Distribution

In this section we test how the cross-correlation matrix will change (with respect to observed characteristics) if we let the distribution of the data be Gaussian. We generate new data of size $(N \times T)$ (the same as original) from multivariate normal distributions with zero means and the empirical correlation matrix C as covariance matrix. Next we apply our method, defined by formulas (4)–(7), to compute new correlation matrix and its characteristics. We repeat this routine $10,000$ times for each considered market.

Figures 11, 12, 13, and 14 present histograms of analyzed characteristics. The results of the analysis show that this approach keeps the main characteristics the same except one. All characteristics saved their observed values in average and estimated shape of distribution is similar with the one provided by bootstrapping for most of the characteristics in each market. The number of eigenvalues above the edge hasn't saved its observed value in American and Russian markets but it appeared to be less than for bootstrapping in average (11 and 8, respectively) with very small probability for other values (Fig. 14a, d).

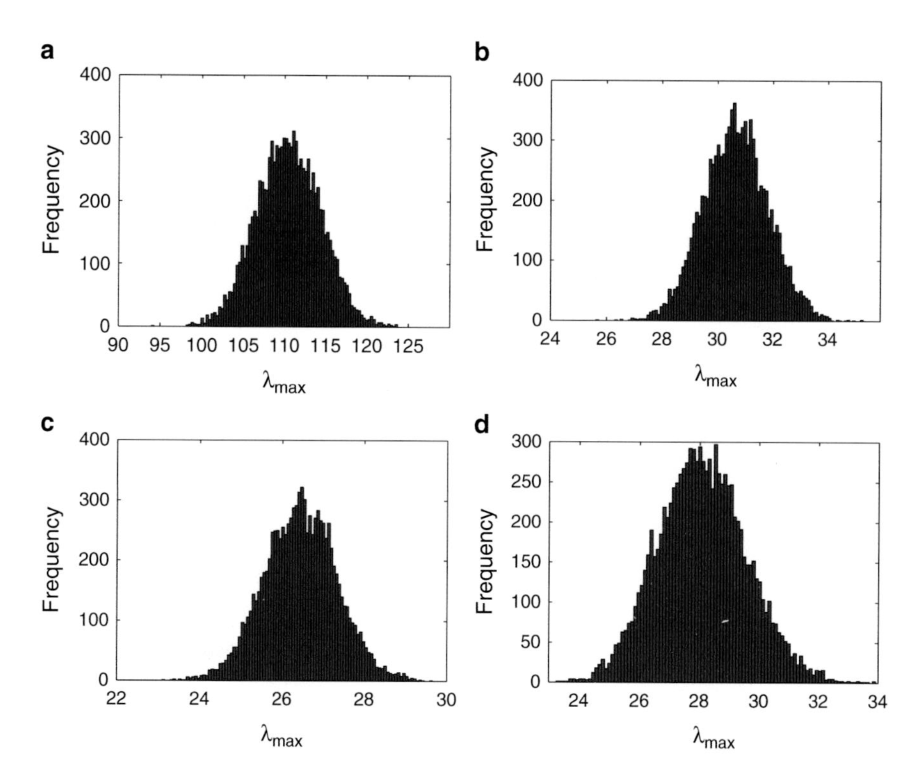

Fig. 8 Estimated distribution of λ_{max} for (**a**) American, (**b**) German, (**c**) Chinese, and (**d**) Russian markets

5.3 Multivariate Student Distribution

As in the previous section, we simulate our data ($N \times T$ time series), but this time using multivariate Student distribution with the empirical correlation matrix C as covariance matrix and 3 degrees of freedom. Then again we apply our method, defined by formulas (4)–(7), to compute new correlation matrix and its characteristics. We repeat this routine $10,000$ times for each considered market.

The histograms on Figs. 15, 16, and 17 show that again the characteristics saved their observed values in average in each market. But this time variance is much less and estimated shape of distribution is not reminiscent of the one provided by bootstrapping. For the number of eigenvalues above λ_+ the picture is completely different from previous two tests. The average value is significantly higher for all markets and estimated shape of distribution also differs (Fig. 18). It means that this characteristic is sensitive to distribution of returns.

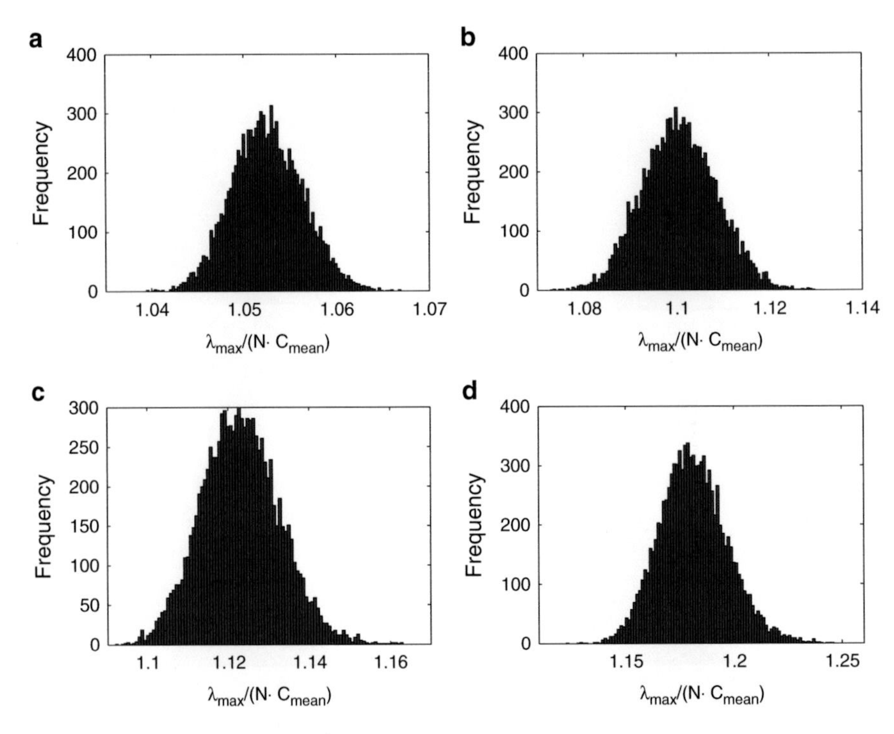

Fig. 9 Estimated distribution of $\frac{\lambda_{max}}{N \cdot \overline{C}}$ for (**a**) American, (**b**) German, (**c**) Chinese, and (**d**) Russian markets

6 Concluding Remarks

Four different stock markets (Russian, American, German, and Chinese) are compared with respect to deviation of spectral properties of correlation matrix to predictions provided by RMT. It is observed that (like in the previous studies), there is one largest eigenvalue significantly higher than upper bound λ_+ of RMT range, and it is very close to the product $N \cdot \overline{C}$, where N denotes the number of stocks and $\overline{C}$—the average value of correlation. Average value of correlation is about 0.3 for all markets, except Russian, which is surprisingly high. In contrast, the number of eigenvalues above λ_+, and the value of these numbers, differs from one market to another one. It can be related with sectors interconnections in different markets. Stability of observed phenomena was tested using bootstrapping method to see whether they are specific for considered markets or not. The analysis showed that the most of observed deviations from RMT are stable, the exception is the number of eigenvalues above the upper bound in American and Russian markets. This characteristic is not stable with respect to distribution of returns too.

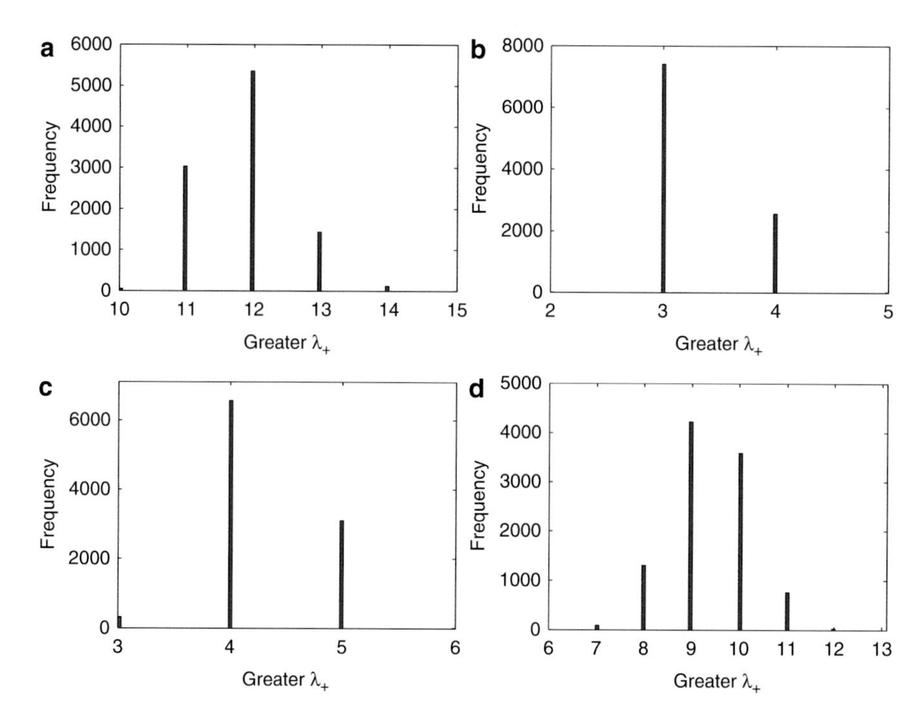

Fig. 10 Estimated distribution of the number of λ above λ_+ for (**a**) American, (**b**) German, (**c**) Chinese, and (**d**) Russian markets

Acknowledgements This work is partly supported by LATNA Laboratory, NRU HSE, RF government grant 11.G34.31.0057 and RFHR grant 15-32-01052.

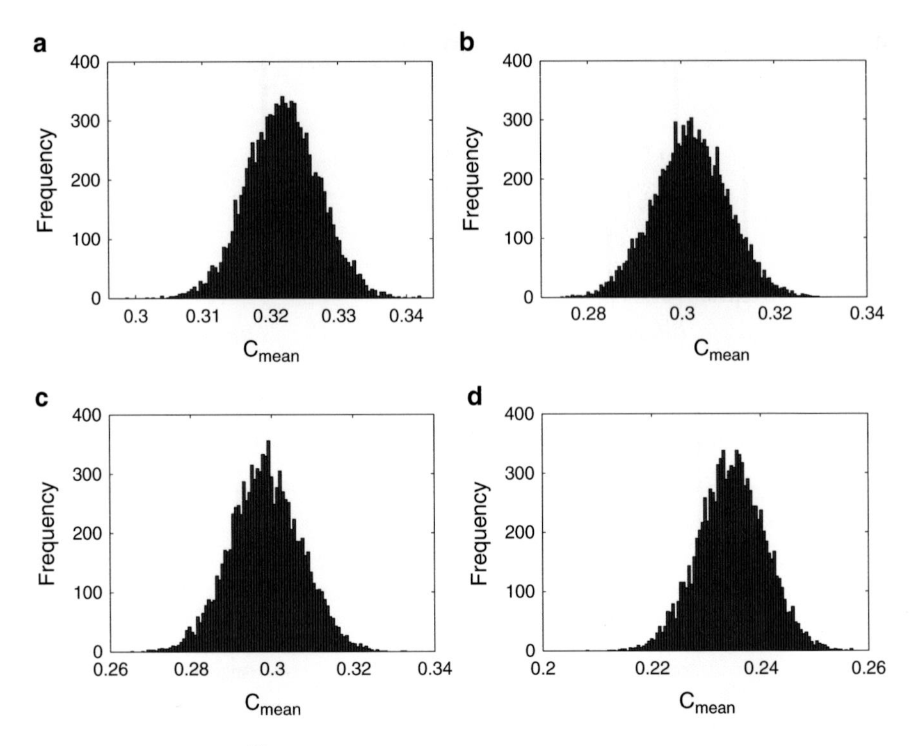

Fig. 11 Distribution of $\overline{C}$ for (**a**) American, (**b**) German, (**c**) Chinese, and (**d**) Russian markets from data generated by MVN

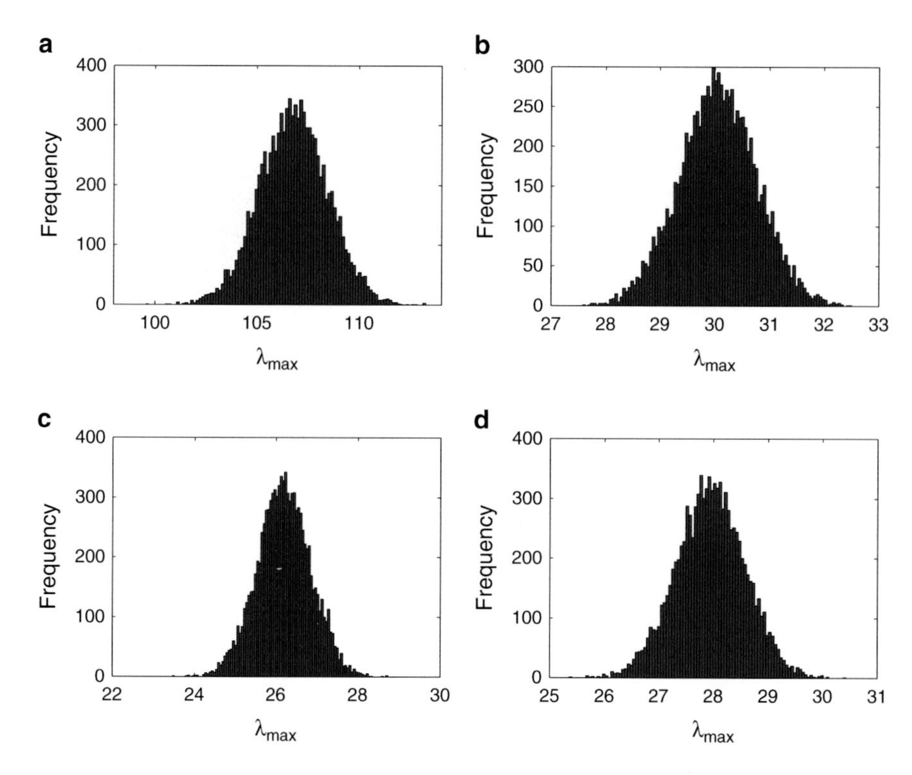

Fig. 12 Distribution of λ_{max} for (**a**) American, (**b**) German, (**c**) Chinese, and (**d**) Russian markets from data generated by MVN

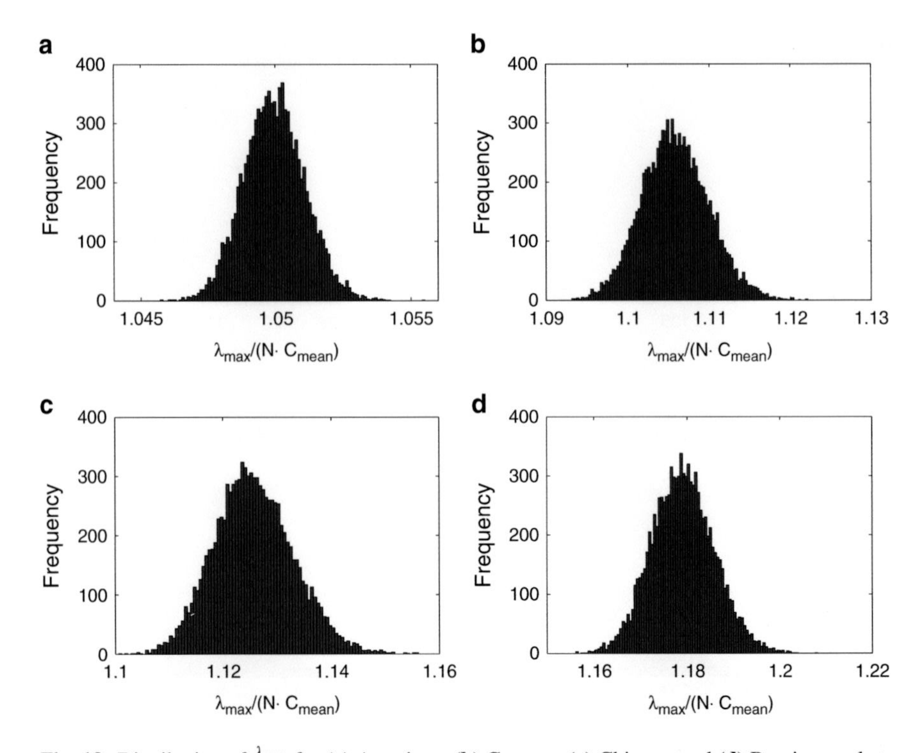

Fig. 13 Distribution of $\frac{\lambda_{max}}{N \cdot C}$ for (**a**) American, (**b**) German, (**c**) Chinese, and (**d**) Russian markets from data generated by MVN

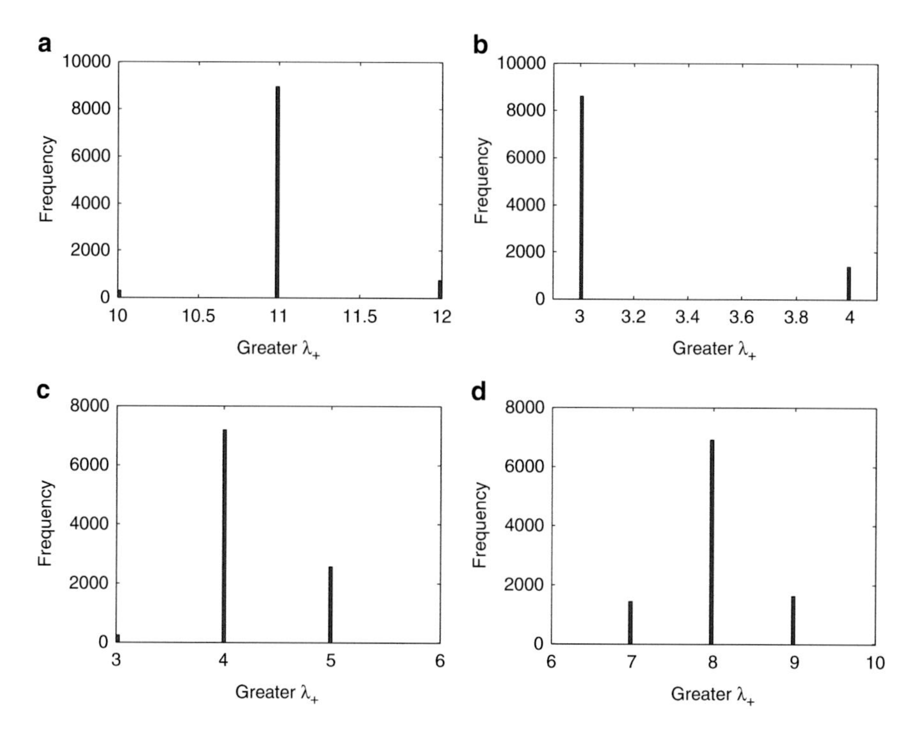

Fig. 14 Distribution of the number of λ above λ_+ for (**a**) American, (**b**) German, (**c**) Chinese, and (**d**) Russian markets from data generated by MVN

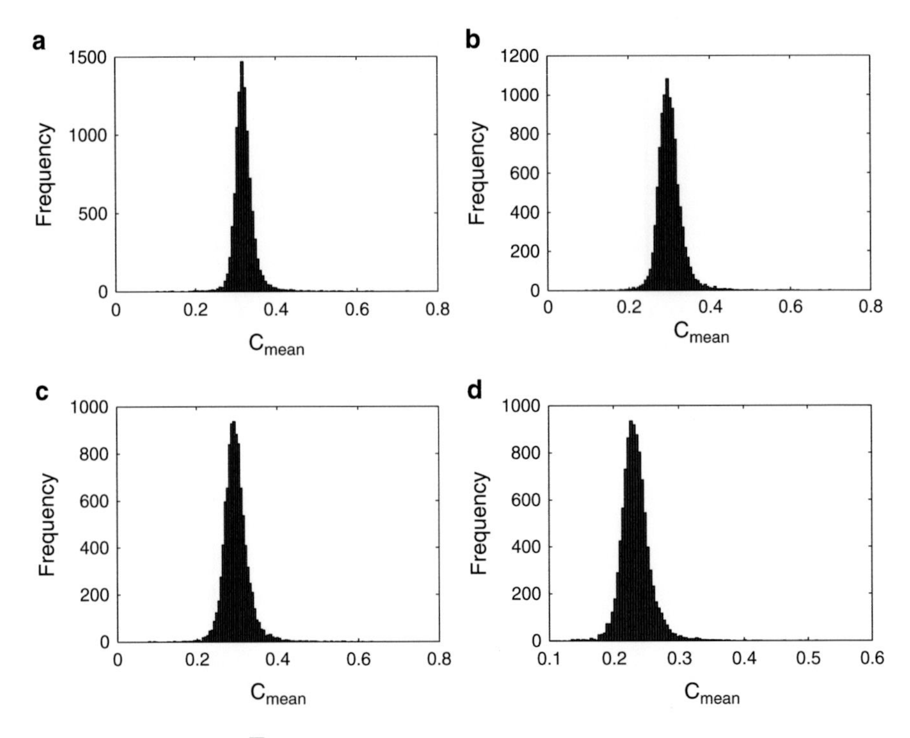

Fig. 15 Distribution of $\overline{C}$ for (**a**) American, (**b**) German, (**c**) Chinese, and (**d**) Russian markets from data generated by MVStudent

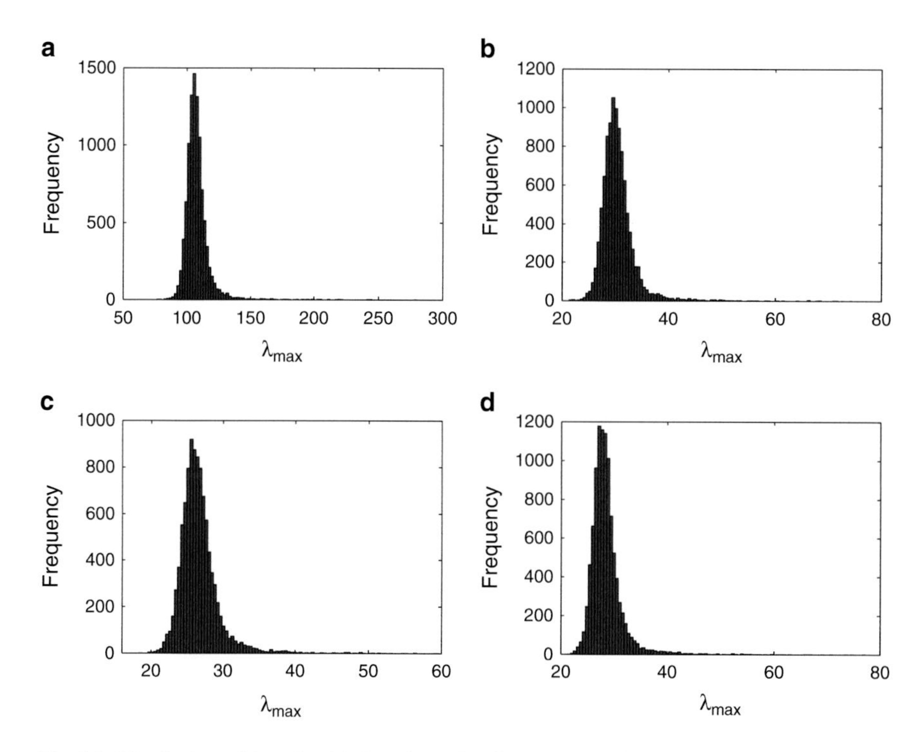

Fig. 16 Distribution of λ_{max} for (**a**) American, (**b**) German, (**c**) Chinese, and (**d**) Russian markets from data generated by MVStudent

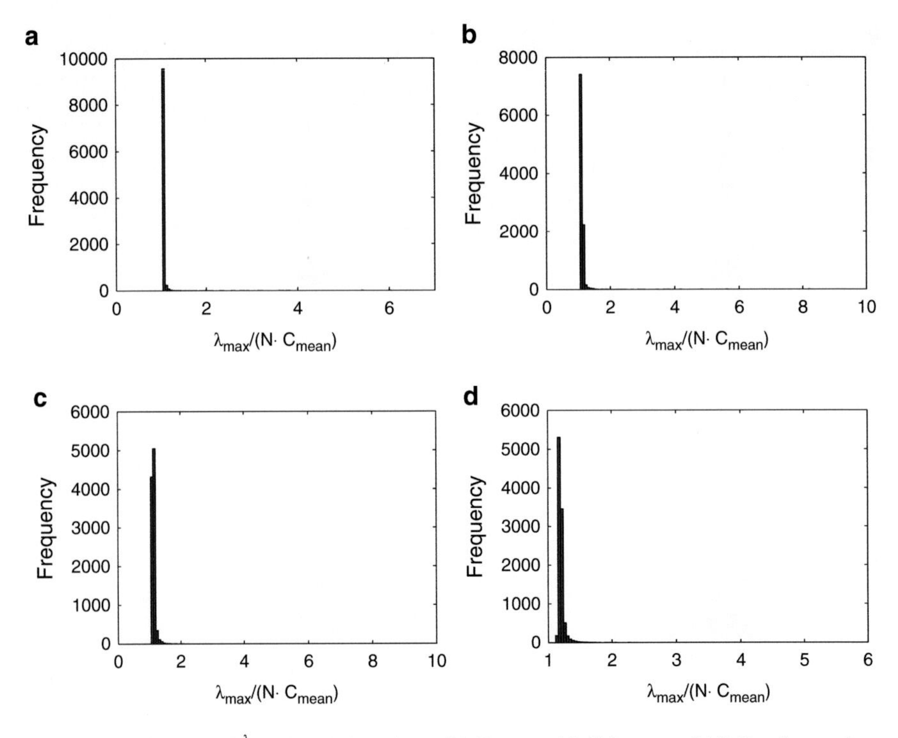

Fig. 17 Distribution of $\frac{\lambda_{max}}{N \cdot \overline{C}}$ for (**a**) American, (**b**) German, (**c**) Chinese, and (**d**) Russian markets from data generated by MVStudent

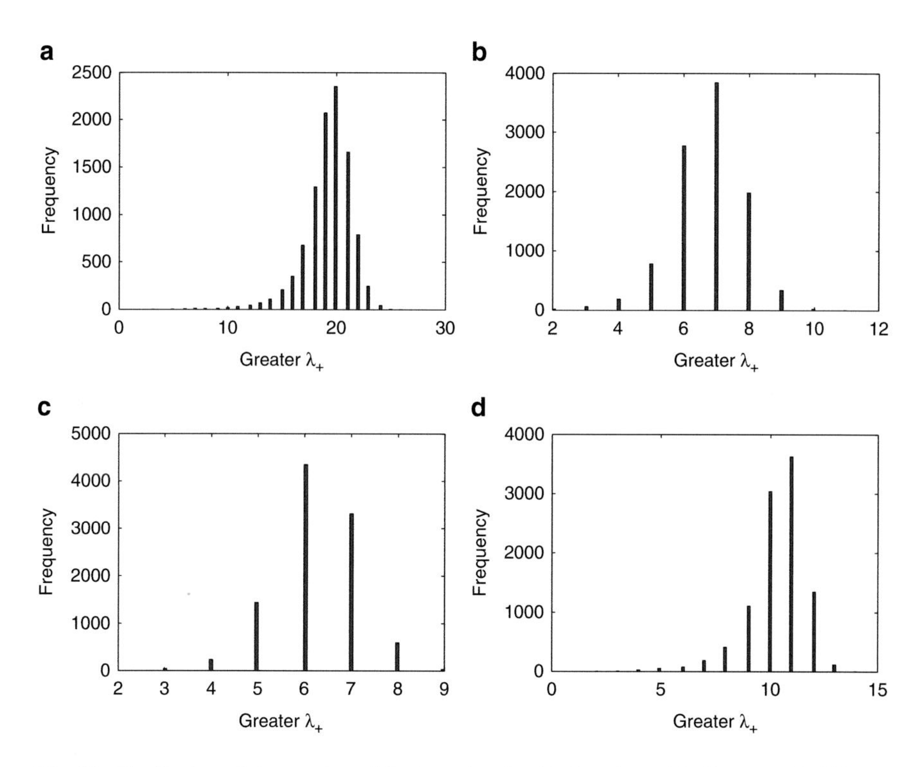

Fig. 18 Distribution of the number of λ above λ_+ for (**a**) American, (**b**) German, (**c**) Chinese, and (**d**) Russian markets from data generated by MVStudent

References

1. Boginski, V., Butenko, S., Pardalos, P.M.: On structural properties of the market graph. In: Innovations in Financial and Economic Networks (Anna Nagurney, Editor), pp. 29–45, Edward Elgar Publishing, (2003)
2. Boginski, V., Butenko, S., Pardalos, P.M.: Mining market data: a network approach. Comput. Oper. Res. **33**(11), 3171–3184 (2006)
3. Conlon, T., Ruskin, H.J., Crane, M.: Cross-correlation dynamics in financial time series. Physica A **388**(5), 705–714 (2009)
4. Cukur, S., Eryigit, M., Eryigit, R.: Cross correlations in an emerging market financial data. Physica A **376**, 555–564 (2007)
5. Edelman, A.: Eigenvalues and condition numbers of random matrices. SIAM J. Matrix Anal. Appl. **9**(4), 543–560 (1988)
6. Elton, E.J., Gruber, M.J.: Modern Portfolio Theory and Investment Analysis. Wiley, New York (1995)
7. Kwapien, J., Oswiecimka, P., Drozdz, S.: The bulk of the stock market correlation matrix is not pure noise. Physica A **359**, 589–606 (2006)
8. Laloux, L., Cizeau, P., Bouchaud, J.P., Potter, M.: Noise dressing of financial correlation matrices. Phys. Rev. Lett. **83**(7), 1467–1470 (1999)
9. Marchenko, V., Pastur, L.: Distribution of some sets of random matrices. Math. USSR-Sb **1**, 457–483 (1967)
10. Markowitz, H.M.: Portfolio Selection: Efficient Diversification of Investments. Wiley, New York (1959)
11. Nguyen, Q.: One-factor model for the cross-correlation matrix in the Vietnamese stock market. Physica A **392**(13), 2915–2923 (2013)
12. Oh, G., Eom, C., Wang, F., Jung, W.S., Stanley, H.E., Kim, S.: Statistical properties of cross-correlation in the Korean stock market. Eur. Phys. J. B **79**, 55–60 (2011)
13. Plerou, V., Gopikrishnan, P., Rosenow, B., Amaral, L.A.N., Stanley, H.E.: Universal and nonuniversal properties of cross correlations in financial time series. Phys. Rev. Lett. **83**(7), 1471–1474 (1999)
14. Sengupta, A.M., Mitra, P.P.: Distributions of singular values for some random matrices. Phys. Rev. E **90**(3), 3389–3392 (1999)

Statistical Uncertainty of Minimum Spanning Tree in Market Network

Anastasia Komissarova and Petr Koldanov

Abstract The paper deals with uncertainty in market network analysis. The main problem addressed is to investigate statistical uncertainty of Kruskal algorithm for the minimum spanning tree in market network. Uncertainty of Kruskal algorithm is measured by the probability of q incorrectly included edges. Numerical experiments are conducted with the returns of a set of 100 financial instruments traded in the US stock market over a period of 250 days in 2014. Obtained results help to estimate the reliability of minimum spanning tree in market network analysis.

Keywords Market network • Market network analysis • Minimum spanning tree • Kruskal algorithm • Statistical procedures • Uncertainty of statistical procedures

1 Introduction

Financial market is a complex system that can be studied in the framework of network analysis. Usually, network structures of the financial market are based on Pearson correlations of stock's returns. Each stock represents the node of network and similarities between stocks (weight of edges) are measured by a Pearson correlations between them [5]. Data mining on the market network can be conducted using different filtering techniques applied to the complete weighted graph (network). One of the filtering procedures is the construction and analysis of the minimum spanning tree (MST). The MST is a tree that includes every vertex and the total weight of all included edges is maximized. In order to construct the MST, Kruskal's algorithm can be applied [2]. Following this algorithm the list of edges is sorted in decreasing order according to the weight and an edge is added to the MST if it does not create a cycle. MST was proved to provide a useful

---

A. Komissarova • P. Koldanov (✉)
National Research University Higher School of Economics, 136, Rodionova Str., Nizhny Novgorod, Russia
e-mail: pkoldanov@hse.ru

V.A. Kalyagin et al. (eds.), *Models, Algorithms and Technologies for Network Analysis*, Springer Proceedings in Mathematics & Statistics 156,
DOI 10.1007/978-3-319-29608-1_10

information for market analysis, in particular it defines a hierarchical structure on financial market [4].

Correct interpretation of MST in market network cannot avoid to take into account uncertainty of its identification. Statistical uncertainty of different filtration techniques for market network analysis was studied in [1]. This approach was applied for analysis of statistical uncertainty of network structures for different markets. The experimental study showed that market graphs, maximum cliques, and maximum independent sets are more reliable with respect to statistical uncertainty than MST. Uncertainty was measured in [1] by a total number of statistical errors of first and second kind. In the present paper we take a different point of view and measure the statistical uncertainty of MST by q-FWER (Family Wise Error Rate) introduced in [3]. The main idea is to allow to have no more than q errors of the first kind and to analyze the statistical uncertainty according to this condition. This analysis is complementary to [1] and it helps to estimate the reliability of MST in market network analysis.

The paper is organized as follows. In Sect. 2 we introduce reference and sample MST and describe the estimation of uncertainty for additive loss function. In Sect. 3 we introduce a q-loss function and q-FWER (Family Wise Error Rate). In Sect. 4 we give and discuss the results of numerical experiments to evaluate the statistical uncertainty of MST by q-FWER for the US stock market. In Sect. 5 we give concluding remarks.

2 Statistical Uncertainty

We model the stocks returns on financial market by random variables $R_i(t)$, $t = 1, \ldots, n, i = 1, 2, \ldots, N$, where N is the number of stocks on the market, and n is the number of days of observations. In our study we assume that $R_i(t)$, $t = 1, \ldots, n$, are independent identically distributed random variables. Denote by R_i the distribution of $R_i(t)$. We assume that the random vector $R = (R_1, R_2, \ldots, R_N)$ has multivariate normal distribution $R \sim N(a, \Sigma)$, where $a = (a_1, \ldots, a_N)$ is the vector of means and Σ is the covariance matrix. *Reference MST* is defined as maximum spanning tree in the complete weighted graph with N nodes and weights given by the matrix of correlations $||\rho_{i,j}||$, $\rho_{i,j} = \sigma_{i,j}/\sqrt{\sigma_{i,i}\sigma_{j,j}}$. To construct the reference MST we use the Kruskal algorithm. Initially, the current set of edges is set empty. Then, among all remaining edges, an edge of maximal weight is selected and added to the existing set if this operation does not create a cycle. If there is a cycle, the edge is skipped. If there are no more edges, the algorithm stops.

Let $r_i(t)$, $k = 1, \ldots, N$, $t = 1, \ldots, n$, be observations of random variables $R_i(t)$. Vectors $r(t) = (r_1(t), r_2(t), \ldots, r_N(t))$, $t = 1, 2, \ldots, n$, represent a sample of the size n from distribution $R = (R_1, R_2, \ldots, R_N)$. One can define the sample covariances by

$$s_{i,j} = \frac{1}{n-1} \sum_{t=1}^{n} (r_i(t) - \overline{r}_i)(r_j(t) - \overline{r}_j), \quad \overline{r}_i = \frac{1}{n} \sum_{t=1}^{n} r_i(t).$$

Sample correlations are then defined by

$$r_{i,j} = \frac{s_{i,j}}{\sqrt{s_{i,i} s_{j,j}}}.$$

Sample MST is defined as maximum spanning tree in the complete weighted graph with N nodes and weights given by the matrix of sample correlations $||r_{i,j}||$. To construct the sample MST we use the same Kruskal algorithm. Initially, the current set of edges is set empty. Then, among all remaining edges, an edge of maximal weight is selected and added to the existing set if this operation does not create a cycle. If there is a cycle, the edge is skipped. If there are no more edges, the algorithm stops.

To measure statistical uncertainty one can compare the sample MST with the reference MST. This approach was developed in [1]. Such comparison is based on risk function connected with possible losses. For MST we introduce a set of hypothesis:

- $h_{i,j}$: edge between vertices i and j is not included in the reference MST and
- $k_{i,j}$: edge between vertices i and j is included in the reference MST.

Let us consider two types of errors:

- Type I error: edge is included in the sample MST when it is absent in the reference MST and
- Type II error: edge is not included in the sample MST when it is present in the reference MST.

Let $a_{i,j}$ be the loss associated with the error of the first kind and $b_{i,j}$ the loss associated with the error of the second kind for the edge $(i;j)$. Risk for additive loss function is defined by

$$R_{add}(MST, n) = \sum_{1 \leq i \leq j \leq N} (a_{i,j} P_n(k_{i,j}|h_{i,j}) + b_{i,j} P_n(h_{i,j}|k_{i,j})) \tag{1}$$

where $P_n(k_{i,j}|h_{i,j})$ is the probability of rejecting hypothesis $h_{i,j}$ when it is true and $P_n(h_{i,j}|k_{i,j})$ is the probability of accepting hypothesis $h_{i,j}$ when it is false. Risk function $R_{add}(MST, n)$ measures the statistical uncertainty of MST. It depends on the choice of the losses $a_{i,j}$ and $b_{i,j}$ and on the number of observations n.

3 Statistical Uncertainty of Kruskal Algorithm for MST

Statistical uncertainty of MST for different markets was investigated in [1] for additive loss functions $R_{add}(MST, n)$. In the present paper we take a different point of view and measure the statistical uncertainty of MST by the probability to have not less than q Type I errors. Usually, in multiple hypotheses testing significance level of multiple testing statistical procedure is defined by the probability to have not less than 1 error of Type I. Our approach is a generalization of this. Define the loss function:

$$Loss_q = \begin{cases} 1, \text{ if there is not less than } q \text{ incorrectly included edges} \\ 0, \text{ otherwise} \end{cases} \quad (2)$$

The risk function is then $R_q(MST, n) = E(Loss_q)$. This is connected with q-FWER (Family Wise Error Rate) introduced in [3]. In order to define the reliability of MST obtained by Kruskal algorithm, we investigate by simulations the risk function $R_q(MST, n)$ as function of parameter q and numbers of observations.

4 Numerical Experiments

We use as a weight matrix $(||\rho_{i,j}||)$ for the reference MST of the correlations of returns of a set of 100 financial instruments traded in the NASDAQ100 over a period of 250 trading days in 2014. Then we simulate multivariate normal distribution $N((0, \ldots, 0), ||\rho_{i,j}||)$ n times to get a sample of the size n. To evaluate the risk function $R_q(MST, n)$ (probability to have not less than q Type I errors) we replicate the experience 500 times, and use the frequency estimation for probability. To measure uncertainty we solve the equation $R_q(MST, n) = P_0$, i.e. we calculate the number of observations n needed to achieve the risk level P_0. The results are summarized in Tables 1, 2, 3, 4, and 5. One can see that an increase in q leads to a considerable decrease in numbers of observations needed to reach the given level of risk. For instance, the probability of not less than 1 incorrectly included edge is equal to 0.1 with 700,000 observations, but only 1200 observations are required when $q = 10$.

Figures 1, 2, and 3 give the number of observations needed to reach a given risk level as a function of risk P_0 for three values of q: 1, 5, and 10. One can see, that dependence on q is not linear. This fact gives a possibility to control uncertainty of MST by a choice of q.

Table 1 Number of observations needed to reach the value of risk $R_q(MST, n) = 0.1$

n	7×10^5	4×10^5	3.5×10^5	1.25×10^5	0.1×10^5	5000	3500	2500	2300	1000
q	1	2	3	4	5	6	7	8	9	10

Table 2 Number of observations needed to reach the value of risk $R_q(MST, n) = 0.3$

n	2.5×10^5	1.5×10^5	1.3×10^5	0.5×10^5	4000	2100	1600	1200	1000	600
q	1	2	3	4	5	6	7	8	9	10

Table 3 Number of observations needed to reach the value of risk $R_q(MST, n) = 0.5$

n	1.5×10^5	0.36×10^5	0.1×10^5	5000	3500	2000	1500	1100	600	550
q	1	2	3	4	5	6	7	8	9	10

Table 4 Number of observations needed to reach the value of risk $R_q(MST, n) = 0.7$

n	50,000	13,000	4000	2500	2000	1200	900	800	500	450
q	1	2	3	4	5	6	7	8	9	10

Table 5 Number of observations needed to reach the value of risk $R_q(MST, n) = 0.9$

n	40,000	9000	2500	1500	1000	640	540	460	430	350
q	1	2	3	4	5	6	7	8	9	10

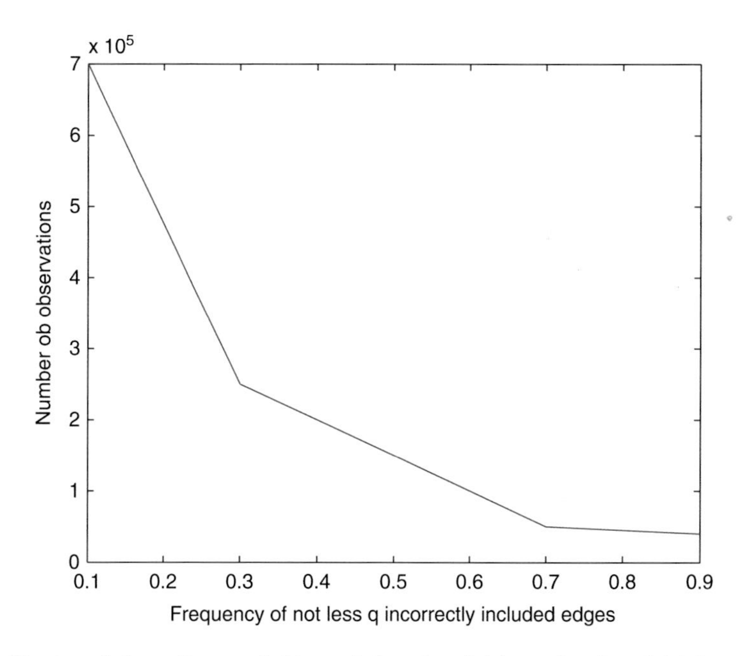

Fig. 1 Number of observations needed to reach the value of risk as a function of risk for $q = 1$

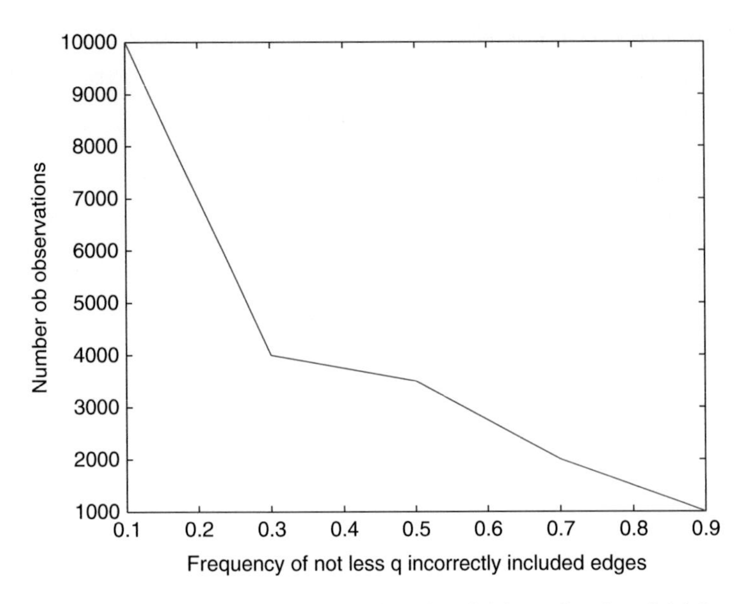

Fig. 2 Number of observations needed to reach the value of risk as a function of risk for $q = 5$

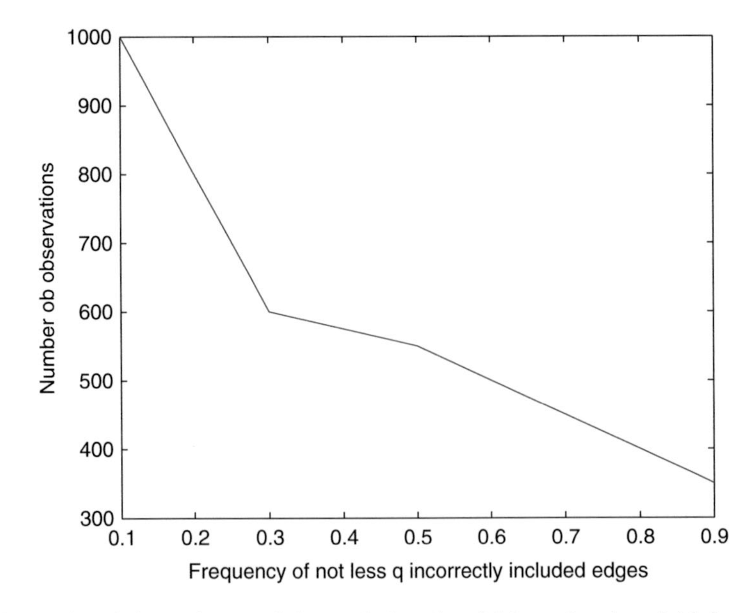

Fig. 3 Number of observations needed to reach the value of risk as a function of risk for $q = 10$

5 Concluding Remarks

The most important conclusion from our study is that the MST cannot be regarded as a stable structure even if we use a very light risk q-FWER. Moreover, we need a huge amount of observations to reach the 90 % confidence of not at all, at most one, or at most two incorrectly included edges. The increase in numbers of observations does not cause a significant reduction in probability of not less than $q = 1, 2, 3$ incorrectly included edges. However, nonlinearity of dependence on q of risk of MST gives a possibility to control uncertainty of MST by a choice of q.

Acknowledgements This work is partly supported by Russian Federation Government grant, N. 11.G34.31.0057, RFBR grant 14-01-00807, and HSE Scientific Fund 15-09-0239.

References

1. Kalyagin, V.A., Koldanov, A.P., Koldanov, P.A., Pardalos, P.M., Zamaraev, V.A.: Measures of uncertainty in market network analysis. Physica A **413**(1), 59–70 (2014)
2. Kruskal, J.B.: On the shortest spanning subtree of a graph and the traveling salesman problem. Proc. Am. Math. Soc. **7**(1), 48–50 (1956)
3. Lehmann, E.L., Romano, J.: Generalizations of the familywise error rate. Ann. Stat. **33**, 1138–1154 (2005)
4. Mantegna, R.N.: Hierarchical structure in financial markets. Eur. Phys. J. B **11**, 93–97 (1999)
5. Mantegna, R., Stanley, E.: An Introduction to Econophysics: Correlations and Complexity in Finance, Cambridge University Press, Cambridge (2000)

Identification of Cliques and Independent Sets in Pearson and Fechner Correlations Networks

Oleg Kremnyov and Valery A. Kalyagin

Abstract We use two market network models for the analysis of stock market: Pearson correlation network and Fechner correlation network. The main goal is to estimate statistical uncertainty of identification of maximum cliques and maximum independent sets. It is shown that identification of maximal cliques and maximal independent sets in Fechner correlation network is distribution free in a certain class of distributions which is not true for Pearson correlation network. This fact gives advantage for Fechner correlation network in the case where no prior information is known about distribution of returns.

Keywords Market network • Market graph • Cliques • Independent sets • Pearson correlation network • Fechner correlation network • Statistical procedures • Risk function • Uncertainty of identification procedures

1 Introduction

Market network is a complete weighted graph. Every stock in a network corresponds to a vertex in a graph and weights of edges between vertices correspond to values of correlations between stock's returns. Market graph is a subgraph of the market network obtained from the network by removal of all edges whose weights do not exceed a given threshold. Maximum cliques and maximum independent sets in the market graph provide a valuable information about the market structure [1–3].

Due to the stochastic nature of stocks returns there is an uncertainty in identification of maximum cliques and maximum independent sets in the market graph. A general approach to handle statistical uncertainty of market network structures was proposed in [4]. In the present paper we investigate statistical uncertainty of cliques and independent sets for two market network models: Pearson correlation network and Fechner correlation network. We show that identification of maximal cliques and maximal independent sets in Fechner correlation network is distribution

O. Kremnyov (✉) • V.A. Kalyagin
National Research University Higher School of Economics, Nizhny Novgorod, Russia
e-mail: oyukremnyov@edu.hse.ru; vkalyagin@hse.ru

V.A. Kalyagin et al. (eds.), *Models, Algorithms and Technologies for Network Analysis*, Springer Proceedings in Mathematics & Statistics 156,
DOI 10.1007/978-3-319-29608-1_11

free in a certain class of distributions which is not true for Pearson correlation network. This fact gives advantage for Fechner correlation network in the case where no prior information is known about distribution of returns.

The paper is organized as follows. In Sect. 2 we describe the overall model of the market network and define maximum cliques and independent sets in the market graph. In Sect. 3 we give a description of statistical procedures used for identification of cliques and independent sets. In Sect. 4 we describe statistical uncertainty of identification procedures. In Sect. 5 we present the results of numerical experiments and give comments on it. In the last Sect. 6 we summarize the main finding of the paper.

2 Random Variables Network

All graphs in this paper are simple, i.e. finite, undirected, without loops or multiple edges. Clique in a graph is a set of vertices where any two vertices are connected by an edge. Independent set in a graph is a set of vertices where any two vertices are not connected by an edge. To model the market network we use the model of *random variables network*. Random variables network is given by a random vector $R = (R_1, R_2, \ldots, R_N)$ and a measure of association γ. In market network random variable R_i represents return of the stock i. Weight of the edge between nodes i and j is given by a measure of association $\gamma(R_i, R_j)$ which characterizes pair-wise connections between random variables R_i and R_j.

One can identify threshold graph in random variables network. It is set by a certain value of threshold γ_0. An edge between vertices R_i and R_j is included in the threshold graph if and only if the measure of the association $\gamma(R_i, R_j)$ exceeds the value of threshold γ_0. One can search for the maximum clique, i.e. the clique of the greatest size, and the maximum independent set, i.e. the independent set of the greatest size, in the threshold graph. Market graph (MG) is a special case of the threshold graph, where the random variables are the stock's returns. In this paper we identify the maximum weighted maximal clique (MCMW), the maximum clique with the greatest total weight which is a sum of weights of all edges in a clique and the minimum weighted maximum independent set (MISMW), a maximum independent set with the lowest total weight.

Inputs of maximum clique identification procedure are observations of stock returns and output is the MCMW in the market graph. Similarly, inputs of maximum independent set identification procedure are observations of stock returns and output is the maximum independent set with the lowest total weight MISMW. We model observations of returns of the stock i by random variables $R_i(t)$, $t = 1, 2, \ldots, n$. We assume that random variables $R_i(t)$ are all independent, identically distributed as R_i. In this setting observations $R(t)$, $t = 1, 2, \ldots, n$, form a sample from distribution R.

3 Identification Statistical Procedures

In Pearson correlation network weight of edges between nodes i and j is given by Pearson correlation between random variables R_i and R_j:

$$\gamma_{i,j}^{P} = \frac{\mathrm{COV}(R_i, R_j)}{\sqrt{\mathrm{VAR}(R_i)\mathrm{VAR}(R_J)}}$$

Let N be the number of stocks and n be the number of days of observations of stock returns. Let $r_i(t)$ be the observed value of the return of stock i for the day t. Sample Pearson correlation is defined by

$$\hat{\gamma}_{i,j}^{P} = \frac{\sum_{t=1}^{n}(r_i(t) - \overline{r}_i)(r_j(t) - \overline{r}_j)}{\sqrt{\sum_{t=1}^{n}(r_i(t) - \overline{r}_i)^2 \sum_{t=1}^{n}(r_j(t) - \overline{r}_j)^2}}$$

For a given threshold γ_0^P an edge (i,j) is included in the market graph if $\gamma_{i,j}^P > \gamma_0^P$. We will call the obtained graph *reference market graph*. To identify the MCMW and MISMW in the reference market graph from observations we use the following statistical procedure: first we identify the market graph by including the edge (i,j) in the market graph if $\hat{\gamma}_{i,j}^P > \gamma_0^P$. This graph will be called *sample market graph*. Once sample market graph is identified we calculate the MCMW and MISMW. Two types of errors can occur with this identification: error of the first kind of the false noninclusion of an edge in the structure and error of the second kind of the false inclusion of an edge in the structure. Quality of identification statistical procedure is measured by these errors.

In Fechner correlation network weight of edges between nodes i and j is given by Fechner correlation between random variables R_i and R_j:

$$\gamma_{i,j}^{FH} = -1 + 2P((R_i - E(R_i))(R_j - E(R_j)) > 0)$$

Sample Fechner correlation is defined by

$$\hat{\gamma}_{i,j}^{FH} = -1 + \frac{2}{n}\sum_{t=1}^{n} \mathrm{sign}((r_i(t) - \overline{r_i}) * (r_j(t) - \overline{r_j})).$$

For a given threshold γ_0^{FH} an edge (i,j) is included in the reference market graph if $\gamma_{i,j}^{FH} > \gamma_0^{FH}$. To identify the MCMW and MISMW in the reference market graph from observations we use the following statistical procedure: first we identify the market graph by including the edge (i,j) in the market graph if $\hat{\gamma}_{i,j}^{FH} > \gamma_0^{FH}$. Thus we obtain sample market graph in Fechner correlation network. Once sample market graph is identified we calculate the MCMW and MISMW. As above two types of errors can occur with this identification: error of the first kind of the false noninclusion of an edge in the structure and error of the second kind of the false inclusion of an edge in the structure.

4 Uncertainty of Statistical Procedures

To measure statistical uncertainty we propose to compare sample network structure with the reference network structure. This comparison is based on risk function. For a given network structure S (market graph, clique, and independent set) we define a set of hypotheses: $h_{i,j}$: edge (i,j) is included in the reference structure S and $k_{i,j}$: edge (i,j) is not included in the reference structure S. Two types of errors can occur:

- Type I error (error of the first kind): edge (vertex) is not included in the sample structure when it is present in the reference structure and
- Type II error (error of the second kind): edge (vertex) is included in the sample structure when it is absent in the reference structure.

To evaluate the losses from Type I and Type II errors in MC and MIS identification we use the following algorithm. As we can have many maximal cliques in a market graph, we choose one with the maximal weight, MCMW. We can also have many maximal independent sets, we also choose one with the minimal weight, MISMW. Let S_{MCR} be a structure MCMW in the reference market graph, and S_{MCS} be MCMW in the sample market graph. Similarly let S_{MISR} be the structure MISMW in the reference market graph, and S_{MISS} be the structure MISMW in the sample market graph. To measure the number of errors (losses) of the Type I for MC, we will compare S_{MCS} with its preimage in reference market graph S_{MCR}. The number of non-vertices in S_{MCR} is the number of errors of Type I for maximal cliques. For MIS, we define S_{MISS} is a direct image in sample market graph of S_{MISR}. The number of vertices in S_{MISS} is the number of errors of Type II for MIS. To measure the number of errors of Type II for MC, we will compare S_{MCR} with its direct image in sample market graph S_{MCS}. The number of non-vertices in S_{MCS} is the number of errors of Type II for maximal cliques. To measure the number of errors of Type II for MIS, we will compare S_{MISS} with its preimage in reference market graph S_{MISR}. The number of vertices in S_{MISR} is the number of errors of the second type for MIS. Let M_1 be the maximal number of errors of Type I and M_2 be the maximal number of errors of Type II. One has for MC, $M_1=\binom{2}{K}$,where K is the size of S_{MCS}. Similarly, $M_1 = \binom{2}{K}$ for MIS, where K is the size of S_{MISR}. Similarly, one has $M_2 = \binom{2}{K}$, for MC, where K is the size of S_{MCR}.

Let X_1 be the number of Type I errors, X_2 be the number of Type II errors. Define the random variable X by

$$X = \frac{1}{2}\left(\frac{X_1}{M_1} + \frac{X_2}{M_2}\right)$$

The risk function is now defined as

$$R(S,n) = E(X) = \frac{1}{2}E\left(\frac{X_1}{M_1} + \frac{X_2}{M_2}\right) \tag{1}$$

This function will measure statistical uncertainty of MC and MIS identification procedures. Note that $0 \leq R(S,n) \leq 1$, and $R(S,n) = 0$ means that there are no errors in identification of the structure S, while $R(S,n) = 1$ gives a maximal number of errors.

5 Results

We investigate statistical uncertainty of identification of cliques and independent sets by simulations. In our experiments distribution of random vector $R = (R_1.R_2,\ldots,R_N)$ is a mixture of multivariate Gaussian and Student distributions. To generate observations $r_i(t)$ we use multivariate Gaussian distribution with probability ν and multivariate Student distribution with 3 degrees of freedom with probability $(1-\nu)$. Both distributions have the same correlation matrix taken from the real US market. Using simulations we compare the risk functions of identification procedures in Pearson and Fechner correlations networks. To make a correct comparison we need to have the same reference structure (market graph, maximum clique, and maximum independent set) in both networks. This is possible because there is a connection between Pearson and Fechner correlations for a mixture of Gaussian and Student multivariate distributions [5]:

$$\gamma^{FH} = \frac{2}{\pi}\arcsin(\gamma^{P}).$$

Therefore if we choose $\gamma_0^{FH} = \frac{2}{\pi}\arcsin(\gamma_0^{P})$, then maximum clique and independent set in the market graph for Pearson correlation network with the threshold γ_0^P will be the same as maximum clique and independent set in the market graph for Fechner correlation network with the threshold γ_0^{FH}.

Figs. 1 and 2 represent the number of observations necessary to achieve the level of risk 0.5 as a function of the parameter ν, i.e. the number of observations n satisfying the equation $R(S,n) = 0.5$. One can see that identification of maximum clique (Fig. 1) and maximum independent set (Fig. 2) is not sensitive to distribution in Fechner correlation network but it is very sensitive to distribution in Pearson correlation network. Moreover uncertainty in Pearson correlation network is lower only in a neighborhood of $\nu = 1$ (Gaussian distribution).

Figs. 3 and 4 represent the number of observations necessary to achieve the level of risk 0.3 as a function of the parameter ν, i.e. the number of observations n satisfying the equation $R(S,n) = 0.3$. One can see that the situation is similar to the level of risk 0.5. Identification of maximum clique (Fig. 3) and maximum independent set (Fig. 4) is not sensitive to distribution in Fechner correlation network but it is very sensitive to distribution in Pearson correlation network. Uncertainty in Pearson correlation network is lower only in a neighborhood of $\nu = 1$ (Gaussian distribution).

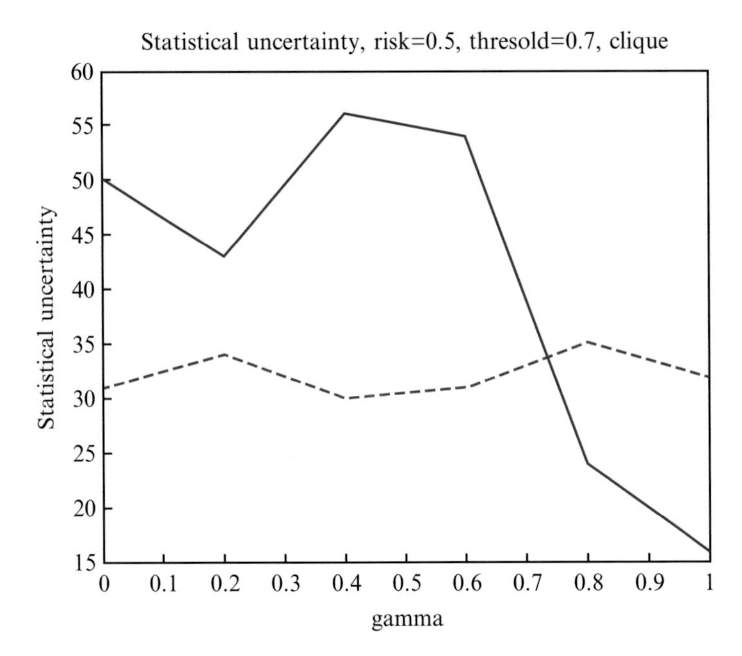

Fig. 1 Number of observations necessary to achieve the level of risk 0.5 for identification of maximum clique in the market graph for Pearson (*solid line*) and Fechner (*dashed line*) correlations networks. Threshold $\gamma_0^P = 0.7$

Figs. 5 and 6 represent the number of observations necessary to achieve the level of risk 0.1 as a function of the parameter ν, i.e. the number of observations n satisfying the equation $R(S, n) = 0.1$. One can see that the situation is similar to the levels of risk 0.5 and 0.3. Identification of maximum clique (Fig. 5) and maximum independent set (Fig. 6) is not sensitive to distribution in Fechner correlation network but it is very sensitive to distribution in Pearson correlation network. Uncertainty in Pearson correlation network is lower only in a neighborhood of $\nu = 1$ (Gaussian distribution).

6 Concluding Remarks

Identification of maximum cliques and maximum independent set in the market graph is investigated for two types of random variables networks: Pearson correlations network and Fechner correlations network. It is shown by numerical simulations that identification of maximum clique and maximum independent set is not sensitive to distribution in Fechner correlation network but it is very sensitive to distribution in Pearson correlation network. This fact has a practical meaning for identification of network structures in real stock markets.

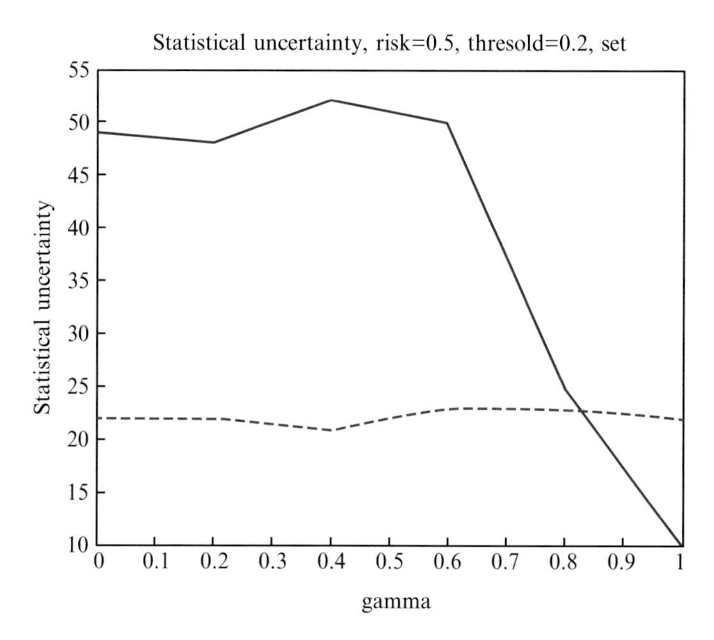

Fig. 2 Number of observations necessary to achieve the level of risk 0.5 for identification of maximum independent set in the market graph for Pearson (*solid line*) and Fechner (*dashed line*) correlations networks. $\gamma_0^P = 0.2$

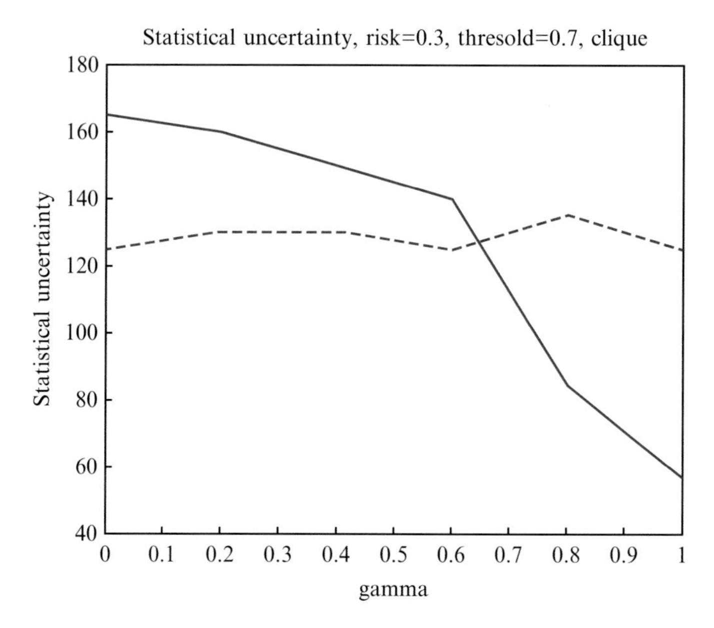

Fig. 3 Number of observations necessary to achieve the level of risk 0.3 for identification of maximum clique in the market graph for Pearson (*solid line*) and Fechner (*dashed line*) correlations networks. $\gamma_0^P = 0.7$

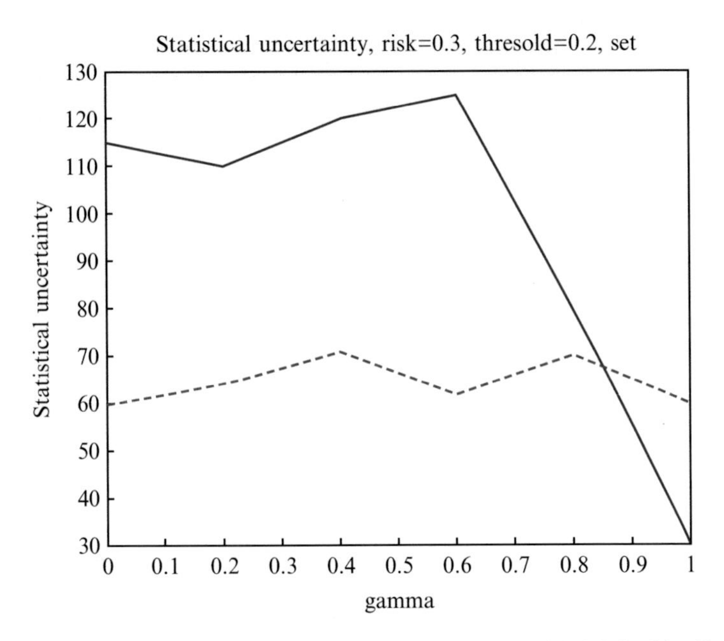

Fig. 4 Number of observations necessary to achieve the level of risk 0.3 for identification of maximum independent set in the market graph for Pearson (*solid line*) and Fechner (*dashed line*) correlations networks. $\gamma_0^P = 0.2$

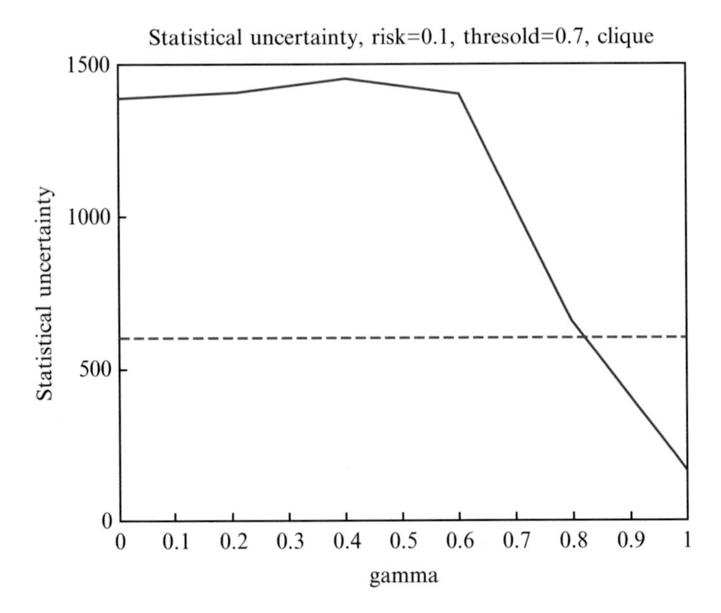

Fig. 5 Number of observations necessary to achieve the level of risk 0.1 for identification of maximum clique in the market graph for Pearson (*solid line*) and Fechner (*dashed line*) correlations networks. $\gamma_0^P = 0.7$

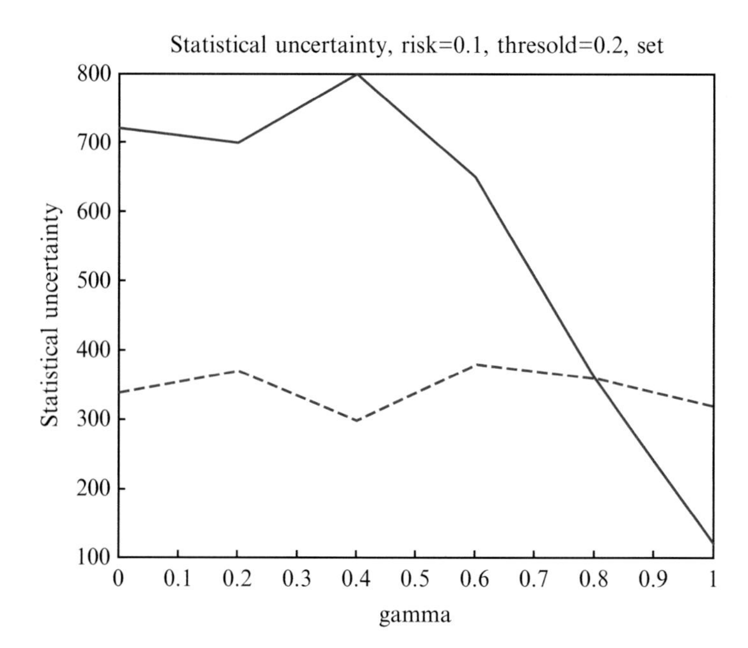

Fig. 6 Number of observations necessary to achieve the level of risk 0.1 for identification of maximum independent set in the market graph for Pearson (*solid line*) and Fechner (*dashed line*) correlations networks. $\gamma_0^P = 0.2$

Acknowledgements This work is partly supported by LATNA laboratory, Russian Federation Government grant 11.G34.31.0057, and RFHR grant 15-32-01052.

References

1. Boginski, V., Butenko, S., Pardalos, P.M.: On structural properties of the market graph. In: Nagurnay, A. (ed.) Innovations in Financial and Economic Networks, vol. 48, pp. 29–35. Edward Elgar, Cheltenham (2003)
2. Boginski, V., Butenko, S., Pardalos, P.M.: Mining market data: a network approach. Comput. Oper. Res. **33**, 3171–3184 (2006)
3. Vizgunov, A., Goldengorin, B., Kalyagin, V., Koldanov, A., Koldanov, P., Pardalos, P.M.: Network approach for the Russian stock market. Comput. Manag. Sci. **11**, 44–55 (2014)
4. Kalyagin, V.A., Koldanov, A.P., Koldanov, P.A., Pardalos, P.M., Zamaraev, V.A.: Measures of uncertainty in market network analysis. Phys. A: Stat. Mech. Appl. **413**, 59–70 (2014)
5. Kalyagin, V.A., Koldanov, A.P., Koldanov, P.A.: Robust identification in random variables network (submitted)

Investigation of Connections Between Pearson and Fechner Correlations in Market Network: Experimental Study

Andrey Latyshev and Petr Koldanov

Abstract Network models for stock market attract a great attention in last decades. Different measures of similarity between stocks attributes are largely used. In general, different measures of similarity can generate different network structures. For instance, Fechner and Pearson correlations networks can have different minimum spanning trees, market graphs, maximum cliques, and maximum independent sets. At the same time it is known that Fechner and Pearson correlations are connected by a monotonic transformation for bivariate Gaussian distributions. This connection can be generalized to bivariate elliptically contoured distributions. In this case it can be shown that network structures are connected too. This fact can be used for data mining in market network. In the present paper we study the connection between Fechner and Pearson correlations for the real market data.

Keywords Stock market • Stock returns • Market network • Pearson correlation • Fechner correlation

1 Introduction

Stock market can be investigated as a complex system [8]. Market network is a complete weighted graph where the vertexes of the graph represent the stocks and weights of edges are given by some measure of similarity between stocks attributes. Network structure is a subgraph of the complete weighted graph obtained by some filtration technique. The popular network structures are MST (minimum spanning tree) [7] and market graph [3].

A. Latyshev (✉)
National Research University Higher School of Economics, Bolshaya Pecherskaya 25, Nizhny Novgorod 603155, Russia
e-mail: andlatyshe@mail.ru

P. Koldanov
National Research University Higher School of Economics, 136, Rodionova Str., Nizhny Novgorod, Russia
e-mail: pkoldanov@hse.ru

V.A. Kalyagin et al. (eds.), *Models, Algorithms and Technologies for Network Analysis*, Springer Proceedings in Mathematics & Statistics 156,
DOI 10.1007/978-3-319-29608-1_12

Traditional measure of similarity used in market network analysis is Pearson correlation. Recently it was shown that sign similarity of stocks returns known as Fechner correlation can be useful for data mining in market network [2]. In general different measures of similarity can generate different network structures. Surprisingly for some class of distributions (for example, for elliptically countered distributions) the popular network structures such as MST and market graphs are connected in Pearson and Fechner correlations networks. This fact is a consequence of the connection between Pearson and Fechner correlations for elliptically countered distributions [1, 5]. The main goal of the present paper is to study this connection for the real market data. This study gives us some arguments to answer the following question: whether distributions of stocks returns can be modeled by elliptically contoured distributions?

2 Pearson and Fechner Correlations

Let N be the number of financial assets and n be the number of observations. Let $P_i(t)$ be the price of stock i on day t ($i = \overline{1,N}$; $t = \overline{1,n}$). The log-return of the stock i per 1-day period from (t-1) until t can be defined as follows:

$$R_i(t) = ln\left(\frac{P_i(t)}{P_i(t-1)}\right)$$

We assume that random variables $R_i(t)$, $(i = 1, N) = \overline{(i = 1, N)}$, are independent when i is fixed, and identically distributed as R_i, $i = \overline{1,N}$), while random variables $R_1,\ R_2,\ \ldots,\ R_N$ have a joint distribution with the Pearson correlation matrix $\|\rho^P_{i,j}\|$, where

$$\rho^P_{i,j} = \frac{\mathrm{COV}(R_i, R_j)}{\sqrt{\mathrm{VAR}(R_i)\mathrm{VAR}(R_j)}}$$

The sample Pearson correlation coefficient between returns of the stocks i and j is defined by

$$r_{i,j} = \frac{\sum (R_i(t) - \overline{R_i})\ (R_j(t) - \overline{R_j})}{\sqrt{\sum (R_i - \overline{R_i})^2}\sqrt{\sum (R_j - \overline{R_j})^2}}$$

Here $\overline{R_i} = \frac{1}{n}\sum_{t=1}^{n} R_i(t)$ is a sample expectation of the ith return.

For any pair of random variables $R_i(t)$ and $R_j(t)$ we can define the pairwise similarity that is based on the probability that their deviations from the corresponding expectations $a_i = E(R_i)$ and $a_j = E(R_j)$ have the same sign:

$$p_{i,j} = P(R_i \geq a_i, R_j \geq a_j \vee R_i < a_i, R_j < a_j) \tag{1}$$

The measure defined in this way is linearly connected with the sign correlation of Fechner which is defined by

$$\rho_{i,j}^F = -1 + 2p_{i,j}$$

Sample Fechner correlation is defined by

$$s_{i,j} = \frac{1}{n}\sum_{t=1}^{n} sign(R_i(t) - \overline{R_i})sign(R_j(t) - \overline{R_j}) \tag{2}$$

where

$$sign(x) = \begin{cases} 1, & x \geq 0 \\ -1, & x > 0 \end{cases}$$

One has for the sample Fechner correlation

$$s_{i,j} = \frac{1}{n}\sum_{t=1}^{n} sign(R_i(t) - \overline{R_i})sign(R_j(t) - \overline{R_j}) = \frac{e_{ij} - d_{ij}}{e_{ij} + d_{ij}}$$

where e_{ij} is the number of the pairs of corresponding returns R_i and R_j that have the same signs and d_{ij} is the number of such pairs that have different signs. Then

$$E(s_{i,j}) = -1 + 2p_{i,j} = \rho_{i,j}^F$$

The interpretation of the proposed measures (1) and (2) can be described in the following way. We suppose that there were 100 observed days. Then the percent of the days when the signs of the deviations of two stock returns are coincident corresponds to the probability measure $p_{i,j}$. The transformation $s_{i,j} = 2p_{i,j} - 1$ gives the sample Fechner correlation. For example, if there were 80 days when the signs of the deviations of two stock returns were the same, then we can find that $p_{i,j} = 0.8$. Hence $s_{i,j} = 0.6$. If $s_{i,j} > 0$, then the stocks i and j had more days when their prices changed in the same directions. If $s_{i,j} < 0$, then the stocks had more days when their prices changed in the different directions. When $s_{i,j} \approx 0$ the prices of stocks i and j fluctuate independently.

3 Connection Between Pearson and Fechner Correlations

This connection is known for bivariate normal distribution [4]. For the sake of completeness we reproduce here the main arguments of the proof. Let X and Y be normally distributed random variables with zero expectations, unit variances, and Pearson correlation $\rho = \rho(X, Y)$. Let U and V be two independent normally distributed random variables with zero expectations and unit variances. The random variables X and Y can be expressed as

$$X = \alpha U + \beta V, \quad Y = \gamma U + \delta V$$

Define the variables U and V from the equations

$$\begin{gathered} E(U) = E(V) = 0, \;\; D(U) = D(V) = 1, \\ E(V^2) = E(U^2) = D(U) + \{E(U)\}^2 = 1 + 0 = 1 \end{gathered}$$

Taking into account that U and V are independent we have

$$E(UV) = E(U)E(V) = 0$$

One has

$$\begin{gathered} D(X) = D(\alpha U + \beta V) = \alpha^2 D(U) + \beta^2 D(V) = \alpha^2 + \beta^2 = 1 \\ D(Y) = D(\gamma U + \delta V) = \gamma^2 D(U) + \delta^2 D(V) = \gamma^2 + \delta^2 = 1 \\ \rho(X, Y) = E(XY) = E((\alpha U + \beta V)(\gamma U + \delta V)) = \\ = \alpha\gamma E(U^2) + \alpha\delta E(UV) + \beta\gamma E(UV) + \beta\delta E(V^2) = \\ = \alpha\gamma + \beta\delta = \rho \end{gathered}$$

Finally we get the system of three equations connecting four parameters α, β, γ, and δ.

$$\begin{cases} \alpha^2 + \beta^2 = 1 \\ \gamma^2 + \delta^2 = 1 \\ \alpha\gamma + \beta\delta = \rho \end{cases}$$

Any solution of the system correspond to the pair of two-dimenision vector with coordinates (α, β), (γ, δ). The length of each vector is equal to one and the scalar product of the vectors is equal to ρ. Let we fix one solution. Then it is obvious that any other solution can be obtained from fixed solution by rotating the plane by arbitrary angle. Assume the following solution

$$\begin{cases} \alpha = 1 \\ \beta = 0 \\ \gamma = \rho \\ \delta = \sqrt{1 - \rho^2} \end{cases}$$

Now we can state the connection between the Pearson and Fechner correlations. The probability of the coincidence of the signs P^S for two bivariate normal variables X and Y can be expressed in the following way

$$\begin{aligned} P^S &= P(X \geq 0, Y \geq 0 \vee X < 0, Y < 0) = \\ &= P(X \geq 0, Y \geq 0) + P(X < 0, Y < 0) = \\ = P(U \geq 0, \rho U &+ \sqrt{1-\rho^2}V \geq 0) + P(U < 0, \rho U + \sqrt{1-\rho^2}V < 0) \end{aligned} \quad (3)$$

If we use the Cartesian plane with the axes U and V, then the proposed sum of probabilities is equal to the measure of the sum of two angles between the lines $U = 0$ and $\rho U + \sqrt{1-\rho^2}V$ divided by the measure of full sector 2π because U and V are independent. The sum of angles can be found using the vectors described below. This sum is equal to $2\pi - \arccos\rho$. Then

$$P^S = \frac{2\pi - 2\arccos\rho}{2\pi} = 1 - \frac{\arccos\rho}{\pi}$$

Now we can establish the connection between Fechner correlation $\rho^F(X, Y)$ and Pearson correlation $\rho^P(X, Y)$ considering $\rho^F_{X,Y} = 2P^S - 1$

$$\rho^F_{X,Y} = 1 - \frac{2}{\pi}\arccos\rho^P(X, Y) \quad (4)$$

It can be shown that the Eq. (4) can be applied to the pairs of stocks that have elliptically contoured joint distribution of the returns.

4 Statistical Framework for Testing Relation Between Pearson and Fechner Correlations

To test the statistical hypothesis of connections between Pearson and Fechner correlations we use the following probabilistic model for the market returns: random variables $R_i(t)$, $i = \overline{1, N}$, are independent when i is fixed, and identically distributed as $(i = 1, N)$, $i = \overline{(i = 1, N)}$, and random variables $R_1, R_2, \ldots, R_N$ have a joint distribution with the Pearson correlation matrix $\|\rho^P_{i,j}\|$. Let us introduce hypotheses:

$$H_{X,Y} : \rho^F_{X,Y} = 1 - \frac{2}{\pi}\arccos\rho^P(X, Y) \text{ vs } K_{X,Y} : \rho^F_{X,Y} \neq 1 - \frac{2}{\pi}\arccos\rho^P(X, Y) \quad (5)$$

We would like to test these hypotheses using the observations from financial market.

According to [6] test for (5) has the form:

$$\varphi(x, y) = \begin{cases} 0, & c_1 < T(x, y) < c_2 \\ 1, & \text{else} \end{cases} \quad (6)$$

where constants c_1 and c_2 are defined from the equation

$$P(c_1 < T(x, y) < c_2 | \rho^F_{X,Y} = 1 - \frac{2}{\pi}\arccos(\rho^P_{X,Y}) = 1 - \alpha$$

To define statistic $T(x, y)$ note that maximum likelihood estimation of (3) is the frequency $h(x, y)$ of the coincidence of the signs of stocks X and Y. This frequency can be written as

$$h(x, y) = \sum_{i=1}^{n} I_i$$

where

$$I_i = \begin{cases} 1, \ sign(x_i) = sign(y_i) \\ 0, \ sign(x_i) \neq sign(y_i) \end{cases}$$

Maximum likelihood estimation of $\frac{2}{\pi} \arccos \rho^P_{X,Y}$ is $\frac{2}{\pi} \arccos(r_{X,Y})$. Thus statistic $T(x, y)$ can be defined as

$$T(x, y) = h(x, y) - \frac{2}{\pi} \arccos(r_{X,Y}) \tag{7}$$

Constant c_1 from (6) is $\frac{\alpha}{2}$—quantile of distribution $T(x, y)$ and constant c_2 is $1 - \frac{\alpha}{2}$—quantile of distribution $T(x, y)$. Distribution of statistic $T(x, y)$ being unknown one can estimate the constants c_1 and c_2 by numerical simulations of bivariate distribution of (X, Y). We consider two types of distributions (both from the class of elliptically contoured distributions)

- The first joint bivariate distribution of (X, Y) is normal with the means equal to zero, the variances equal to one, the period n of observation is 250, 500, or 1000 days optionally, and the correlation between the variables ρ varies from –0.8 to 0.8 with an increment 0.1.
- The second joint bivariate distribution of (X, Y) is Student distribution with three degrees of freedom, the period n of observation is 250, 500, or 1000 days optionally, and the correlation between the variables ρ varies from –0.8 to 0.8 with an increment 0.1.

The number of replications is 100,000 for every set of parameters. After the described operations we determine the bounds (c_1, c_2) on the significance levels $\alpha = 0.05$ and $\alpha = 0.01$ for known n and ρ using the following equations

$$P(T(x, y) < c_1(\alpha, \rho, n)) = \frac{\alpha}{2}$$

$$P(T(x, y) > c_2(\alpha, \rho, n)) = \frac{\alpha}{2}$$

Next the test for the hypotheses of the connection between sign and classical correlations on the real market data is developed. For every pair of stocks i and j we compute their returns R_i and R_j. Then we calculate the sample Pearson correlation $r_{i,j}$ and the frequency of the coincidence of the signs $s_{i,j}$ and define

$$\Delta_{i,j} = s_{i,j} - (1 - \frac{\arccos r_{i,j}}{\pi})$$

After that we compare the resulting $\Delta_{i,j}$ with the boundary values $c_1(\alpha, \rho, n)$ and $c_2(\alpha, \rho, n)$ and make a decision.

The shown procedure has to be completed for all the possible C_N^2 pairs of stocks $\{(i,j)|\ i,j = 1, 2, \ldots, N,\ i \neq j\}$. The analysis of the results (the percentage of acceptances and rejections of the hypotheses) lets us make some conclusions about the nature of the distribution of the market stocks returns.

5 Testing Connection Between Pearson and Fechner Correlations for the Real Market Data

In this section an analysis of the connection between Pearson and Fechner correlation for the American stock market is presented. We selected $N = 30$ stocks that are included in the DJI index. The period of observation is from January 01, 2009 until July 19, 2013 (1000 trading days in total). We divided our observation period in three different ways: into four periods (250 days each), into two periods (500 days each) by grouping the periods 1–2 and 3–4, and the whole period of 1000 days. Totally we analyze 435 joint distributions of each pair of DJI stocks. The results of the described tests are represented in the tables below. In Table 1 we use the critical values c_1 and c_2 obtained from normal distribution.

One can see from Table 1 that the average of proportion of rejected hypotheses is many times larger than corresponding significance level. It means that the conjecture of connection between Fechner and Pearson correlations for the set of DJI stocks is not confirmed if the critical values c_1 and c_2 are calculated using normal distribution.

In Table 2 we use the critical values c_1 and c_2 obtained from Student distribution.

One can see from Table 2 that the average of proportion of rejected hypotheses is lower than corresponding significance level. It means that the conjecture of connection between Fechner and Pearson correlations for the set of DJI stocks is confirmed if the critical values c_1, c_2 are calculated using Student distribution.

Table 1 Proportion of rejected hypotheses for different periods of observations. Critical values of tests are taken from normal distribution

α	Period 1	Period 2	Period 3	Period 4	Periods 1–2	Periods 3–4	Periods 1–4
0.05	0.137	0.128	0.119	0.282	0.168	0.262	0.301
0.01	0.071	0.039	0.036	0.160	0.067	0.181	0.186

Table 2 Proportion of rejected hypotheses for different periods of observations. Critical values of tests are taken from student distribution

α	Period 1	Period 2	Period 3	Period 4	Periods 1–2	Periods 3–4	Periods 1–4
0.05	0.002	0.004	0.002	0.030	0.002	0.075	0.059
0.01	0.000	0.000	0.000	0.000	0.000	0.000	0.000

6 Concluding Remarks

First experimental study on connections between Pearson and Fechner correlations for the real market conducted in the paper shows that the conjecture of connection can be accepted at least for a part of the market. A deeper analysis of the rejections of hypothesis of connection between Pearson and Fechner correlations shows that the hypothesis is rejected for the pairs of stocks with two hubs KO (Coca-Cola) and NKE (Nike). If we remove these stocks, the conjecture of connection is confirmed for both choice of critical values c_1 and c_2. This phenomenon needs a further investigation.

Acknowledgements This work is partly supported by LATNA laboratory, Russian Federation Government grant 11.G34.31.0057, and RFHR grant 15-32-01052.

References

1. Anderson, T.W.: An Introduction to Multivariate Statistical Analysis, 3rd edn. Wiley, New York (2003)
2. Bautin, G.A., Kalyagin, V.A., Koldanov, A.P., Koldanov, P.A., Pardalos, P.M.: Simple measure of similarity for the market graph construction. Comput. Manag. Sci. **10**, 105–124 (2013)
3. Boginski, V., Butenko, S., Pardalos, P.M.: Statistical analysis of financial networks. J. Comput. Stat. Data Anal. **48**(2), 431–443 (2005)
4. Cramer, H.: Mathematical Methods of Statistics. Princeton Mathematical Series, vol. 9. Princeton University Press, Princeton (1945)
5. Gupta, F.K., Varga, T., Bodnar, T.: Elliptically Contoured Models in Statistics and Portfolio Theory. Springer, New York (2013). ISBN:978-1-4614-8153-9
6. Lehmann, E.L., Romano, J.P.: Testing Statistical Hypotheses. Springer, New York (2005)
7. Mantegna, R.N.: Hierarchical structure in financial markets. Eur. Phys. J. B. **11**, 93–97 (1999)
8. Mantegna, R.N., Stanley, H.E.: An Introduction to Econophysics: Correlations and Complexity in Finance. Cambridge University Press, Cambridge (2000)

Testing the Stationarity of Sign Coincidence in Market Network

Dmitry E. Mozokhin and Alexander P. Koldanov

Abstract The problem of stationarity of sign coincidence of returns is considered. Stationarity of sign coincidence for a pair of stocks is tested by two-sample Kolmogorov–Smirnov and Chi-Square tests. Multiple comparison procedures, such as Bonferroni and Holm procedures, are employed to test stationarity of sign coincidence in market network and to control the family wise error rate (FWER). The method is validated for testing stationarity of stock's prices and returns. It is shown that the hypothesis of stationarity is rejected for prices and it is not rejected for returns and their sign coincidence on some significance level.

Keywords Stock market • Stock price • Stock return • Correlations • Sign coincidence • Stationarity • Multiple comparison • Multiple testing statistical procedures • Banferroni procedure • Holm procedure • Family wise error rate

1 Introduction

Probabilistic models for stock prices and their returns were introduced in [1, 5]. Modern development of this topic is presented in [9]. It is known that prices are not stationary [6] and returns of stocks can be considered as stationary (in [8] the authors point out that returns are serially independent). Sign coincidence (sign similarity) of returns was introduced in market network analysis in [2] as a measure of similarity between stocks. This measure has some interesting properties with

D.E. Mozokhin (✉)
National Research University Higher School of Economics, Bolshaya Pecherskaya 25, Nizhny Novgorod 603155, Russia
e-mail: dim_moz@mail.ru

A.P. Koldanov
Lab LATNA, National Research University Higher School of Economics, Bolshaya Pecherskaya 25, Nizhny Novgorod 603155, Russia,
e-mail: akoldanov@hse.ru

V.A. Kalyagin et al. (eds.), *Models, Algorithms and Technologies for Network Analysis*, Springer Proceedings in Mathematics & Statistics 156,
DOI 10.1007/978-3-319-29608-1_13

respect to classical Pearson correlation [3]. Therefore it is important to investigate general properties of this measure of similarity, and, in particular, its stationarity. As far as we know there is no work that is devoted to this problem.

Our method for testing stationarity has two main parts. First, it is stationarity verification for quantitative characteristics of stocks that is implemented, thanks to two-sample homogeneity test. Second, we apply a multiple comparison technique and use family wise error rate (FWER) to control the risk. Multiple comparison is conducted with Bonferroni and Holm multiple testing statistical procedures, comparisons of binomial proportions, and Chi-square test of homogeneity. The method is validated for testing stationarity of stock's prices and returns. Using this method we confirm the previous study (prices are not stationary, returns are stationary) and obtain our main result: sign coincidences pass the stationarity tests for some significance levels.

This paper is organized as follows. In Sect. 2 the problem statement with main definitions and notations is provided. Section 3 contains description of statistical procedures that are employed in the paper. In Sect. 4 the conducted experiments and the analysis of the result observed are presented. Section 5 emphasizes the main result of the paper.

2 Problem Statement

Let N and n be the number of stocks in the financial market and number of observations, respectively. Define by $p_i(t)$ the price of stock i for the day t, where $i = 1..N, t = 1..n$. Also, denote by $r_i(t)$ the daily return of stock i for the period from $(t-1)$ to t that can be calculated by

$$r_i(t) = ln\frac{p_i(t)}{p_i(t-1)} \tag{1}$$

Sign coincidence $c_{i,j}$ of returns i and j is defined by

$$c_{i,j}(t) = \begin{cases} 1, \ r_i(t)r_j(t) \geq 0 \\ 0, \ otherwise \end{cases}, \tag{2}$$

where $r_i(t)$ and $r_j(t)$ are stock returns i and j, respectively $(i \neq j), t = 1 \dots n$. Note that the total number of $c_{i,j}$ is $\binom{N}{2}$.

For testing stationarity we use the following pre-processing technique. The vectors of prices (p_i) and returns (r_i) are divided into k intervals of the same length. It is assumed that $r_i(1) \dots r_i(\frac{n}{k})$ (also $p_i(1) \dots p_i(\frac{n}{k})$ are the observations of a random variable $R_i^{(1)}$ $(P_i^{(1)})$. In this case, $R_i^{(t)}$ $(P_i^{(t)})$ describes the behavior of daily returns(prices) of the stock i for the period from $((\frac{n}{k}-1)*t+1)$ to $(\frac{n}{k})*t$.

By analogy with returns and prices, vectors of sign coincidence are divided into k parts which are descried by the relevant random variables. It means that $c_{ij}(1).\therefore.c_{ij}(\frac{n}{k}))$ $(i \neq j)$ are observations of random variable $C_{ij}^{(t=1)}$ and so on for $t = 1..k$.

With considering all definitions and the assumption that the stationarity of the characteristics is equivalent to pairwise testing the null hypothesis of homogeneity, the problem can be interpreted in the following way: Firstly, for prices and returns it is needed to test the fact that random variables $R_i^{(t)}$ $(P_i^{(t)})$ $(t = 1, 2, \ldots, k)$ are independent and identically distributed for fixed stock i. The total number of hypotheses is equal to $N\binom{k}{2}$ and we study the stationarity problem as a multiple decision problem. Secondly, for measures of sign coincidence between returns the following hypothesis has to be tested:

$$H_0^i : F_l^i(r) = F_m^i(r),$$

where $l, m = 1 \ldots k; l \neq m; i = 1 \ldots N$.

This hypothesis tests the homogeneity of two distribution functions that correspond to parts l and m of stock i.

It should be noted that there are several procedures to solve multiple comparisons problem and control FWER. For instance, Bonferroni and Holm procedures, which are described below, can be utilized for it.

3 Statistical Procedures

3.1 Multiple Comparisons

By definition, testing of each statistical hypothesis contains the possibility of error of the first kind. The more we check the hypotheses on the same data, the more likely to allow at least one such error. This phenomenon is called the effect of multiple comparisons (*multiple comparisons* or *multiple testing*). Let V be the number of errors of the first kind, then in a simultaneous test of a set of statistical hypotheses, the goal is to minimize the number of false rejections. Since the probability to make a mistake in at least one of these M (in our case M is equal to $\binom{k}{2}$) comparisons is equal to $1-(1-\alpha)$, which substantially exceeds the original value of the significance level (for example, $\alpha = 0.05$), then a further increase in the number of testable hypotheses will be to lead to an inevitable increase in the error of the first kind. So, if $V \geq 1$, we will make at least one error of the first kind and the likelihood of such errors is called "*group error probability (family wise error rate)*". By definition, FWER $= P(V \geq 1)$. Accordingly, when we want to control an error of first kind at a certain significance level α, we must have FWER $\leq \alpha$. Finally, in order to make adjustments in the significance level, there are a number of methods, such as the Bonferroni procedure and Holm procedure.

3.2 Bonferroni Procedure

It is one of the simplest and most well-known methods of control over the group error probability. This method states that in order to achieve the level of significance α it is enough to reject the hypothesis H_0^i, for which $p \leq \frac{\alpha}{M}$, where M—total number of hypotheses, and p is a p-value. The formal definition of the Bonferroni procedure can be presented as follows. Let $H_1, H_2, \ldots, H_M$—a family of hypotheses, and $p_1, p_2, \ldots, p_M$—corresponding p-values. It should be noted that p-value is a function of the observed sample results which is calculated as the lowest α for which the null hypothesis is rejected for a given set of observations. Denote by I an unknown subset of true null hypotheses that has power m_0. Then FWER is the probability of rejection of at least one hypothesis of I. Bonferroni correction method argues that the rejection of all $p_i \leq \frac{\alpha}{M}$ provides an FWER $\leq \alpha$.

$$FWER = P(V \geq 1) \leq P\left(\bigcup_{i=1}^{m_0} \widetilde{P_i} \leq \alpha\right) \leq \sum_{i=1}^{m_0} P(\widetilde{P_i} \leq \alpha) \leq \sum_{i=1}^{m_0} \frac{\alpha}{m} = m_0 \frac{\alpha}{m} \leq \alpha$$

3.3 Holm Procedure

Holm (Holm–Bonferroni) procedure is a method of controlling a group of probability of errors. Holm procedure is based on an algorithm that includes the following steps:

- The initial p-values are arranged in nondecreasing order: $p_1 \leq p_2 \leq \ldots \leq p_m$. These values correspond to family of hypotheses which are being tested $H_1, H_2, \ldots, H_m$.
- If $p_1 \geq \frac{\alpha}{m}$, then hypotheses $H_1, H_2, \ldots, H_m$ are non-rejected and the algorithm stops. Otherwise, if $p_1 < \frac{\alpha}{m}$, we have to reject the hypothesis H_1 and continue checking the remaining hypotheses at significance level $\frac{\alpha}{m-1}$.
- If $p_2 \geq \frac{\alpha}{m-1}$, then hypotheses $H_2, H_3, \ldots, H_m$ are non-rejected and the algorithm stops. Otherwise, if $p_2 < \frac{\alpha}{m-1}$, we have to reject the hypothesis H_2 and continue checking the remaining hypotheses at significance level $\frac{\alpha}{m-2}$, etc.

It should be noted that Bonferroni and Holm procedures are not the best available procedures. There are many different methods that control the FWER and are more powerful than Holm method. For instance, Hochberg and Hommel procedures can be considered as more precise. However, the results observed show that in our case it is enough to employ Bonferroni and Holm approaches to detect stationarity.

3.4 Global Test for Comparisons of k Binomial Proportions

In order to test stationarity for sign coincidence of returns the comparisons of k binomial proportions are utilized. This approach is described in more detail in [7]. The general idea is the following: Let $X_i \sim BIN(m, p_i)$, where m is the length of each part of division and equal to $\frac{n}{k}$; $\widehat{p_i} = \frac{X_i}{m}$. Naturally, it is stated that proportions are independent.

For testing $H_0 : p_1 = \ldots = p_k$ an inverse-sine transformation that gives a constant variance is done since standard errors of $\widehat{p_i}$ depend on unknown p_i.

$$\breve{p}_i = \sin^{-1}\sqrt{\frac{X_i + \frac{3}{8}}{n_i + \frac{3}{4}}} \sim N\left(\sin^{-1}\sqrt{p_i}, \frac{1}{4n_i}\right)$$

Using these transformed values, it is analyzed that the statistics G has an asymptotic chi-square distribution with $k-1$ degrees of freedom.

$$G = \frac{\sum_{i=1}^{k} n_i(\breve{p}_i - \breve{p})^2}{\frac{1}{4}}, \breve{p}_i = \sin^{-1}\sqrt{\frac{\sum_{i=1}^{k} X_i + \frac{3k}{8}}{\sum_{i=1}^{k} n_i + \frac{3k}{4}}}$$

Thus, the approach is reduced to calculation of the statistics G and comparison of it with critical value for some significance level.

3.5 Chi-Square Test of Homogeneity

According to [4] we have k sequences of observations and in each of them some event E (compliance of a sign) occurs $v_1, \ldots, v_k$ times, respectively. The question is, is there any reason to believe that the event E has the same constant but unknown probability p in all cases. Obviously, estimation of p should be the frequency of the event E in the total data-set.

$$p^* = 1 - q^* = \frac{1}{n}\sum_j v_{j*}$$

After that p^* can be substituted in Chi-square statistic that has $k-1$ degrees of freedom in order to test the hypothesis that was mentioned above.

$$\chi^2 = \sum_j \frac{(v_j - n_j p^*)^2}{n_j p^* q^*} = \frac{1}{p^* q^*}\sum_j \frac{v_j^2}{n_j} - n\frac{p^*}{q^*}$$

4 Experiments

The conducted experiments have the following objectives:

- to check the null hypothesis of stationarity for prices of stocks and their returns and confirm the fact that returns are stationary and prices are not.
- to investigate the stationarity of sign coincidence of returns

In the first part the two-sample Kolmogorov–Smirnov (KS) test is utilized for testing homogeneity of distributions. Let $F_{1,l}(x)$ and $F_{1,m}(x)$ be empirical distribution functions which characterize the distributions of l and m parts, respectively, that are obtained, thanks to division of the first stock into k parts. It is necessary to note that these parts are considered as independent. Thus, KS test checks the homogeneity in pairs for each stock. One needs a multiple comparisons in this case.

The second part is aimed to investigate stationarity of sign coincidence using global comparisons test and Chi-square test of homogeneity which are described above. Also, in both cases we change number of parts for division ($k = 2, 5, 10, 20$) and significance level ($\alpha = 0.05, 0.5$) in order to conduct the comparison analysis.

4.1 Data-Set Description

For the experiments, information concerning prices of stocks of German financial market from 2010 to 2011 has been collected. Namely, we take 85 tickers from DAX, SDAX, and TECDAX indexes. Each ticker has 500 observations which correspond to closing prices.

4.2 Stationarity of Prices

First of all, we investigate prices which according to [6] are non-stationary. Let $k = 10$, it means that we study stationarity on time frame within 50 days (approximately 2 months). In this case the total number of homogeneity hypotheses is equal to 45 and the largest number of non-rejected hypotheses corresponds to the stronger stationarity. In other words, if all 45 hypotheses are not rejected for some stock, then it will mean that its prices or returns are stationary. On the assumption of adjustments, the results observed are presented in Fig. 1. Analyzing it, we can state that prices of stocks are definitely non-stationary, since on the average the null hypothesis of stationarity is not rejected in 4 cases for Bonferroni procedure. As for the other approaches, the utilization of the single step procedure (with constant significance level α) leads to complete absence of stationarity because 4 % of hypotheses are not rejected only.

Ticker	α = const	Bonferroni	Holm
ADSDE	1	2	1
ALVDE	2	7	6
BASDE	2	5	3
BAYNDE	4	6	6
BEIDE	8	14	13
BMWDE	2	3	3
CBKDE	3	8	6
CONDE	2	2	2
DAIDE	1	5	1
DB1DE	2	8	7
DBKDE	1	6	5
DPWDE	5	14	8
DTEDE	2	2	2
EOANDE	0	0	0
FMEDE	1	2	2
FREDE	1	6	2
HEIDE	3	5	4
HEN3DE	2	6	3
IFXDE	3	6	6
LHADE	0	3	0
LINDE	2	4	3
LXSDE	3	6	6
MRKDE	1	7	6
MUV2DE	1	1	1
RWEDE	1	1	1
SAPDE	3	9	8
SDFDE	5	9	7
SIEDE	3	4	3
TKADE	3	6	3
VOW3DE	4	5	5
AOXDE	2	5	4
B5ADE	2	5	3
BDTDE	0	1	0
BVBDE	1	1	1
CAPDE	1	4	2
COMDE	3	7	4
EVDDE	2	2	2
DBANDE	1	2	1
DEZDE	3	4	4
DICDE	2	4	2
GSC1DE	1	1	1
GFKDE	1	3	1
GMMDE	2	2	2
GLJDE	2	5	4
HABDE	3	4	3
HHFADE	2	6	2
HBMDE	1	1	1
KWSDE	3	5	3
P1ZDE	3	5	3
PUMDE	6	12	9
SLTDE	0	1	0
SIX3DE	0	4	0
SURDE	1	4	3
TTKDE	2	4	3
VIB3DE	1	2	1
ZO1DE	1	4	3
AFXDE	2	5	4
AIXADE	0	6	3
BBZADE	1	3	1
BC8DE	2	2	2
COKDE	1	4	1
COPDE	2	6	3
DLGDE	2	7	5
DRIDE	1	4	1
DRW3DE	2	3	3
EVTDE	2	4	2
FNTNDE	3	14	13
JENDE	2	4	2
KBCDE	0	3	0
LPKDE	2	5	5
M5ZDE	2	7	4
MORDE	1	5	1
NDX1DE	1	3	1
NEMDE	4	4	4
O1BCDE	2	3	2
PFVDE	3	4	4
PSANDE	1	4	3
QIADE	3	5	4
QSCDE	0	2	0
S92DE	1	2	1
SBSDE	0	6	0
SOWDE	2	4	4
SRT3DE	2	2	2
UTDIDE	3	5	5
WDIDE	3	4	3
Mean	1,9647	4,5882	3,1529
Std	1,3754	2,7915	2,5612

Fig. 1 Stationarity of prices. Integer value in cells determines the number of accepted hypotheses of homogeneity, $k = 10$, $\alpha = 0.05$, the total number of hypotheses is 45

4.3 Stationarity of Returns

To investigate the stationarity of returns the same methodology is employed. We divide given vectors of returns into 10 and 20 parts which correspond to 2 and 1 months, respectively.

At first, for $\alpha = 0.05$ and $k = 10$ the results observed are shown in Fig. 2. It is necessary to note that the total number hypotheses is 45. Hence, we assume the stationarity of returns, since approaches that control FWER do not reject on average 44.7 hypotheses. However, for some tickers such as "*ALVDE*" and "*CBKDE*" the single step procedure rejects approximately the half of hypotheses and at the result, it means that in general the hypothesis of stationarity for returns can be rejected.

Figure 3 shows that Bonferroni and Holm procedures still accept the hypothesis even if the initial probability of error is very large ($\alpha = 0.5$). Also, it should be noted that in both cases the number of non-rejected hypotheses is matched completely when Bonferroni and Holm approaches are utilized.

In addition to this, we test the hypotheses of stationarity for $k = 5$ and $\alpha = 0.05$. The results are shown in Fig. 4. It should be noted that majority of hypotheses are not rejected also, since we have 190 hypotheses and on the average 189 are supported from them. Therefore we can state that returns can be considered as stationary. What is more, tickers "*ALVDE*" and "*CBKDE*" have non-stationary returns and this fact is supported by calculations that are presented in Fig. 4.

4.4 Stationarity of Sign Coincidence

To test the stationarity of sign coincidence the Global test for comparisons of k binomial proportions and Chi-square test of homogeneity are employed with $\alpha = 0.05$. The number of hypotheses is $\binom{N}{2}$, where N is the number of stocks. Therefore we take the first ten tickers and test 45 hypotheses, the results are presented in Figs. 5 and 6. If a cell contains 0, it means that the sign coincidence between $TICKER_i\_TICKER_j$ is non-stationary for some value k.

The last row is sum of columns. It allows to conclude that sign coincidence are stationary. Moreover, the sign coincidence between tickers "*ALVDE*" and "*CBKDE*" which is of great interest is stationary because the null hypothesis is rejected once.

5 Concluding Remarks

A new method is proposed to test stationarity of stocks characteristics (prices, returns, sign coincidence). The method confirms the previously obtained results for prices and returns. Besides, new problem of testing the hypothesis of stationarity for sign coincidence of returns is considered. It is shown that sign coincidence passes the stationarity test. It is interesting to investigate stationarity of other measures of similarity between stocks, such as Pearson correlation and partial correlation. It will be a subject of further researches.

Ticker	α = const	Bonferroni	Holm
ADSDE	42	45	45
ALVDE	36	45	45
BASDE	43	45	45
BAYNDE	44	45	45
BEIDE	45	45	45
BMWDE	44	45	45
CBKDE	24	44	44
CONDE	37	45	45
DAIDE	41	45	45
DB1DE	42	45	45
DBKDE	36	45	45
DPWDE	38	44	44
DTEDE	36	45	45
EOANDE	33	42	42
FMEDE	40	45	45
FREDE	40	45	45
HEIDE	35	42	42
HEN3DE	41	45	45
IFXDE	44	45	45
LHADE	35	44	44
LINDE	44	45	45
LXSDE	40	45	45
MRKDE	42	45	45
MUV2DE	34	45	45
RWEDE	27	42	42
SAPDE	39	45	45
SDFDE	37	45	45
SIEDE	45	45	45
TKADE	38	45	45
VOW3DE	42	45	45
AOXDE	43	45	45
B5ADE	34	45	45
BDTDE	39	45	45
BVBDE	34	44	44
CAPDE	42	45	45
COMDE	41	45	45
EVDDE	45	45	45
DBANDE	45	45	45
DEZDE	39	45	45
DICDE	42	45	45
GSC1DE	37	45	45
GFKDE	39	45	45
GMMDE	40	45	45
GLJDE	43	45	45
HABDE	40	45	45
HHFADE	37	45	45
HBMDE	43	45	45
KWSDE	44	45	45
P1ZDE	45	45	45
PUMDE	45	45	45
SLTDE	45	45	45
SIX3DE	42	45	45
SURDE	34	45	45
TTKDE	43	45	45
VIB3DE	37	45	45
ZO1DE	41	45	45
AFXDE	43	45	45
AIXADE	42	45	45
BBZADE	31	41	41
BC8DE	44	45	45
COKDE	45	45	45
COPDE	45	45	45
DLGDE	40	45	45
DRIDE	45	45	45
DRW3DE	42	44	44
EVTDE	39	45	45
FNTNDE	40	45	45
JENDE	43	45	45
KBCDE	43	45	45
LPKDE	42	45	45
M5ZDE	38	45	45
MORDE	44	45	45
NDX1DE	42	45	45
NEMDE	44	45	45
O1BCDE	31	41	41
PFVDE	39	45	45
PSANDE	43	45	45
QIADE	37	45	45
QSCDE	37	44	44
S92DE	40	45	45
SBSDE	44	45	45
SOWDE	40	43	43
SRT3DE	45	45	45
UTDIDE	42	45	45
WDIDE	45	45	45
Mean	40,2118	44,7059	44,7059
Std	4,2709	0,8567	0,8567

Fig. 2 Stationarity of returns. Integer value in cells determines the number of accepted hypotheses of homogeneity, $k = 10$, $\alpha = 0.05$, the total number of hypotheses is 45

Ticker	α = const	Bonferroni	Holm
ADSDE	70	190	190
ALVDE	48	186	186
BASDE	66	190	190
BAYNDE	66	190	190
BEIDE	100	190	190
BMWDE	71	190	190
CBKDE	48	187	187
CONDE	58	190	190
DAIDE	70	189	189
DB1DE	70	190	190
DBKDE	54	188	188
DPWDE	56	190	190
DTEDE	55	190	190
EOANDE	56	189	189
FMEDE	79	190	190
FREDE	84	190	190
HEIDE	67	189	189
HEN3DE	64	190	190
IFXDE	65	190	190
LHADE	57	189	189
LINDE	109	190	190
LXSDE	70	190	190
MRKDE	72	190	190
MUV2DE	60	188	188
RWEDE	50	187	187
SAPDE	74	190	190
SDFDE	75	190	190
SIEDE	101	190	190
TKADE	54	189	189
VOW3DE	67	189	189
AOXDE	90	190	190
B5ADE	47	189	189
BDTDE	62	190	190
BVBDE	55	186	186
CAPDE	104	190	190
COMDE	61	190	190
EVDDE	80	190	190
DBANDE	107	190	190
DEZDE	68	190	190
DICDE	90	190	190
GSC1DE	65	187	187
GFKDE	75	190	190
GMMDE	54	190	190
GLJDE	74	189	189
HABDE	92	190	190
HHFADE	78	190	190
HBMDE	114	190	190
KWSDE	123	190	190
P1ZDE	72	190	190
PUMDE	108	190	190
SLTDE	105	190	190
SIX3DE	85	190	190
SURDE	64	186	186
TTKDE	104	190	190
VIB3DE	104	190	190
ZO1DE	77	190	190
AFXDE	99	190	190
AIXADE	90	190	190
BBZADE	82	165	165
BC8DE	107	189	189
COKDE	102	190	190
COPDE	120	190	190
DLGDE	67	190	190
DRIDE	107	190	190
DRW3DE	95	190	190
EVTDE	67	189	189
FNTNDE	78	190	190
JENDE	90	190	190
KBCDE	98	190	190
LPKDE	93	189	189
M5ZDE	62	190	190
MORDE	92	190	190
NDX1DE	74	190	190
NEMDE	84	190	190
O1BCDE	55	190	190
PFVDE	61	190	190
PSANDE	72	190	190
QIADE	88	189	189
QSCDE	69	188	188
S92DE	82	181	181
SBSDE	134	189	189
SOWDE	73	188	188
SRT3DE	113	190	190
UTDIDE	93	190	190
WDIDE	109	190	190
Mean	79,4118	189,1059	189,1059
Std	20,1319	2,9722	2,9722

Fig. 3 Stationarity of returns. Integer value in cells determines the number of accepted hypotheses of homogeneity, $k = 20$, $\alpha = 0.5$, the total number of hypotheses is 190

Ticker	α = const	Bonferroni	Holm
ADSDE	9	10	10
ALVDE	5	9	7
BASDE	9	10	10
BAYNDE	7	10	10
BEIDE	10	10	10
BMWDE	8	10	10
CBKDE	3	5	5
CONDE	6	10	10
DAIDE	8	10	10
DB1DE	9	10	10
DBKDE	7	9	9
DPWDE	6	10	10
DTEDE	7	9	9
EOANDE	6	6	6
FMEDE	9	10	10
FREDE	10	10	10
HEIDE	7	9	8
HEN3DE	10	10	10
IFXDE	10	10	10
LHADE	7	9	9
LINDE	9	10	10
LXSDE	10	10	10
MRKDE	8	10	10
MUV2DE	6	9	9
RWEDE	4	6	6
SAPDE	10	10	10
SDFDE	7	10	10
SIEDE	10	10	10
TKADE	7	9	9
VOW3DE	10	10	10
AOXDE	8	10	10
B5ADE	7	10	10
BDTDE	4	9	9
BVBDE	5	7	6
CAPDE	9	10	10
COMDE	8	10	10
EVDDE	10	10	10
DBANDE	10	10	10
DEZDE	6	10	10
DICDE	7	10	10
GSC1DE	5	10	10
GFKDE	6	10	10
GMMDE	7	10	10
GLJDE	7	10	10
HABDE	6	9	9
HHFADE	7	10	10
HBMDE	10	10	10
KWSDE	10	10	10
P1ZDE	10	10	10
PUMDE	9	10	10
SLTDE	10	10	10
SIX3DE	8	10	10
SURDE	7	10	10
TTKDE	10	10	10
VIB3DE	10	10	10
ZO1DE	7	10	10
AFXDE	10	10	10
AIXADE	8	10	10
BBZADE	6	6	6
BC8DE	10	10	10
COKDE	10	10	10
COPDE	10	10	10
DLGDE	9	10	10
DRIDE	10	10	10
DRW3DE	10	10	10
EVTDE	10	10	10
FNTNDE	10	10	10
JENDE	9	10	10
KBCDE	9	10	10
LPKDE	8	10	10
M5ZDE	6	10	10
MORDE	10	10	10
NDX1DE	8	9	9
NEMDE	9	10	10
O1BCDE	5	7	7
PFVDE	7	10	10
PSANDE	8	10	10
QIADE	7	10	10
QSCDE	8	10	10
S92DE	7	10	10
SBSDE	10	10	10
SOWDE	7	9	8
SRT3DE	10	10	10
UTDIDE	10	10	10
WDIDE	9	10	10
Mean	8,0824	9,6	9,5412
Std	1,7942	1,0259	1,1186

Fig. 4 Stationarity of returns. Integer value in cells determines the number of accepted hypotheses of homogeneity, $k = 5$, $\alpha = 0.05$, the total number of hypotheses is 10

Connection	k = 20	k = 10	k = 5	k = 2
ADSDE_ALVDE	1	1	0	1
ADSDE_BASDE	1	1	1	1
ADSDE_BAYNDE	1	1	1	1
ADSDE_BEIDE	1	1	1	1
ADSDE_BMWDE	1	1	1	0
ADSDE_CBKDE	1	0	1	1
ADSDE_CONDE	1	1	1	1
ADSDE_DAIDE	1	1	1	0
ADSDE_DB1DE	1	1	1	1
ALVDE_BASDE	1	1	1	1
ALVDE_BAYNDE	1	1	1	1
ALVDE_BEIDE	1	1	1	1
ALVDE_BMWDE	1	1	1	1
ALVDE_CBKDE	1	1	1	1
ALVDE_CONDE	1	1	1	1
ALVDE_DAIDE	1	1	0	1
ALVDE_DB1DE	1	1	1	1
BASDE_BAYNDE	1	1	1	1
BASDE_BEIDE	1	1	1	1
BASDE_BMWDE	1	1	1	1
BASDE_CBKDE	1	0	0	1
BASDE_CONDE	1	1	1	1
BASDE_DAIDE	1	1	0	1
BASDE_DB1DE	1	1	1	1
BAYNDE_BEIDE	1	1	1	1
BAYNDE_BMWDE	1	1	1	1
BAYNDE_CBKDE	0	0	0	1
BAYNDE_CONDE	1	1	1	1
BAYNDE_DAIDE	1	1	1	1
BAYNDE_DB1DE	1	1	1	1
BEIDE_BMWDE	0	1	1	1
BEIDE_CBKDE	1	1	1	1
BEIDE_CONDE	1	1	1	1
BEIDE_DAIDE	1	1	1	1
BEIDE_DB1DE	0	1	1	1
BMWDE_CBKDE	1	1	1	1
BMWDE_CONDE	1	1	1	1
BMWDE_DAIDE	1	1	1	1
BMWDE_DB1DE	1	1	1	1
CBKDE_CONDE	1	1	1	1
CBKDE_DAIDE	1	1	1	1
CBKDE_DB1DE	1	1	1	1
CONDE_DAIDE	1	1	1	1
CONDE_DB1DE	1	1	1	1
DAIDE_DB1DE	1	1	1	1
Sum	42	42	40	43

Fig. 5 Global test

	Connection	k = 20	k = 10	k = 5	k = 2
1	ADSDE_ALVDE	1	0	0	1
2	ADSDE_BASDE	1	1	0	0
3	ADSDE_BAYNDE	1	1	1	1
4	ADSDE_BEIDE	1	1	1	1
5	ADSDE_BMWDE	1	0	0	0
6	ADSDE_CBKDE	1	0	1	1
7	ADSDE_CONDE	0	0	1	1
8	ADSDE_DAIDE	1	0	0	0
9	ADSDE_DB1DE	1	1	1	1
10	ALVDE_BASDE	0	0	0	1
11	ALVDE_BAYNDE	0	1	1	1
12	ALVDE_BEIDE	1	1	1	1
13	ALVDE_BMWDE	1	1	1	1
14	ALVDE_CBKDE	1	0	1	1
15	ALVDE_CONDE	0	0	0	1
16	ALVDE_DAIDE	1	0	0	1
17	ALVDE_DB1DE	0	0	1	0
18	BASDE_BAYNDE	1	1	1	0
19	BASDE_BEIDE	1	1	1	1
20	BASDE_BMWDE	1	1	1	1
21	BASDE_CBKDE	1	0	0	1
22	BASDE_CONDE	1	1	1	1
23	BASDE_DAIDE	0	0	0	1
24	BASDE_DB1DE	1	1	1	1
25	BAYNDE_BEIDE	1	1	1	1
26	BAYNDE_BMWDE	1	1	1	1
27	BAYNDE_CBKDE	0	0	0	1
28	BAYNDE_CONDE	1	1	1	1
29	BAYNDE_DAIDE	1	1	0	1
30	BAYNDE_DB1DE	1	1	1	1
31	BEIDE_BMWDE	0	1	1	1
32	BEIDE_CBKDE	1	1	1	1
33	BEIDE_CONDE	0	1	1	1
34	BEIDE_DAIDE	1	1	1	1
35	BEIDE_DB1DE	1	1	1	0
36	BMWDE_CBKDE	1	1	0	1
37	BMWDE_CONDE	1	1	1	1
38	BMWDE_DAIDE	1	1	1	0
39	BMWDE_DB1DE	1	1	1	1
40	CBKDE_CONDE	1	0	1	1
41	CBKDE_DAIDE	1	0	0	1
42	CBKDE_DB1DE	1	1	1	1
43	CONDE_DAIDE	1	1	1	1
44	CONDE_DB1DE	1	1	1	1
45	DAIDE_DB1DE	1	1	1	1
	Sum	36	30	32	38

Fig. 6 Chi-square test

Acknowledgements This work is partly supported by LATNA laboratory, Russian Federation Government grant 11.G34.31.0057, RFBR grant 14-01-00807 and HSE Scientific Fund 15-09-0239.

References

1. Bachelier, L.: Theorie la speculation. Annales de l'Ecole Normale Superieure. **17**, 21–86 (1900)
2. Bautin, G.A., Kalyagin, V.A., Koldanov, A.P., Koldanov, P.A., Pardalos, P.M.: Simple measure of similarity for the market graph construction. Comput. Manag. Sci. **10**, 105–124 (2013)
3. Bautin, G.A., Kalyagin, V.A., Koldanov, A.P.: Comparative analysis of two similarity measures for the market graph construction. In: Springer Proceedings in Mathematics and Statistics, vol. 59, pp. 29–41. Springer, New York (2013)
4. Cramer, H.: Mathematical Methods of Statistics. Princeton Mathematical Series, vol. 9. Princeton University Press, Princeton (1945)
5. Kendall, M.G.: The analysis of economic time-series. Part 1. Prices. J. R. Stat. Soc. **96**, 11–25 (1953)
6. Michael, P.: Purchasing power parity yet again: evidence from spatially separated commodity markets. In: Nobay, R., Peel, D.A. (eds.) Liverpool Research Papers in Economics and Finance. ASIN:B0018TE6IO (1992)
7. Nashimoto, K.: Multiple comparison of k binomial proportions. In: Haldeman, K.M., Christopher, M.T. (eds.) Computational Statistics and Data Analysis, vol. 68, pp. 202–212. Elsevier, New York (2013)
8. Rosenberg, B., Ohlson, J.A.: The stationary distribution of returns and portfolio separation in capital markets: a fundamental contradiction. J. Financ. Quant. Anal. **11**, 393–402 (1976)
9. Shiryaev, A.N.: Essentials of Stochastic Finance: Facts, Models, Theory. Advanced Series on Statistical Science and Applied Probability. Word Scientific Publishing Co., Singapore (2003)

Synchronization and Network Measures in a Concussion EEG Paradigm

Ioannis Pappas, Gianluca Del Rossi, John Lloyd, Joseph Gutmann, James Sackellares, and Panos M. Pardalos

Abstract The objective of this work is to characterize the neurophysiologic changes in a patient who suffered a concussion during a football practice via quantitative and graph-theoretic measures and to evaluate the results with respect to the pre-concussion state. We report on a high school athlete who sustained a self-reported concussion while wearing a 16-channel portable EEG recorder and a specially instrumented helmet capable of recording biomechanical impact data. This opportune occurrence has enabled a detailed assessment of the type and duration of changes that occur to human brain function immediately following a sports-related concussion. In the post-concussion EEG segment, we observed a significant decrease in both the quantitative and graph-theoretic measures. More specifically, we observed a significant decline in the cross-mutual information measure between certain pairs of electrodes as well as in the global efficiency of the corresponding brain network. The deviations in the selected quantitative and graph-theoretic measures partially corroborate the usual clinical characteristics of the post-concussion state but further investigation for additional data and evaluation of alternative quantitative measures are needed.

Keywords EEG brain network • Network measures of interconnection • Joint entropy • Cross-mutual information measure • Global efficiency

---

I. Pappas (✉)
University of Florida, Gainesville, FL, USA
e-mail: ioannis.p.pappas@ufl.edu; pardalos@ufl.edu

G. Del Rossi • J. Lloyd • J. Gutmann
University of South Florida, Tampa, FL, USA
e-mail: gdelross@health.usf.edu; drjohnlloyd@tampabay.rr.com; josephgutmann@yahoo.com

J. Sackellares
MD VA Hospital, Gainesville, FL, USA
e-mail: jc.sackellares@gmail.com

P.M. Pardalos
Department of Industrial and Systems Engineering, University of Florida, 303 Weil Hall, Gainesville, FL 32608, USA

National Research University Higher School of Economics, Nizhny Novgorod, Russia

V.A. Kalyagin et al. (eds.), *Models, Algorithms and Technologies for Network Analysis*, Springer Proceedings in Mathematics & Statistics 156,
DOI 10.1007/978-3-319-29608-1_14

1 Introduction

According to the Centers for Disease Control and Prevention (CDC), there are up to 3.8 million concussions occur each year with over 300,000 requiring treatment from health care providers [11] and a significant number of unreported cases [12]. Besides the existence of guidelines provided by the American Academy of Neurology [8], the accurate and timely diagnosis of sports-related concussion remains an elusive challenge for clinicians, because there are currently no reliable biological or imaging markers for the objective detection of a concussion. Detection and diagnosis may be complicated further by an athlete's tendency to mask symptoms when tested using qualitative measures in an effort to return to play quickly.

The need for a means to accurately diagnose acute-sports concussions is underscored by the recognition that repeated concussions could be a significant risk factor for delayed manifestation of neurodegenerative diseases such as chronic traumatic encephalopathy [7], and for conditions such as second-impact syndrome (SIS), a rapid and catastrophic brain swelling following a second trauma occurring minutes, days, and weeks after the initial concussion [1].

EEG was established over 70 years ago as a diagnostic tool for assessing brain function following traumatic head injury on the basis that the moment of head injury, neurons discharge resulting in a release of neurotransmitters followed by neuronal suppression [19]. Most EEG data following traumatic brain injury are from animal models [20]. These models may not be directly applicable to human concussion due to differences in normal EEG patterns among species and difficulty in reproducing human concussions occurring during sports events in an animal model. In humans, the waking EEG has a well-defined pattern, with characteristic waveforms and frequencies in each region of the cerebral cortex. Any pathological process that causes altered consciousness or impairment in concentration and attention can cause diffuse slowing in the EEG background rhythms and changes in spatial organization of normal background rhythms, which can be visually detected by a trained electroencephalographer. Most published EEG studies following traumatic brain injury have examined EEG changes in the days to months following the injury [16]. Reports of EEG changes immediately after head injury are limited, with only a few studies describing EEG findings immediately after head injuries [6] and related to football head injuries [10]. These latter studies describe diffuse slowing in the EEG background, usually due to an increase in theta range frequencies (5–8 Hz). The findings are most prominent immediately after the injury, tend to vary with the severity of the injury, and may persist from minutes to days [17].

The utility of EEG for detecting acute concussion in sports is limited by the challenges of reliably obtaining an interpretable signal, and the fact that most sports physicians are not trained in EEG interpretation. In 2014, Del Rossi et al. [5], conducted a study that demonstrated the feasibility of acquiring continuous EEG recordings during active play in high school football players. Using data from the Del Rossi study, we explore the use of an automated signal analysis of the EEG, to detect mild concussion, based on a network analysis approach. We chose to use network analysis because it provides information regarding both spatial and temporal aspects of the EEG. Cross-mutual information was used to assess statistical correlations between signals derived from each electrode because the measure does not require assumption of linearity or stationarity of the signals [3, 9, 13, 14].

2 Methods

2.1 Source of EEG Data

A continuous multichannel EEG recording previously performed in a de-identified 17-year-old male high school football player during a practice session was analyzed. The digital recording had been obtained as part of an IRB approved research protocol after obtaining informed written subject assent and parental consent. Changes in the EEG were analyzed in relation to kinematic data obtained simultaneously in the same subject.

2.2 Instrumentation and Materials

EEG recordings were obtained using a 16-electrode StatNet (HydroDot, Inc.) electrode system and a Nicolet (CareFusion Corp., Madison, WI) wireless ambulatory EEG amplifier at 1024 Hz sampling rate. The StatNet system approximates the standard International 10/20 System of electrode placement. Fifteen minutes prior to practice, the EEG was obtained with the subject in the alert relaxed state, with eyes closed. For the remainder of the recording, the patient engaged in active football practice on the field. The total duration of the recording was approximately 3 h and 14 mins. In addition, the subject wore a Riddell Revolution Speed football helmet fitted with the head impact telemetry system (HITS), developed by Simbex, Inc. (Lebanon, NH), which incorporates an array of uni-axial accelerometers in the crown of a player's helmet. Linear head impact acceleration data were acquired and sampled at 1 kHz, from which angular head kinematic is estimated.

After training, the data acquisition file was verified for completeness using the Nicolet One software package, then exported as a CSV file for subsequent processing. The following electrodes, each referenced to an average, were used for this analysis: Fp1, Fp2, F7, F8, T3, T4, T5, T6, O1, O2, C3, C4, CZ, A1, and A2. A high pass digital filter of 1 Hz and low pass digital filter of 35 Hz were applied prior to quantitative analysis, for which custom scripts were developed using MATLAB (The MathWorks, Inc., Natick, MA, USA) and its package EEGLAB [4].

2.3 *Quantitative Analysis of EEG Epochs*

EEG spatiotemporal characteristics were analyzed using network analysis of each 3 s epochs. Epochs were selected by visual inspection by a board certified electroencephalographer. The first segment was selected when the subject sat in the locker-room in an alert relaxed state with eyes closed. The second segment was selected at the beginning of the practice when the athlete was walking onto the field. The third, fourth, and fifth segments were selected when the athlete was actively engaged in practice. Selected resting segments were chosen based on relative absence of artifact in the EEG. This was an arduous effort for the data in their totality were noisy due to the nature of the recording.

2.3.1 Network Construction

Let X and Y be two random discrete variables. According to Shannon, the information that is shared between two signals is connected with their degree of randomness [14]. Entropy is a quantitative measure that can encapsulate the randomness of signal and is defined as [3]

$$H(X) = -\sum_{x \in X} P_X(x) log\left(P(x)\right), \tag{1}$$

where P_X is the probability density function of X. The conditional entropy of X given the information of Y is denoted as $H(X/Y)$ is defined as

$$H(X/Y) = -\sum_{y \in Y}\sum_{x \in X} P_{XY}(x,y) log\left(P_{X|Y}(x|y)\right), \tag{2}$$

where P_{XY} is the joint probability distribution of X and Y and $P_{X|Y}$ is the conditional probability distribution of X with respect to Y. Similarly, the joint entropy of X and Y is denoted as $H(X, Y)$ and is defined as

$$H(X,Y) = -\sum_{y \in Y}\sum_{x \in X} P_{XY}(x,y) log\left(P_{XY}(x,y)\right). \tag{3}$$

The relationship among the entropy, joint entropy, and conditional entropy can be highlighted by the following equation:

$$H(X,Y) = H(X) + H(Y/X) = H(Y) + H(X/Y). \tag{4}$$

Intuitively, the above relation states that the joint entropy of two random variables is nothing but the conditional entropy of one random variable w.r.t. the other random variable, plus the entropy of the other random variable. The joint entropy indicates the total uncertainty of the two random variables. In other words, it means the total amount of information needed to describe the random variables. Notice that the joint entropy is lesser than the individual sum of entropy of each random variable. This reduction is due to the presence of shared knowledge provided by one random variable regarding the other random variable. In fact, total uncertainty as a whole may not provide any meaningful information about the relationship between any two random variables. However, the common knowledge that is shared between the two random variables, i.e., the mutual information, may provide insights into the relationship. For example, if the mutual information between the random variables X and Y is exactly equal to entropy of X, then the random variable Y can completely describe the random variable X. Furthermore, if the mutual information between the random variables X and Y is zero, then the random variables X and Y are completely unrelated.

Mutual information between two random variables X and Y is denoted as I(X,Y) and is defined as [3]

$$I(X,Y) = H(X) - H(X/Y) = H(Y) - H(Y/X) = H(X) + H(Y) - H(X,Y). \quad (5)$$

It can be proved that mutual information can be rewritten as

$$I(X,Y) = \sum_{x \in X} \sum_{y \in Y} P_{XY}(x,y) log \frac{P_{XY}}{P_X P_Y}. \quad (6)$$

The above relationships are very natural, and they indicate the simple yet powerful mechanism to quantify the information that one random variable contains about the other random variable. In other words, mutual information indicates the amount of reduction in uncertainty of one random variable due to the knowledge provided by the other random variable. The mutual information measure is symmetric and non-negative. Furthermore, two random variables have a zero mutual information if and only if they are independent. Mutual information can be seen as measure of dependency between two random variables. When compared to the correlation coefficient, a zero value of mutual information indeed guarantees independence irrespective of the probability distribution of the random variables. In addition, no linear dependency between the random variables is assumed in calculation of the mutual information. Therefore, mutual information is a general similarity measure that is more suitable for EEG signals.

In order to incorporate the temporal effect, we calculated the mean mutual information values (we call it cross-mutual information and abbreviate it as CI) between electrodes over time delays of approximately 0–500 ms. We used 64 bins to construct the histograms for the appropriate probability density functions.

A weighted network was constructed using each electrode as a node. Interelectrode dependence between all electrode (node) pairs was quantified using cross-mutual information and this value was assigned as a weight to each edge.

2.3.2 Network Analysis

Network characteristics were quantified using a measure of global efficiency [15, 18]. Global efficiency is a calculation of the shortest path between networks, determined by averaging the inverse of the shortest paths. We define the length between two nodes as the inverse of their weight, i.e., $l_{ij} = \frac{1}{w_{ij}}$. The shortest path between two nodes i and j is defined as d_{ij} and was calculated using Dijkstra's algorithm [2].

In turn, for the resulting network G with set of nodes N, we calculated the global efficiency as

$$E(G) = \frac{1}{(|N||N| - 1)} \sum_{i,j,\ i \neq j,\ i,j \in N} \frac{1}{d_{ij}}. \tag{7}$$

We observe that the higher the global efficiency, the shorter the paths (on average) in the network.

3 Results

In the following five figures we present the average CMI interdependencies for the selected segments. Electrode placements follow the placement of the StatNet system and are named according to their respective brain regions. By convention odd numbers denote electrodes placed over the left cerebral hemisphere, while even numbers denote electrodes placed over the right cerebral hemisphere and z is used to denote midline electrodes. Abbreviations are as follows: Fp (Frontpolar), F (Frontal), C (Central), P (Parietal), O (Occipital), T (Temporal), and A (Ears).

Next, we present the global efficiency of the selected segments as well as the severity of the head impacts that took place throughout the recording with respect to their rotational accelerations.

The CMI distribution reveals attenuated connections between close and distant areas of the brain during the resting state before the athlete starts to practice (Fig. 1). The distribution of CMI among the edges seems to exhibit uniform behavior in the resting state with insignificant deviations across its spectrum. The global efficiency was calculated as 0.5397 for this state. During the athlete's increased athletic activity, an increase in the totality of the connections of the graph (Figs. 2 and 3) is evident; a fact that is mapped to an increased global efficiency with values 0.9808 and 1.2833, respectively. After the concussive hit, we observe a decrease in the weights of the edges that mainly stems from the occipital region (Fig. 4). Global efficiency is now decreased to a value of 1.0389. As the athlete continues to play under the influence of the concussive hit, the weights of the edges are continuously

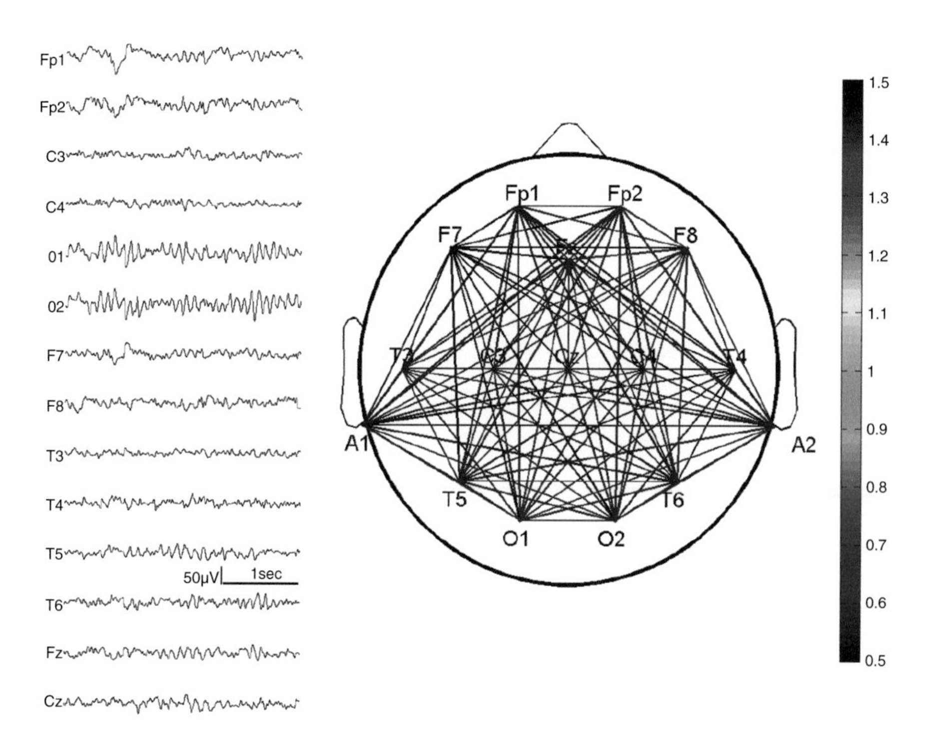

Fig. 1 Network representation of the cross-mutual information interdependencies between the recording electrodes before the athlete starts practice and approximately 1 h and 30 mins before the concussive event

weakened (Fig. 5). In addition the global efficiency tends to reach the pre-concussive resting state with a value of 0.5517 although the athlete is still engaged in the training.

The concussive hit exhibits the greatest rotational acceleration (4026 rad/s^2) recorded throughout the football practice. Following the concussive impact, there were a significant number of subsequent hits and it is unclear what effect, if any, these additional impacts had on the rate of decrease in global efficiency of the CMI network.

Additionally, although the EEG recording was not investigated in its totality, it seems that global efficiency follows a pattern beginning with the resting state value, reaching a significant level during intense athletic activity, then declining during the acute and sub-acute time frame following the concussive event, and eventually returning to a resting state level although the athlete is still active during practice (Fig. 6). In this scheme, the concussive impact seems to affect crucially the global efficiency as it becomes the time point when the global efficiency begins to decrease.

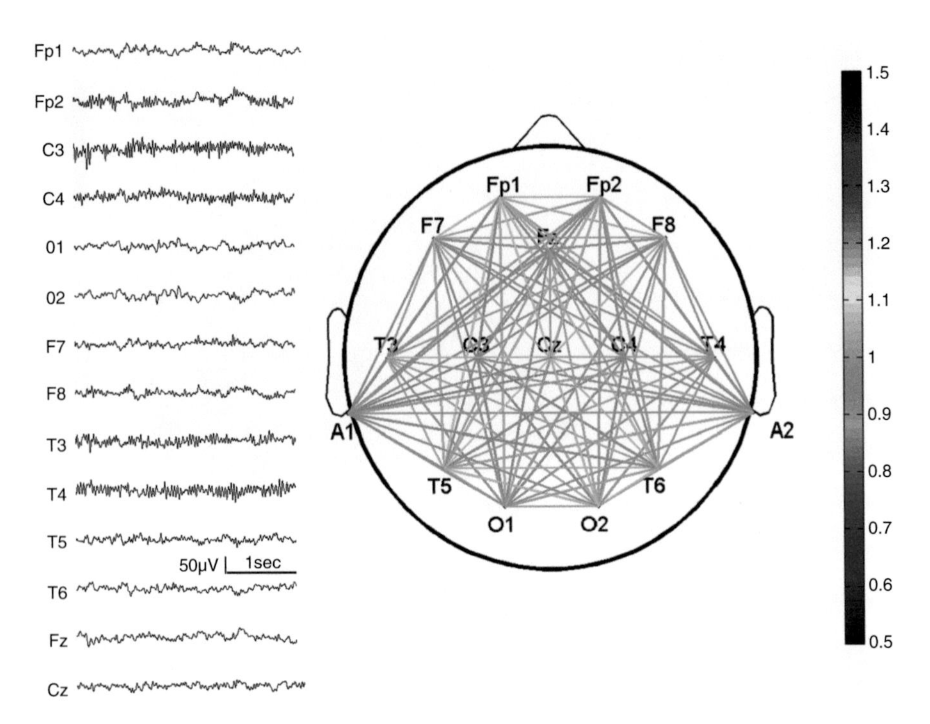

Fig. 2 Network representation of the cross-mutual information interdependencies between the recording electrodes when the athlete starts walking onto the field and approximately 1 h and 26 mins before the concussive event

4 Discussion

The aforementioned observations may not provide a definitive answer with respect to characterizing the concussive impact. The global efficiency of the CMI-based network was decreased after the concussive event; including small declines in the connections between both distantly connected and closely connected areas of the brain, which continued for the duration of time that the athlete remained on the field. According to the definition, this means a decreased information exchange between distant and close parts of the brain—an observed symptom of the concussive brain in the acute period.

From a quantitative point of view, CMI is considered as a statistical measure that represents the exchange of information between signals. Beyond this statistical definition though, we cannot draw conclusions about how those types of connections are established. It is important to keep in mind that CMI is not necessarily intertwined with the exchanging information mechanisms in the brain. Additional nonlinear interdependencies are needed to determine the mechanisms that lead to such deviations.

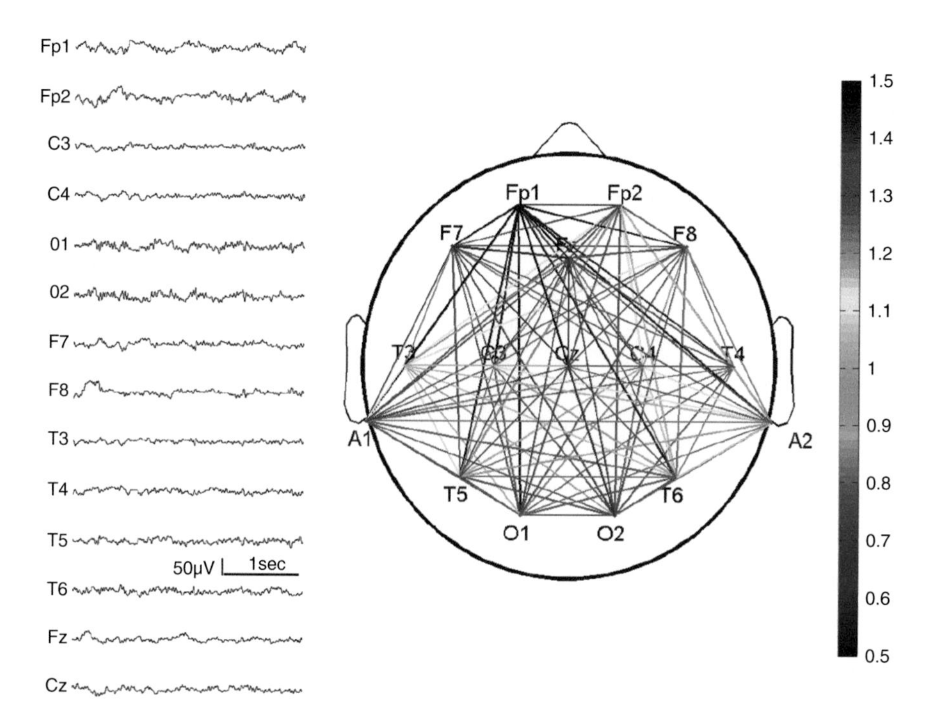

Fig. 3 Network representation of the cross-mutual information interdependencies between the recording electrodes when the athlete is actively engaged in the practice and approximately 26 mins before the concussive event takes place

From a neurophysiological point of view, it is not clear whether this decrease in global efficiency can be attributed to the concussive event and was not due to fatigue due to high frequency of hits in the second half of the practice. Moreover, it was extremely difficult for the recording team to associate each segment of the EEG recording with the behavioral task of the subject in order to further investigate the relationship between global efficiency and the subject's behavior in the practice field.

Finally, the fact that we possess only one dataset of live EEG recording during a reported concussion limits our ability to establish distilled connections between concussion, global efficiency, and other neurophysiological variables.

In conclusion, we analyzed a unique EEG real-time recording of a concussive event that was experienced during an athlete's football practice. We applied a nonlinear measure to examine the correlations between different areas of the brain in artifact-free segments before and after the concussive event. Based on previous remarks, mutual information may stand as a possible discriminant for identifying concussive events. Additional data and measures are needed for analysis to construct a robust framework for identifying concussion via EEG analysis. It appears though

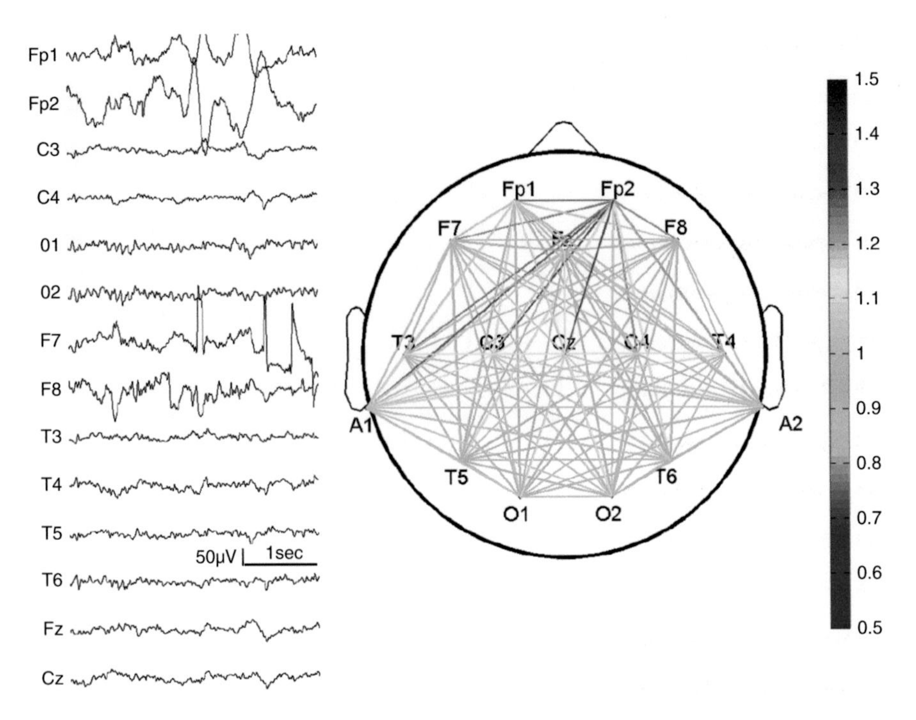

Fig. 4 Network representation of the cross-mutual information interdependencies between the recording electrodes when the athlete is actively engaged in the practice and approximately 6 mins after the concussive event takes place

that quantitative EEG is affected in the post-concussive acute period. However, it remains to be confirmed whether these changes are a recurring pattern in concussions and not a random phenomenon attributed to fatigue or other factors.

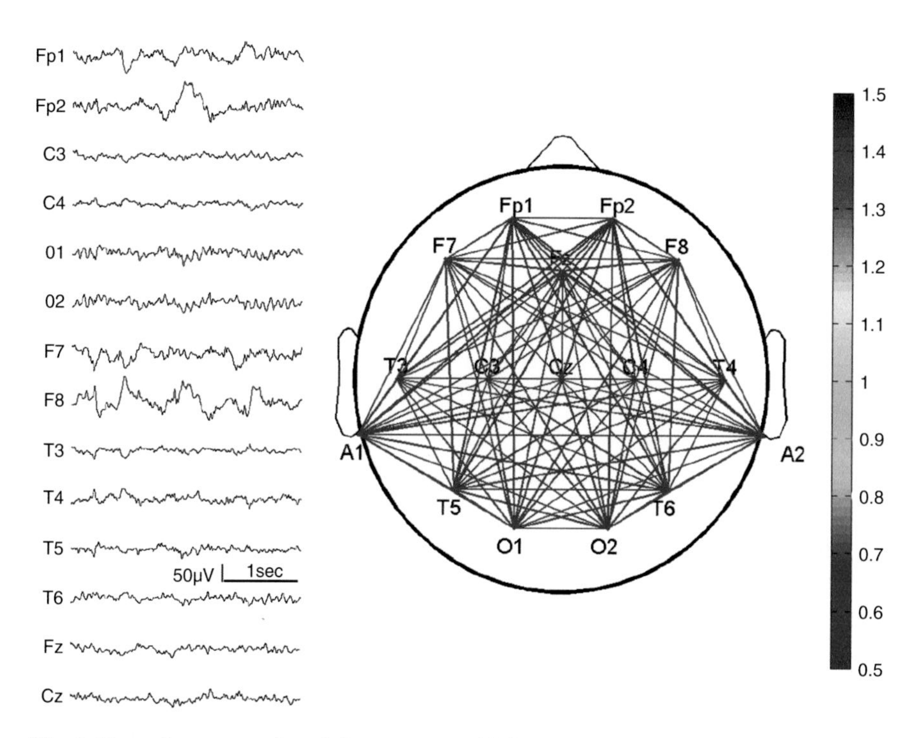

Fig. 5 Network representation of the cross-mutual information interdependencies in an artifact-free epoch where the athlete is actively engaged in the practice and approximately 45 mins after the concussive event takes place

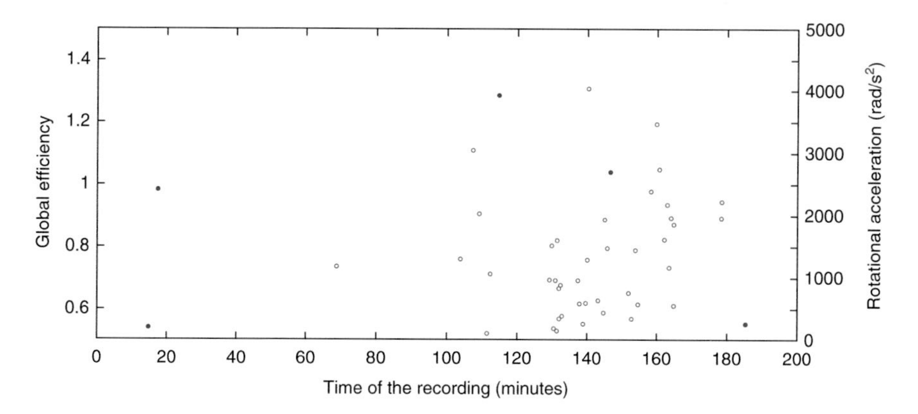

Fig. 6 Evolution of the global efficiency of the five selected segments. The *dots* represent the selected segments and are associated with the global efficiency on the left y-axis. The *circles* represent the hits that were recorded during practice and are associated with their rotational acceleration on the right y-axis

References

1. Cifu, D., Drake, D.: Repetitive head injury syndrome. eMedicine.com (2006)
2. Cormen, T.H., Stein, C., Rivest, R.L., Leiserson, C.E.: Introduction to Algorithms, 2nd edn. McGraw-Hill, New York (2001)
3. Cover, T.M., Thomas, J.A.: Elements of Information Theory. Wiley Series in Telecommunications and Signal Processing. Wiley-Interscience, New York (2006)
4. Delorme, A., Makeig, S.: EEGLAB: an open source toolbox for analysis of single-trial EEG dynamics including independent component analysis. J. Neurosci. Methods **134**(1), 9–21 (2004)
5. Del Rossi, et al.: Real-time changes in human EEG immediately following athletic-related concussion. Manuscript in preparation (2004)
6. Dow, R.S., Ulett, G., Raaf, J.: Electroencephalographic studies immediately following head injury. Am. J. Psychiatr. **101**, 174–183 (1944)
7. Gavett, B.E., Stern, R.A., McKee, A.C.: Chronic traumatic encephalopathy: a potential late effect of sport-related concussive and subconcussive head trauma. Clin. Sports Med. **30**(1), 179-xi (2011)
8. Giza, C.C., Kutcher, J.S., Ashwal, S., Barth, J., Getchius, T.S., Gioia, G.A., et al.: Summary of evidence-based guideline update: evaluation and management of concussion in sports: report of the guideline development subcommittee of the American academy of neurology. Neurology **80**(24), 2250–2257 (2013)
9. Hayes, M.H.: Statistical Digital Signal Processing and Modeling, 1st edn. Wiley, New York (1996)
10. Hughes, J.R., Hendrix, D.E.: Telemetered EEG from a football player in action. Electroencephalogr. Clin. Neurophysiol. **24**(2), 183–186 (1968)
11. Langlois, J.A., Rutland-Brown, W., Wald, M.M.: The epidemiology and impact of traumatic brain injury: a brief overview. J. Head Trauma Rehabil. **21**(5), 375–378 (2006)
12. McCrea, M., Hammeke, T., Olsen, G., Leo, P., Guskiewicz, K.: Unreported concussion in high school football players: implications for prevention. Clin. J. Sport Med. **14**(1), 13–17 (2004). Official Journal of the Canadian Academy of Sport Medicine
13. Pikovsky, A., Rosenblum, M., Kurths, J., Hilborn, R.C.: Synchronization: a universal concept in nonlinear science. Am. J. Phys. **70**(6), 655–655 (2002)
14. Shannon, C.: A mathematical theory of communication. Bell Syst. Tech. J. **27**, 379–423, 623–656 (2009)
15. Skidmore, F., Korenkevych, D., Liu, Y., He, G., Bullmore, E., Pardalos, P.M.: Connectivity brain networks based on wavelet correlation analysis in Parkinson fMRI data. Neurosci. Lett. **499**(1), 47–51 (2011)
16. Teel, E.F., Ray, W.J., Geronimo, A.M., Slobounov, S.M.: Residual alterations of brain electrical activity in clinically asymptomatic concussed individuals: an EEG study. Clin. Neurophysiol. **125**(4), 703–707 (2014)
17. Thatcher, R.: Electroencephalography and mild traumatic brain injury. In: Slobounov, S., Sebastianelli, W. (eds.) Foundations of Sport-Related Brain Injuries, pp. 241–265. Springer, New York (2006)
18. Watts, D.J., Strogatz, S.H.: Collective dynamics of "small-world" networks. Nature **393**(6684), 440–442 (1998)
19. Werner, C., Engelhard, K.: Pathophysiology of traumatic brain injury. Br. J. Anaesth. **99**(1), 4–9 (2007)
20. West, M., Parkinson, D., Havlicek, V.: Spectral analysis of the electroencephalographic response in experimental concussion in the rat. Electromyogr. Clin. Neurophysiol. **53**(2), 192–200 (1982)

Video-Based Pedestrian Detection on Mobile Phones with the Cascade Classifiers

Ksenia G. Shipova and Andrey V. Savchenko

Abstract The paper is devoted to the problem of real-time pedestrian recognition on mobile phones. The insufficient quality of conventional detection methods is highlighted. We propose here a specialized procedure of data gathering and preprocessing to train cascade classifiers. Firstly, automobile video recorder is used to get real pedestrian images. Secondly, the application for training sample preprocessing is designed to prepare positives and negatives by image cutting. Experimental results in testing under real road conditions with several mobile phones reveal that the best quality (3 % of false positives and 19 % of false negatives rate) is achieved with the Haar features. In conclusion we emphasized that sometimes it is necessary to choose faster, but less accurate, object detection algorithm, because in this case it is possible to process more number of frames in a fixed period of time. Hence, the total object detection accuracy can be increased.

Keywords Real-time recognition • Viola–Jones method • AdaBoost cascade classifier • Haar features • Local binary patterns

1 Introduction

Nowadays, the need to use methods of computer vision and image analysis increases dramatically. In particular, the issue of pedestrian detection in the current period is of paramount importance and generates a lot of practical interest because contemporary road conditions cause introduction of such technologies into human life [1]. Despite plenty of methods offered [2, 3], there is no unified approach. The literature review reveals that simple algorithms provide inexcusably high false positives and

K.G. Shipova
National Research University Higher School of Economics, Nizhniy Novgorod, Russia
e-mail: shipovaxenia@gmail.com

A.V. Savchenko (✉)
Laboratory of Algorithms and Technologies for Network Analysis, National Research University Higher School of Economics, Nizhny Novgorod, Russia
e-mail: avsavchenko@hse.ru

V.A. Kalyagin et al. (eds.), *Models, Algorithms and Technologies for Network Analysis*, Springer Proceedings in Mathematics & Statistics 156,
DOI 10.1007/978-3-319-29608-1_15

negatives rate, and complex ones mostly require too much time, energy, and memory resources for implementation [1]. Moreover, though there exist several quite fast algorithms [4], their performance is still insufficient to be implemented in mobile devices. In the paper one of the most popular detection methods suitable for mobile applications, namely, the histogram of oriented gradient (HOG) descriptor [5], is examined to compare algorithms efficiency.

Unfortunately, the quality of the HOG method is sometimes insufficient for practical applications [6]. Thus, this paper is intended to research into the details of the Viola–Jones method [3, 7]: AdaBoost cascade classifier training algorithm [8], Haar features [7], and local binary patterns (LBP) [9] are to be considered as components of the fundamental model, because training procedure needs closer examination to get improved and adopted to pedestrian recognition. The primary practical aim of the investigation is to develop an application for real-time pedestrian detection on mobile phone with the OpenCV library. The proposed approach is expected to reduce problem concerning human representation and environment variability with minimal costs due to the introduction of new solutions in data selection and preprocessing.

The rest of the paper is organized as follows: in Sect. 2, we describe the cascade classifier training procedure. In Sect. 3, we conduct an experimental study of the application created and compare the results obtained with the existing detection method. In Sect. 4, the findings and concluding comments are presented.

2 Materials and Methods

This part of the article is meant to represent the approach used in carrying out the study on pedestrian detection. Pedestrian recognition is a technically challenging task owing to certain problems that are caused by object and environment complexity [1, 5]:

- absence of a standard pedestrian image;
- radical differences in person representation: facial expressions, hairstyles, clothes, various items, posture, height, shape parameters, etc.;
- object configuration diversity, strong dependence on its spatial location, orientation, and scale; and
- variety of weather conditions, landscapes, illumination, etc.

The factors enumerated are admittedly the major reasons for false positives and false negatives. In an attempt to reduce these risks, the HOG method [5] has been developed. At this juncture, the algorithm dominates the pedestrian detection sphere. We can use an idea that the main edges of a pedestrian remain relatively constant, especially around the leg area, and histograms are able to show it. So, with the HOG method we evaluate well-normalized local histograms of image gradient orientations in a dense grid. Despite general recognition of the approach, it has specific drawbacks. For example, traditional HOG method cannot extract the

Fig. 1 Demonstration of the HOG-based pedestrian detection

body local features in comparatively large image region [10]. Figure 1 represents insufficient detection accuracy of the method. That is why in the present research it is decided to investigate the Viola–Jones method and modify it to achieve better detection results than existing algorithms provide.

Traditional Viola–Jones method involves decision trees boosting via primitive functions: Haar features [7] or LBP [9, 11]. Since these functions can be considered as weak classifiers, the rigorous boosting technique provides their merging to create a strong one. As a result, we have a cascade of strong classifiers [12, 13] and can detect objects in video stream by comparing image representations and classifier structure. This idea is implemented as a multistage procedure. We can use its flexibility and modify it to improve detection accuracy.

The first step implies training samples selection to adjust the classifier to only pedestrian detection. The main quality criteria are veracity of data gathered and samples quantity. In order to maximize the benefit at this stage, automobile video recorder is used to get real pedestrian images. Total sample size is 3377 positive images which contain objects in their natural environment and 8515 negative ones without any pedestrian pattern (640x480 resolution to get the real-time detection results). In addition, we used the well-known Daimler Pedestrian Detection Benchmark Dataset [14]. Such a tremendous body of data raises a possibility to detect pedestrians in the most complicated cases: overlap, sudden pose changing, or conditions worsening. Generally, the measures offered are directed at avoiding problems concerning human representation multiplicity, environment variability, and object configuration diversity.

Special samples preparation is supposed to be managed at the second step. The innovative idea is that every positive sample is to contain only one object which matches it in size. It is also proposed that pieces of positive samples which do

Algorithm 1: The proposed algorithm of pedestrian detection

Input: Video frames $\{F(t)\}, t \in \{1, \ldots, T\}$, training set $\{S_i\}$
Output: Video frames $\{F(t)\}$ with detected positions of pedestrians
1: **for** each each training image S_i **do** {training phase}
2:     **while** an object "pedestrian" exists **do**
3:         Detect the borders of the pedestrian
4:         Write file path to the positive sample, object number, coordinates and size into *positive.dat* file
5:     **end while**
6:     **if** the object "pedestrian" does not exist in the area **then**
7:         Cut the area as a negative training sample
8:         Write file path to the negative sample into *negative.dat* file
9:     **end if**
10: **end for**
11: Create a vec-file from positive and negative sets using *opencv_createsamples* utility
12: Train cascade via the *opencv_traincascade* utility
13: Assign $counter = 0$
14: **for** each each video frame $F(t)$ **do** {detection phase}
15:     **if** $F(t)$ contains an object "pedestrian" **then**
16:         Assign $counter := counter + 1$
17:         **if** $counter > C$ **then**
18:             draw a rectangle with coordinates of detected object in frames $F(t - C + 1), \ldots, F(t)$
19:         **end if**
20:     **else**
21:         Assign $counter = 0$
22:     **end if**
23: **end for**
24: **return** the modified set of video frames $\{F(t)\}$

not contain objects can be used as negative ones. A key for appropriate ideas implementation lies in profound software support, because an additional application is to be written to cut positive and negative images. The program allows filling in the description file and looking over the whole dataset at a speed of one image per second at the same time. Finally, samples and the description are processed with *opencv_createsamples* utility to be converted into the vec-file.

At the third step the set of samples is used as a base for the classifier training procedure. The main goal here is the generation of xml-cascades that describe all possible variants of object form and location. To train the classifier, *opencv_traincascade* utility is used and training parameters are chosen experimentally. The proposed method with pedestrian detection at each frame and combining the results with a committee [15] is presented in Algorithm 1. Here parameter C is the number of frames to reliably detect the pedestrian.

This algorithm was implemented in an Android mobile application (Fig. 2).

Examples of usage of the proposed approach with the Haar-based features are shown in Fig. 3. As one can see, here the quality of pedestrian detection is much higher when compared with conventional HOGs (Fig. 1). The next section experimentally supports this claim.

Fig. 2 Graphical user interface of the developed mobile application

3 Experimental Results and Discussion

The system testing is carried out in six stages. To evaluate if cascade classifiers are appropriate for the problem of real-time pedestrian detection, it is decided to compare their performance with the results that the OpenCV-implemented HOG method demonstrates. Tests are conducted to reveal application potential and evaluate performance of Haar-trained classifier, LBP-trained classifier, and HOG method quantitatively. We captured real HD-video with duration 600 s, which contains 4152 pedestrians. The distance from the camera to each pedestrian is between 6 and 10 m. Three mobile phones were used in this experiment. Their descriptions are presented in Table 1.

Each frame of the video stream is scaled to the same resolution 640x480 to obtain comparable results. To estimate the detection quality, we measure the number of true positives (TP), false positives (FP), true negatives (TN), and false negatives (FN).

The results of this experiment are demonstrated in Table 2. Here we compute true positive rate (TPR) and false positive rate (FPR) by using the following equations

Fig. 3 Demonstration of the Haar features-based pedestrian detection

Table 1 Descriptions of the mobile phones used in the experiment

	LG G2 GOLD D802	Philips Xenium W8510	LG Optimus L7 P705
CPU	4-core, 2.26 GHz	4-core, 1.2 GHz	1-core, 1 GHz
Display	5.2", 1980x1080, FULL HD IPS	4.7", 1280x720, TFT IPS	4.3", 800x480, IPS
RAM	2 GB	1 GB	512 Mb
Operating system	Android 5.0.2	Android 4.2.2	Android 4.0.3

Table 2 Experimental results

	LG G2			Philips W8510			Philips W8510		
	Haar	LBP	HOG	Haar	LBP	HOG	Haar	LBP	HOG
TPR (%)	**97.2**	84.1	86.4	**94.4**	82.3	91.6	**85.4**	83.8	85.0
FPR (%)	**19.3**	42.9	20.3	**24.3**	69.3	25.4	45.6	56.8	**41.25**
Frame processing time (ms)	98	**71**	88	132	**107**	145	308	237	**166**

$$TPR = \frac{TP}{TP + FN} \cdot 100\%$$

$$FPR = \frac{FP}{TN + FP} \cdot 100\%$$

In this table we bold the highest values of TPR and the lowest values of FPR and processing time for each mobile phone. Here, firstly, the best quality (2.8 % of false positives and 19.3 % of false negatives rate) is obtained by the classifier trained via Haar features. The HOG method provides similar computational complexity (except the obsolete LG P705 L7), but its TPR is rather worse. The LBP-trained classifier does not provide normal detection quality: the average number of skipped objects per second is inexcusable. Thus, the Haar features seem to be effective for pedestrian detection in actual practice.

Finally, one can notice that the detection quality (TPR and FPR) is not the same for different phones, though the input video was identical. It can be explained by different characteristics of our smartphones. In fact, the more powerful the device is, the more is the processed number of frames in a second. Hence, the quality of pedestrian detection with the most powerful LG G2 is the best in all cases. It is a very remarkable fact. It is widely known, that the higher accuracy is achieved in object detection in a single image with the more complex algorithms. Hence, it is necessary to choose the correct balance between accuracy and average processing time. However, in video-based processing the situation is quite different. Really, if the frame processing time is higher than the frame rate (20 frames per second), the faster algorithm can process more number of frames and increase its quality by combining the results from serial frames. For instance, in the obsolete LG P705 L7 phone, the quality of the HOG and the Haar features is the same as the performance of the HOG here is twice higher, when compared with the performance of the Haar features. However, if the quality of single-image detector is too poor (as in case of the LBP features), even very fast processing does not lead to a superior detection accuracy.

4 Conclusion and Future Work

In this paper we proposed the pedestrian detection method based on the cascade classifier training, which provides 3–10 % higher detection accuracy, than the HOG method. We paid attention to the preparation of the training dataset. Namely, we proposed to use parts of pedestrians as negative images to improve the detection quality. The presented approach allows reducing this time due to the use of an additional application for semi-automatic positive and negative images cutting. We implemented several detection algorithms in an Android application. It was experimentally shown that the best quality is achieved with the Haar features and by using the modern powerful devices. We emphasized the necessity to minimize the detection speed to process more frames in a second.

We have not used the object tracking as we assumed that a new pedestrian can suddenly appear, and the response time for such kind of event should be minimized. However, in future it is important to explore the possibility to combine our approach with the known tracking algorithms, e.g., the Lucas–Kanade method [16]. Another direction for future research is the application of more complex dissimilarity measures with the HOG features, known to improve the recognition accuracy (e.g., see [6] for details). Finally, it is important to explore other image sources, e.g., infrared images [17] and depth-map cameras [18].

Acknowledgements Andrey Savchenko is supported by RSF (Russian Science Foundation) grant 14-41-00039.

References

1. Benenson, R., Omran, M., Hosang, J., Schiele, B.: Ten years of pedestrian detection, what have we learned? In: ECCV 2014 Workshop on Computer Vision. Lecture Notes in Computer Science, vol. 8926, pp. 613–627. Springer, Heidelberg (2015)
2. Gavrila, D.M., Munder, S.: Multi-cue pedestrian detection and tracking from a moving vehicle. Int. J. Comput. Vis. **73**(1), 41–59 (2007)
3. Viola, P., Jones, M., Snow, D.: Detecting pedestrians using patterns of motion and appearance. IEEE Conf. Comput. Vis. **63**(2), 153–161 (2005)
4. Benenson, R., Mathias, M., Timofte, R., Van Gool, L.: Pedestrian detection at 100 frames per second. In: IEEE Conference on Computer Vision and Pattern Recognition (CVPR), pp. 2903–2910 (2012)
5. Dalal, N., Triggs, B.: Histograms of oriented gradients for human detection. In: International Conference on Computer Vision and Pattern Recognition, pp. 886–893 (2005)
6. Savchenko, A.V.: Probabilistic neural network with homogeneity testing in recognition of discrete patterns set. Neural Netw. **46**, 227–241 (2013)
7. Viola, P., Jones, M.: Rapid object detection using a boosted cascade of simple features. In: IEEE Conference on Computer Vision and Pattern Recognition, pp. I-511–I-518 (2001)
8. Freund, Y., Schapire, R.E.: A decision-theoretic generalization of on-line learning and an application to boosting. J. Comput. Syst. Sci. **55**(1), 119–139 (1997)
9. Ahonen, T., Hadid, A., Pietikainen, M.: Face recognition with local binary patterns. In: European Conference on Computer Vision, pp. 469–481 (2004)
10. Felzenszwalb, P., Girshick, R., McAllester, D., Ramanan, D.: Object detection with discriminatively trained part-based models. IEEE Trans. Pattern Anal. Mach. Intell. **32**(9), 1627–1645 (2009)
11. Trefny, J., Matas, J.: Extended set of local binary patterns for rapid object detection. In: Computer Vision Winter Workshop, pp. 1–7 (2010)
12. Felzenszwalb, P.F., Girshick, R.B.: Cascade object detection with deformable part models. In: IEEE Conference on Computer Vision and Pattern Recognition (CVPR), pp. 2241–2248 (2010)
13. Wu, J., Brubaker, S.Ch., Mullin, M.D., Rehg, J.M.: Fast asymmetric learning for cascade face detection. IEEE Pattern Anal. Mach. Intell. **30**(3), 369–382 (2008)
14. www.gavrila.net/Research/Pedestrian_Detection/Daimler_Pedestrian_Benchmark_D/Daimler_Mono_Ped__Detection_Be/daimler_mono_ped__detection_be.html. Accessed 10 July 2015
15. Savchenko, A.V.: Adaptive video image recognition system using a committee machine. Opt. Mem. Neural Netw. (Inf. Opt.) **21**(4), 219–226 (2012)
16. Lucas, B.D., Kanade, T.: An iterative image registration technique with an application to stereo vision. In: International Joint Conference on Artificial Intelligence, pp. 674–679 (1981)
17. Zhang, L., Wu, B., Nevatia, R.: Pedestrian detection in infrared images based on local shape features. In: IEEE Conference on Computer Vision and Pattern Recognition, pp. 1–8 (2007)
18. Savchenko, A.V., Khokhlova, Ya.I.: About neural-network algorithms application in viseme classification problem with face video in audiovisual speech recognition systems. Opt. Mem. Neural Netw. (Inf. Opt.) **23**(1), 34–42 (2014)

Clustering in Financial Markets

Kristina Sörensen and Panos M. Pardalos

Abstract This chapter considers graph partition of a particular kind of complex networks referred to as power law graphs. In particular, we focus our analysis on the market graph, constructed from time series of price return on the American stock market. Two different methods originating from clustering analysis in social networks and image segmentation are applied to obtain graph partitions and the results are evaluated in terms of the structure and quality of the partition. Our results show that the market graph possesses a clear clustered structure only for higher correlation thresholds. By studying the internal structure of the graph clusters we found that they could serve as an alternative to traditional sector classification of the market. Finally, partitions for different time series were considered to study the dynamics and stability in the partition structure. Even though the results from this part were not conclusive we think this could be an interesting topic for future research.

Keywords Complex network • Financial markets • Price returns • Market network • Market graph • Clustering • Data mining

1 Introduction

Financial analysis of today often involves interpretation of very large data sets. One convenient way to represent this large amount of data is in terms of a network. Network theory has been used to analyze many different concepts, examples span from Internet and social networks to biological networks, and recently financial networks.

K. Sörensen (✉)
KTH Royal Institute of Technology, Stockholm, Sweden
e-mail: kristina.sorensen@live.se

P.M. Pardalos
Department of Industrial and Systems Engineering, University of Florida, 303 Weil Hall, Gainesville, FL 32608, USA

National Research University Higher School of Economics, Nizhny Novgorod, Russia
e-mail: pardalos@ise.ufl.edu

V.A. Kalyagin et al. (eds.), *Models, Algorithms and Technologies for Network Analysis*, Springer Proceedings in Mathematics & Statistics 156,
DOI 10.1007/978-3-319-29608-1_16

Despite arising from different fields many of these networks share topological characteristics which cannot be described neither by uniform random graphs nor by regular lattices. Thus to describe the complex topology of these graphs a new field emerged, complex network theory. One feature observed in many of these networks is the occurrence of a heavy tail in the degree distribution. A network showing this characteristic is called a scale free network or a power law graph. Another common feature in these networks is their tendency to form clustered communities in the graph. This introduces new problems to find specific clusters or partitions of the networks into different clusters.

Several models for representing financial networks have been proposed. Results from previous research revealed overall structure of the market as well as introduced a tool for studying market dynamics [2]. Other considered topics involve the grouping of instruments, stock classification, and finding highly influential actors in the market [1]. Many previous studies have also focused on identifying specific substructures in the graph [22]. One such example is the maximum clique problem, i.e. to identify a complete sub-graph of maximal cardinality in the graph. However, as many other network optimization problems this is NP-hard which often makes it impossible to find an exact solution in a reasonable amount of time.

In this chapter a graph partition of the network is studied. The partitions will be obtained by using two different, well-known objective functions for graph partition. The resulting optimization problems will be presented together with heuristic approaches to solve two partition formulations. Finally, the results for the market graph will be analyzed further to interpret the structure of the market.

2 Graph Theory Concepts

Since networks are represented in terms of graphs some notations from basic graph theory are introduced. Concepts related to graph partition and cluster analysis are also included.

2.1 *Definitions and Notations*

Since networks are represented in terms of graphs some notations from basic graph theory are introduced. Let $G = (V, E)$ be an undirected graph consisting of the set V with $|V| = n$ vertices and the set E with $|E| = m$ edges. We say that A_G is the adjacency matrix representing $G(V, E)$, if A_G is an $n \times n$-matrix such that $A_G = [a_{ij}]_{i,j}^{n}$, with $a_{ij} = 1$ if $(i, j) \in E$ and $i \neq j$ and otherwise $a_{ij} = 0$. The *degree* d_i of a vertex i is the number of edges emanating from it. For every $d_i = d$, we can define $n(d)$ as the number of nodes in G with degree d. This gives rise to a degree distribution of a graph G as the fraction of vertices having degree d. The *(open) neighborhood* $T(i)$ of a vertex $i \in G$ is the set of all vertices sharing an edge with i, i.e. $T(i) = \{j|a_{ij} = 1\}$. A *path* in G is a sequence of edges connecting vertices. The *average path length* is the average number of steps along the shortest path for all possible pairs of the network nodes. The *diameter* of the graph is the longest of all the shortest paths in the graph. The graph G is *connected* if there is a path from any vertex $v \in V$ to any vertex $u \in V$. We call G a *complete graph* if there exists an edge $(i,j) \in E$ for every $i \neq j$ and $i,j \in V$. Given a subset $S \subseteq V$, we denote by $G(S)$ the *sub-graph* induced by the set S.

The complementary graph of G, denoted $\bar{G} = (V, \bar{E})$, is defined as follows. If $(i,j) \in E$, then $(i,j) \notin \bar{E}$ and if $(i,j) \notin E$, then $(i,j) \notin \bar{E}$. In words, one obtains the complementary graph of G by removing all the edges (i,j) present in G, and then introducing all the edges not present in G in the graph. The *edge density*, $\delta(G)$, measures the connectivity in the graph, defined by the ratio between the number of edges in the graph and the maximal possible number of edges in the graph. Mathematically we write

$$\delta(G) = \frac{2|E|}{|V|(|V|-1)}. \tag{1}$$

The *cluster coefficient* reveals to what extent the nodes in the graph tend to cluster together. The local clustering coefficient C_i for a vertex i with degree $d_i > 1$ is defined as the ratio of the number of edges among its neighbors divided by the maximal (possible) number of such edges. For $d_i \leq 1$ C_i is undefined. Mathematically we write C_i as

$$C_i = \frac{2E_i}{d_i(d_i-1)}, \quad d_i > 1 \tag{2}$$

where d_i is the degree of node i and E_i is the number of common edges among its neighbors. The global clustering coefficient C of the entire graph is defined as the mean of the local clustering coefficients, i.e. $C = \frac{1}{n}\sum_i^n C_i$.

Generally speaking, a *cluster* in a network is a set of elements that are more similar to each other than to elements not included in the cluster. Studying graph clusters can reveal topological structure of the network as well as information about the particular elements in the clusters. The similarity criterion varies depending on what property the cluster should reveal. Common criteria include vertex degree, vertex distance, or cluster density.

One special case of cluster is called a *clique*. We say that $C \subseteq V$ is a clique if the induced sub-graph $G(C)$ is complete. A clique is *maximal* if it cannot be contained in any larger clique in the graph, and it is called a *maximum clique* if it is a clique of maximal cardinality in the graph. A problem in graph theory is to identify maximum cliques in a graph, called the *maximum clique* (MC.) problem. The size of a maximum clique is called the *clique number*, denoted $\omega(G)$.

Since the strict requirements of cohesiveness in the clique definition often are difficult to fulfill, several relaxations of cliques have been introduced. Examples of clusters being cliques relaxations include *k-clubs*, *k-cores*, *k-communities*, and

γ-quasi clique, all further discussed in [18]. We say that the set $Q \subseteq V$ with $|Q| = p$ is a *γ-quasi clique* $(0 < \gamma < 1)$ if the graph $G(Q)$ induced by Q is connected and satisfies $|E(G(Q))| \geq \gamma\binom{p}{2}$. This means that we impose the requirement that the edge density of the induced graph $G(Q)$ must be greater or equal to the threshold γ. Note that in the case when $\gamma = 1$, then Q corresponds to a clique.

The opposite of a clique is an *independent set*. An independent set is a set $I \subseteq V$ such that the induced graph $G(I)$ has no edges. The problem of finding an independent set of maximal cardinality in a graph is called the *maximum independent set* (MIS.) problem. By $\alpha(G)$ we denote the size of the largest independent set of G. Note the symmetry between the maximum clique problem and the maximum independent set problem. The set Q is a maximum clique in $\bar{G}$ if and only if Q is a maximum independent set in G. Therefore an MIS. can easily be reformulated into an MC. and vice versa, and hence it holds that $\omega(G) = \alpha(\bar{G})$.

2.2 Clustering and Graph Partitions

Clustering involves the task of partitioning the elements of the graph into disjoint clusters. Generally one seeks a partition of the vertices in a way that maximizes the similarity within the clusters and minimizes the similarity between the clusters. A partition where each cluster is a clique is called a *clique partition*. The *minimal clique partition problem* is to find the smallest integer k such that the vertex set V of G can be partitioned into the k disjoint sets $C_1, \ldots, C_k$, where each C_i is a clique. This minimal integer k is called the clique partitioning number $\bar{\chi}(G)$.

A concept closely related to graph partitioning is graph coloring. A proper *k-coloring* of the vertices of G is an assignment of colors to the vertices in G such that no adjacent vertices in G have the same color. If such a coloring exists, we call the graph G *k-colorable*. Seeking a coloring using a minimal number of colors is called the *graph coloring problem*. The smallest integer k for which the graph G is k-colorable is the *chromatic number* of G denoted $\chi(G)$. In a coloring of G the vertices with the same color are all pairwise non-adjacent, making them by definition independent sets. Thus, the graph coloring problem is equivalent to finding a minimal partition of G into pairwise, disjoint independent sets. Due to the symmetry between cliques and independent sets the *graph coloring problem* of $\bar{G}$ can therefore also be formulated as the *minimum clique partition problem* of G. Again, due to the symmetry we have that $\bar{\chi}(G) = \chi(\bar{G})$.

2.2.1 Desirable Cluster Properties

What constitutes a cluster of high quality will of course depend on the application at hand. However, some characteristics are relevant for most structures. First, the cluster must be connected, thus if there is no path between two vertices u and v, they should not be grouped within the same cluster. By classifying edges as internal if they connect vertices within a cluster to each other, the *internal degree* of a vertex v in a cluster $C \subset V$ is $deg_{int}(v, C) = |T(v) \cap C|$, where $T(v)$ is the neighborhood of v in G. Similarly, edges are identified as external if they connect a vertex in a cluster with a vertex outside the cluster. Thus the *external degree* of a vertex v in a cluster C is $deg_{ext}(v, C) = |T(v) \cap (V \setminus C)|$. Note that with these definitions we have $d_v = deg_{int}(v, C) + deg_{ext}(v, C)$.

In general, if $deg_{int}(v, C) = 0$, then v should not be included in cluster C as it is not connected to the other vertices in C. Similarly $deg_{ext}(v, C) = 0$ implies that C could be a good cluster for v as it has no connections outside C. Generally in clustering one seeks to form clusters such that the induced sub-graph is dense and has few connections to the rest of the graph. We therefore introduce two density measures with respect to a cluster C. We call the density of the sub-graph induced by C *internal* or *intra-cluster density* if it is defined by

$$\delta_{int}(C) = \frac{|\{(u, v) \in E | v \in C, u \in C\}|}{|C|(|C|-1)} = \frac{1}{|C|(|C|-1)} \sum_{v \in C} deg_{int}(v, C). \tag{3}$$

Given a clustering of a graph G into k clusters $\bar{C} = (C_1, C_2 \ldots, C_k)$ we define the *intra-cluster density* of the clustering $\bar{C}$ as the average of the intra-cluster densities of the included clusters.

$$\delta_{int}(G|C_1, C_2 \ldots, C_k) = \frac{1}{k} \sum_{i=1}^{k} \delta_{int}(C_i). \tag{4}$$

Similarly, we introduce the *external* or *inter-cluster density* of a clustering as the ratio of the number of external edges and the maximal possible number of external edges.

$$\delta_{ext}(G|\{C_1, C_2 \ldots, C_k) = \frac{|(u, v)| v \in C_i, u \in C_j, i \neq j\}|}{n(n-1) - \sum_{l=1}^{k} |C_l|(|C_l|-1)} \tag{5}$$

Employing the introduced density measures above a good clustering should have an internal density significantly higher than that of the overall graph, $\delta(G)$, and an external density much lower than $\delta(G)$. Depending on how strict these density constraints are imposed different cluster types can be obtained with the loosest possible definition being a connected component and the strictest being a maximal clique. However, in practice most interesting structures can be found somewhere in between. Computation of connected components can be done in $O(n+m)$ time with a breadth-first search while identifying maximal cliques is NP-complete [20].

2.2.2 Clustering Structure

An important characteristic in a clustering structure is whether the clusters $C_1, C_2 \ldots, C_k$ must be disjoint or if cluster overlap is allowed. In the former case we talk about a *graph partition*, or a "hard" clustering, $C_i \cap C_j = \emptyset, \forall i \neq j$. When clusters overlap, we call this a *graph cover* or a "soft" clustering. In this paper we will focus on the former structure and we will use the term clustering and partition exchangeable, always referring to the hard clustering.

Another distinction for a clustering structure is the one between *flat* versus *hierarchical* clustering. If the partition consists of a set of clusters without any explicit structure that would relate clusters to each other, we talk about a flat clustering. On the other hand, we say that a clustering is hierarchical if it contains several levels of clusters where each top level cluster consists of clusters from lower levels. Which type of clustering that is preferred depends on the network topology. If it is known that the data contains a hierarchical structure, then this should be preferred. However, if the number of clusters is known prior, then a flat clustering approach is preferred over a hierarchical structure [20].

Hierarchical clustering can be separated further into two types, depending on whether the partition is refined or coarsened between each level. In the first type, called top-down or *divisive* hierarchical clustering the graph is recursively spilt into smaller and smaller pieces. In the second version, bottom-up or *agglomerative* clustering, smaller clusters are iteratively merged into larger ones.

2.2.3 Measures to Identify Clusters

Clusters are usually identified with two different approaches, using vertex similarities or a fitness measure. In the former approach one computes a set of similarity values for all vertices and then classifies them into clusters according to their overall score. In the latter case one computes a fitness function over the set of possible clusters and then chooses among the set of clusters that optimize the chosen fitness measure. An extensive overview of clustering techniques can be found in [9, 20].

Density Based Measures

Some approaches use a density based fitness measure to identify maximal subgraphs with a density higher than a certain threshold. As Schaeffer [20] mentions, finding clusters based on their edge density can essentially be considered as special cases of the following decision problem:

> ***Instance:*** *Given an undirected graph $G = (V, E)$, with a density measure $\delta(\cdot)$ over the vertex subsets $S \subseteq V$, a positive integer $k \leq |V|$, and a rational number $\xi \in [0, 1]$.*
> ***Question:*** *Does it exist a subset $S \subseteq V$ such that $|S| = k$ and the density $\delta(S) \geq \xi$?*

Note that if the density measure used is the overall graph density, the problem is NP-complete since for $\xi = 1$ it coincides with the NP-complete maximum clique problem. Many variants and relaxations of this problem have been proposed and studied during the years. Matsuda et al. proposed a model that considers γ-quasi cliques as clusters [14]. They showed that it is NP-complete to determine whether a given graph has a $\frac{1}{2}$ quasi clique of order at least k.

Cut Based Measures

Instead of focusing on the internal density of the cluster one can also measure how connected the cluster is to the rest of the graph. These measures are usually based on cut sizes. Given a graph $G = (V, E)$ and two subsets $S_1 \subseteq V$ and $S_2 \subseteq V$ we define the *cut size*, $c(S_1, S_2)$, of S as the number of edges between nodes in S_1 and nodes in S_2. Mathematically, we write this as

$$c(S_1, S_2) = |\{(u, v) \in E | u \in S_1, v \in S_2\}|. \tag{6}$$

The definition can be extended to a collection of clusters $\Pi = (V_1, \ldots, V_K)$ as the sum of all edges with end nodes in different clusters. We define the cut of a collection of clusters $\Pi = (V_1, \ldots, V_K)$ as

$$C(\Pi) := \frac{1}{2} \sum_{i=1}^{K} c(V_i, \bar{V}_i) \tag{7}$$

where $\bar{V}_i$ is the complement of V_i in V, as $\bar{V}_i = V \setminus V_i$.

A normalization of this metric was introduced by Shi and Malik [21], called the *normalized cut*, $C_N(\Pi)$. They defined it as the ratio between the cut size and the degrees of the vertices.

$$C_N(\Pi) := \frac{1}{2} \sum_{i=1}^{K} \frac{c(V_i, \bar{V}_i)}{vol(V_i)} \tag{8}$$

where $vol(V_i) = \sum_{j \in V_i} d_j$, i.e. the sum over the degrees of the vertices in V_i.

Modularity

Another common measure to identify graph clusters is the metric *modularity*, introduced by Newman and Girvan in [17]. The metric modularity, denoted Q, is defined as

$Q(\Pi) =$ (*the number of the edges that fall within a cluster*) $-$ (*the expected such number if edges were distributed at random*)

The meaning of the first term is clear. However, the second term requires some comments. Expressing this term necessitate choosing a null model for the network, a question we will address soon. First, we introduce P_{ij} as the probability that there is an edge between vertex i and j. Thus, the actual minus the expected number of edges between i and j can be written $A_{ij} - P_{ij}$ and the modularity is proportional to the sum of this quantity over all pairs of vertices in the same cluster. Thus, the modularity can be expressed as

$$Q = \frac{1}{2m} \sum_{ij} [A_{ij} - P_{ij}] \delta(C_i, C_j) \tag{9}$$

where $\delta(C_i, C_j) = 1$ if $C_i = C_j$ and zero otherwise.

Returning to the question of choosing a null model. A possible choice could be to consider a standard uniform random graph, in which edges appear random with equal probability $P_{ij} = p$. However, this model turns out to be a bad representation for many real life graphs. In particular the model often fails to reflect the degree distribution of the graph. One way to deal with this in practice is to approximate the expected degree of each vertex within the model with the actual degree, d_i, of the corresponding vertex i in the real network. The expected degree of i is given by $\sum_j P_{ij}$, giving us the relation

$$\sum_j P_{ij} = d_i \tag{10}$$

A null model in this class, is the one in which edges are distributed at random subject to the constraint. This implies that the expected number of edges between i and j, P_{ij}, can be expressed as a product of separate functions of the degrees.

$$\sum_j P_{ij} = f(d_i) \sum_j f(d_j) = d_i \tag{11}$$

Hence, $f(d_i) = Cd_i$, for some constant C. Furthermore, since $\sum_i d_i = 2m$ (m being the number of edges in the graph) we can write

$$2m = \sum_i \sum_j P_{ij} = C^2 \sum_i \sum_j d_i d_j = (2mC)^2 \tag{12}$$

which gives $C = \frac{1}{\sqrt{2m}}$, and hence $P_{i,j} = \frac{d_i d_j}{2m}$.

Thus, the modularity can be rewritten as

$$Q = \frac{1}{2m} \sum_{ij} \left[A_{ij} - \frac{d_i d_j}{2m} \right] \delta(C_i, C_j) \tag{13}$$

2.3 *Random Graph Models*

2.3.1 Uniform Random Graph Model

The theory of random graphs was introduced in 1959 in the work of Erdös and Rényi [7]. In the context of their probabilistic method a random graph can be described in the following way. Consider the situation where we try to study the existence of graphs G_P with a specific property P. Let the existence of such a graph be represented by the random variable X. Then, one can construct a probability space such that the appearance of G_P with property P can be described by the event E. Showing that the probability of observing this event E is larger than zero, i.e. showing that $P(X = E) > 0$ implies that such a graph G_P with property P in fact can exist. By studying the distributions of probability spaces of these kind random graphs are introduced.

In their first paper Erdös and Rényi introduced two formulations for the uniform random graph model. The first version, $G(n, m)$ assigns a uniform probability to all graphs with n nodes and m edges. By setting $N = \binom{n}{2}$ we can see that $G(n, m)$ has $\binom{N}{m}$ elements, all with probability $\binom{N}{m}^{-1}$. In the second formulation denoted $G(n, p)$, a graph is constructed by introducing edges between nodes with an independent probability p, where $0 < p < 1$. One can easily identify similarities between the two formulations since all graphs with n nodes and m edges will have the same probability $p^m(1-p)^{\binom{n}{2}-m}$ in the $G(n, p)$ model. From now on we will continue working with the second formulation of the model.

With the notation above a graph in $G(n, p)$ has $\binom{n}{2} \cdot p$ expected number of edges. Therefore the degree distribution of a particular vertex v is given by the *Binomial distribution*, and we have

$$P(d_v = k) = \binom{n-1}{k} p^k (1-p)^{n-1-k}. \tag{14}$$

Letting $n \to \infty$ we get that for the case $np = constant$ the degree distribution tends to the *Poisson distribution* OBS1.

$$P(d_v = k) = \frac{(np)^k e^{-np}}{k!} \tag{15}$$

Many properties of the $G(n, p)$ model have been studied, some fundamental results cover graph connectivity, emergence of a giant connected component, as well as results about graph diameter, independent sets, cliques, and colorings. The interested reader is referred to [4] for a more comprehensive review of the different properties of random graphs.

Two characteristics worth mentioning in this context are the degree distribution and cluster coefficient of a random graph $G(n, p)$. First, as stated above the degree distribution for $G(n, p)$ tends to the Poisson distribution as n grows large. This is the first drawback when using this model to represent real-life graphs. Many real life graphs have instead shown to exhibit a degree distribution with a heavy right end tail [6, 8]. This kind of degree distribution is often referred to as a *power law distribution*. Secondly, the clustering coefficient of a random graph $G(n,p)$ is given by $C_R = \frac{<k>}{n} = p$. This is a second indication that $G(n,p)$ is not suitable for modelling real life networks since it has been shown that in many real life graphs the clustering coefficient highly exceeds this number [19].

2.3.2 Power Law Random Graph Model

Following the discoveries that the topology of many real life networks could not be accurately modelled by the classical uniform random graph theory new models for describing these scale free networks have been presented. A common feature of these models is the occurrence of a power law in the right end tail of their degree distribution. This section will therefore introduce the power law distribution and its specific properties. We then move on and discuss some proposed graph models for generating networks with a power law degree distribution.

One says that the random variable $X > 0$ follows a *power law* if it has the probability density function.

$$f(x)_X = \frac{\alpha}{x^{\beta}}, \quad x \in S \tag{16}$$

where S is the support of x, α is a normalization constant, and β is the power law exponent.

A characteristic of power law distributions is their scale invariance property. That a function $f(x)$ is scale invariant means that scaling x with a constant c is equivalent to scaling the function itself with a constant, that is

$$f(x) = \alpha x^{\beta} \Rightarrow f(cx) = \alpha(cx)^{\beta} = c^{\beta} f(x) \tag{17}$$

The concept of a *power law graph* arises when the degree distribution of the vertices in a graph G follows (or closely approximates) some power law, i.e. when the number of vertices y with degree x in the graph can be described by the relation $y = \frac{e^{\alpha}}{x^{\beta}}$.

2.4 *The Market Graph Model*

By employing the market graph model introduced by Boginski et al. [1] we construct the market graph by representing traded instruments by vertices and introducing edges if the Pearson cross-correlation between two instruments exceeds a certain threshold, θ. This can be expressed in terms of the graph adjacency matrix $A = [a_{i,j}]_{i,j=1}^{n}$ as

$$a_{ij} = \begin{cases} 1, & \text{if } C_{i,j} \geq \theta \\ 0, & \text{if } C_{i,j} < \theta \end{cases} \tag{18}$$

where $\theta \in [-1, 1]$. The cross-correlation between i and j is given by

$$C_{i,j} = \frac{E(R_i R_j) - E(R_i)E(R_j)}{\sqrt{Var(R_i)Var(R_j)}} \tag{19}$$

where $R_i(t)$ is the daily return of instrument i at time t.

$$R_i(t) = \frac{P_i(t)}{P_i(t-1)} \tag{20}$$

and $P_i(t)$ is the closing price of instrument i at time t.

This results in an undirected, unweighted graph, represented with an adjacency matrix $A(\theta) = [a_{i,j}]_1^n$, where $a_{i,j}$ is 1 if there is an edge between i and j and 0 otherwise.

Graph characteristics such as edge distribution, cluster coefficient, maximum cliques, and independent sets can be examined to study the structure of the market. Many previous studies have shown that above a certain threshold the degree distribution of the market graph will follow a power law [2, 3, 10, 22].

In this paper we focus on the problem of partitioning the market graph into disjoint clusters. In terms of the market graph this can be interpreted as a division into different, strongly connected segments of the market.

3 Graph Partitioning Methods

In this section two formulations for graph partition are presented based on two different fitness measures, the normalized cut (8) and modularity (13). Both formulations result in integer programs which turn out to be NP-hard problems.

3.1 Minimizing Normalized Cut

The first formulation seeks a partition of V into (a fixed number of) k disjoint subsets such that the normalized cut (8) of the partition is minimized. This approach was introduced in [21] for image segmentation and is solved using a spectral relaxation of the problem. That approach was further studied in [13]. The objective function in this case is given by

$$\underset{(A_1,\ldots,A_k)}{\text{minimize}} \quad C_N(A_1, \ldots, A_k) \tag{21}$$

Following the methodology presented in [13] this can be written as a min trace problem by introducing the indicator variables

$$h_{i,j} = \begin{cases} \frac{1}{\sqrt{vol(A_j)}}, & \text{if } v_i \in A_j \\ 0 & \text{otherwise.} \end{cases} \quad (22)$$

Next we set the matrix H to be the matrix with the k indicator vectors as its columns, i.e. $H = \{h_j\}_{j=1}^k$. Now, since $HH' = I$, $h_i'Dh_i = 1$, and that $h_i'Lh_i = \frac{cut(A_i,\bar{A}_i)}{vol(A_i)}$, the *k-way* $C_N(\Pi)$ minimization problem (21) can be reformulated as

$$\begin{aligned} \underset{A_1,\ldots,A_k}{\text{minimize}} \quad & Tr(H'LH) \\ \text{subject to} \quad & H \quad \text{as in} \quad (22) \\ & H'DH = I. \end{aligned} \quad (23)$$

Now, relaxing the discreteness condition on h_j, and introducing T by $T = D^{1/2}H$, we can write the relaxed problem in the following way

$$\begin{aligned} \underset{T \in \mathbf{R}^{n \times k}}{\text{minimize}} \quad & Tr(T'D^{-1/2}LD^{-1/2}T) \\ \text{subject to} \quad & TT' = I. \end{aligned} \quad (24)$$

System (24) is a standard trace minimization problem and its solution is obtained by choosing the matrix T to contain the k first eigenvectors of L_{sym} as columns. Substituting back $H = D^{-1/2}T$, we see that H will consists of the first k eigenvectors of the matrix L_{rw}, or equivalent to the first k generalized eigenvectors of $Lu = \lambda Du$. This results in the normalized spectral algorithm from [21] for arbitrary k.

3.2 *Maximizing Modularity*

Several graph partition formulations with modularity maximization have been proposed. Here we only present the integer formulation introduced in [5]. Other commonly used formulations include the spectral relaxation presented by Newman [15], this has great similarities with the relaxed spectral formulation presented for the normalized cut.

The formulation used here, first presented in [5], results in a linear integer program. The objective is to find a partition Π of V that maximizes the modularity. Note that in this formulation the number of clusters k in the partition is not fixed.

First we introduce the variable f_{ij} for each pair (i, j) of vertices in the graph, where

$$f_{ij} = \begin{cases} 1, & \text{if } i \text{ and } j \text{ belong to the same cluster} \\ 0, & \text{otherwise.} \end{cases} \quad (25)$$

These variables can be interpreted as an equivalence relation over V and thus form a partition by its equivalence classes. To ensure consistency we must impose the following constraints on the relation.

$$\begin{aligned} &\text{reflexivity} && \forall i : f_{ii} = 1 \\ &\text{symmetry} && \forall i, j : f_{ij} = f_{ji} \\ &\text{transitivity} && \forall i, j, l : f_{ij} + f_{jl} - 2f_{il} \leq 1. \end{aligned} \tag{26}$$

Using the introduced decision variables f_{ij} the objective function can be expressed as

$$Q = \frac{1}{2m} \sum_{ij} \left[A_{ij} - \frac{d_i d_j}{2m} \right] f_{ij} \tag{27}$$

where as before m is the total number of edges in the graph and d_i denotes the degree of vertex i.

The modularity maximization problem is then given by

$$\begin{aligned} &\underset{f_{ij}}{\text{maximize}} && Q \\ &\text{subject to} && f_{ij} \text{ as in } (26), \quad f_{ij} \in [0, 1]. \end{aligned} \tag{28}$$

Since we consider undirected graphs, we have $f_{ij} = f_{ji}$, so it is enough to introduce $\binom{n}{2} = O(n^2)$ optimization variables f_{ij} for $i < j$. However, there are $\binom{n}{3}$ constraints from (26). Brandes et al. showed [5] several characteristics of modularity maximization, including a proof that the decision version of the problem is NP-complete.

3.3 *Algorithms*

Considering the complexity of both the formulations in the previous section the algorithms presented here will be heuristic. When solving (28) we will use a greedy agglomerative approach, similar to the ones presented in [12, 16], while (23) will be solved using the spectral relaxation (24) and the approach described in [13].

3.3.1 Spectral Algorithm for the Normalized Cut

A partition from minimizing the normalized cut will be found by considering the relaxed problem (24). This problem is computed by solving the generalized eigenvalue problem for L. The obtained relaxed solution must then be made feasible for the original problem, taking the discrete constraints into consideration. Several approaches have been proposed for this including directional cosine method, randomized projection heuristic, and clustering rounding. We will adapt the method suggested in [13], using k-means algorithm on the eigenvectors of the normalized Laplacian L_{rw} to obtain a feasible solution.

3.3.2 Greedy Algorithm for Modularity

Problem (28) is solved by using a greedy agglomerative hierarchical heuristic that follows a scheme similar to [12]. The algorithm is based on an aggregation process with two different phases. The first phase performs small changes by shifting nodes between clusters and a second phase merges entire clusters, resulting in larger changes. Starting from singleton clusters the algorithm evaluates the modularity change in every phase, ΔQ, of each possible move/merging and then performs the action that would result in the largest modularity increase. The algorithm will alternate between these two actions as long as an improvement in the modularity is possible.

4 Simulation Result

4.1 Cluster Properties of the Market Graph

By considering the closing prices of stocks on the New York Stock market (comprising of NYSE, Nasdaq, and AMEX) a market graph was created. The original data consisted of 504 observations of 6330 stocks taken from *Yahoo Finance* with observations made between January 4th 2012 and December 31 2013.

In order to obtain more reliable results two pre-processing procedures were applied on the original data. First, all illiquid instruments were removed. This was done by removing all instruments that had no trading volume for more than 20 % of the observations. The second filtering procedure was introduced due to the large amount of Exchange traded funds (ETFs) present on the American market. The ETFs were removed since they often aim to track the market itself making them highly correlated with most stocks in the market. Their presence adds a noise of highly correlated instruments, not reflecting the overall behavior of the market. After applying these two procedures 4519 instruments remained, these time series were used to construct the market graph and its adjacency matrix A_{ij} by using Eqs. (18) and (19).

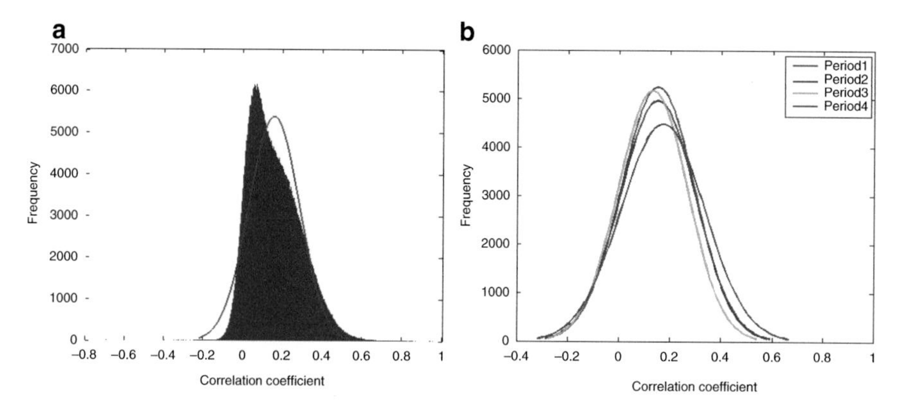

Fig. 1 (**a**) Correlation distribution and fitted distribution for entire period. (**b**) Fitted correlation distribution for different time periods

4.1.1 Correlation Distribution

The correlation distribution represents the fundamental structure of the market. A plot of the correlation distribution for the entire time period can be found in the left-hand graph in Fig. 1 together with a fitted normal distribution, with $\mu = 0.1532$ and $\sigma = 0.1264$. One can see that the correlation distribution of the US market does not seem to fit perfect with the normal distribution. Even though both tails of the distribution are covered the shape of the fitted curve is not consistent with the data. However, it is interesting to note that stocks seem to mainly exhibit positive correlation, suggesting that stock prices will often move in the same direction. This has been observed before and has then been interpreted as a sign of globalization with the motivation that more and more stock effect each other positively [1]. The graph on the right in Fig. 1 shows fitted distributions for different, shorter time periods, each period consisting of 100 observations. Even though there are some differences between the different periods the correlation distribution of the market remains stable over the considered time intervals.

4.1.2 Edge Density

The density of the market graph will of course depend on the correlation threshold. Varying the threshold θ generates graphs of different degrees of correlation. Figure 2 shows the edge density for different thresholds, as expected the density will decrease with increasing threshold.

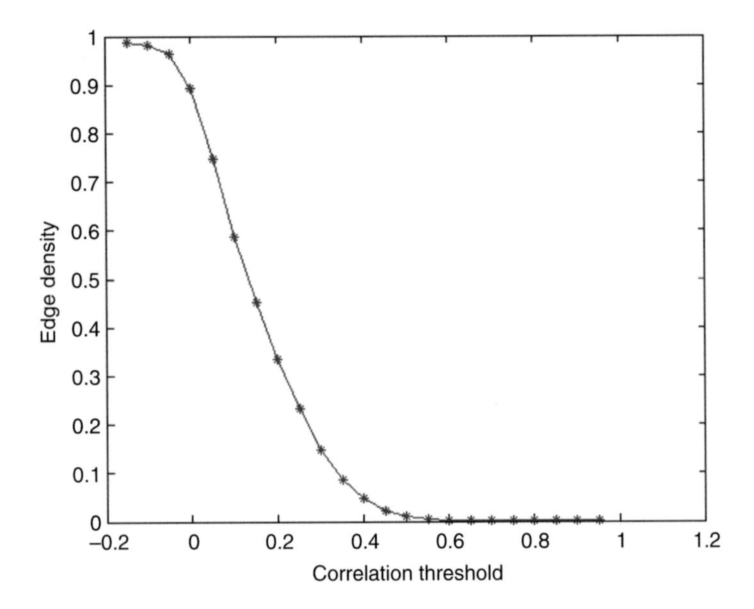

Fig. 2 Edge density as a function of correlation threshold

4.1.3 Clustering Coefficient

By computing the global clustering coefficient for graphs of different θ we found that the cluster coefficient was larger among positively correlated stocks than for negatively correlated stocks. As an example, the edge density of the graph obtained with threshold 0.6 is very close to that of the complementary graph with threshold -0.05. However, the corresponding global clustering coefficients of the two graphs are $C = 0.76$ and $C = 0.19$, respectively. Hence, one can suspect that positively correlated stocks tend to cluster more in the graph than negatively correlated stocks. This feature has been observed previously for other market graphs [1, 2].

For higher positive thresholds the global clustering coefficient appears almost constant. Figure 3 shows the graphs clustering coefficient for positive θ. For all $\theta \in [0.2, 0.9]$ the clustering coefficient of the graph remains in the interval $[0.70, 0.82]$. It should be noted that this is significantly higher than what would be expected from a uniform random graph with the same edge density. A high clustering coefficient is a common feature among many real life graphs, indicating that the market graph could possess a community structure.

4.1.4 Degree Distribution

By fixing the threshold a specific market graph is obtained. For this graph a degree distribution can be studied. As was also found in [1] the degree distribution is

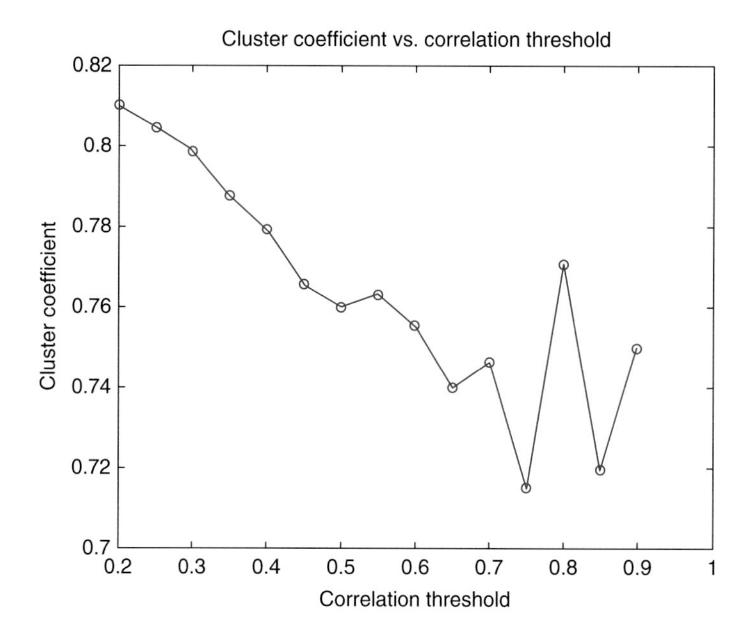

Fig. 3 Cluster coefficient as a function of correlation threshold

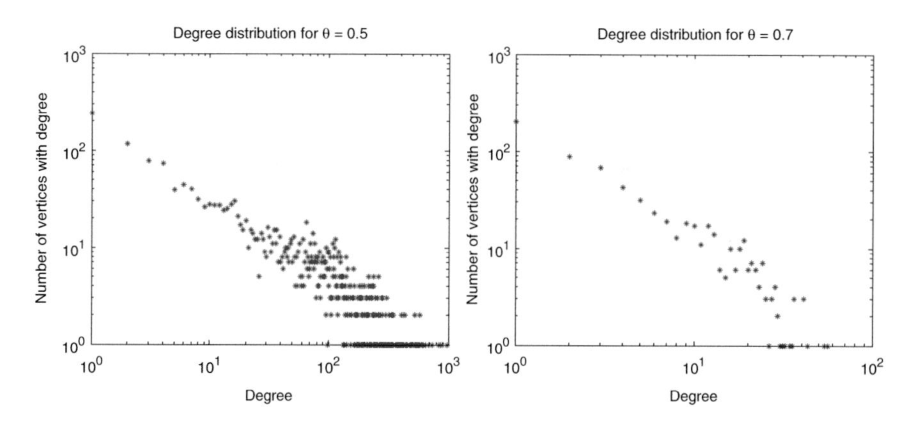

Fig. 4 Degree distribution for $\theta = \{0.5, 0.7\}$

filled with noise for lower thresholds, however for higher values the power law behavior becomes more clear. Figure 4 shows the degree distribution for thresholds $\theta = \{0.5, 0.7\}$ in a loglog plot. From the figure one can notice that the noise in the graph decreases as the threshold is increased. Also, it is interesting to note that the slope is lower compared to the edge distribution of many other real life graphs. For instance, the Web graph has been estimated to follow a power law with slope 2.18 [8]. The small exponent suggests that there could exist many vertices with high degree in the graph implying that there could exist larger clusters in the graph.

4.2 *Partitions of the Market Graph*

Partitions of different instances of the market graph are obtained with the two formulations presented in Sect. 3. The simulation setup is described in detail below.

4.2.1 Simulation Setup

By solving the two partition formulations for market graph instances partitions were obtained. In a first attempt to evaluate the result of the two approaches we ran both algorithms N times on each studied graph. For each of these N partitions the following metrics were computed.

- **Number of clusters found**—Remember that this is a free variable for the Modularity based formulation while it is fixed for the normalized cut approach.
- **Internal clustering density**—computed from Eq. (4)
- **Max internal cluster density**—computed from Eq. (3)
- **Min internal cluster density**—computed from Eq. (3)
- **External cluster density**—computed from Eq. (5)
- **Min cluster size**
- **Max cluster size**

Then, the average over all N values were taken, and the results are reported in Table 1 and in the Appendix. Since the formulation based on normalized cut requires a fixed number of clusters, k, as input, this algorithm was applied with three different $k = 10, 20, 30$ for each graph.

Additionally, to evaluate the consistency of the partitions for each algorithm we compute the adjusted rand index (ARI) [11] between some obtained partitions. The ARI measures the overlap between two partitions and is defined in the following way. Given a set $V = (1, 2, \ldots, n)$ and the partitions $X = (X_1, \ldots, X_s)$ and $Y = (Y_1, \ldots, Y_t)$ of V we can define the following quantities

a —the number of pairs of elements in V that are in the same set in X and in the same set in Y.

b —the number of pairs of elements in V that are in different sets in X and in different sets in Y.

c —the number of pairs of elements in V that are in the same set in X and in different sets in Y.

d —the number of pairs of elements in V that are in different sets in X and in the same set in Y.

Using these quantities the ARI is defined as

$$ARI = \frac{\binom{n}{2}(a+d) - [(a+b)(a+c) + (c+d)(b+d)]}{\binom{n}{2}^2 - [(a+b)(a+c) + (c+d)(b+d)]}. \tag{29}$$

ARI has expected value 0 and maximal value 1, corresponding to identical partitions. It can be used to compare both partitions obtained by the same approach, to evaluate the method's consistency as well as comparing the results obtained from different formulations. Computed ARI for different pairs of partitions can be found in Table 2 in the Appendix.

4.2.2 Partition Structure

The result from the simulations shows that for the lower thresholds $\theta \in [0.4, 0.5]$, both approaches produce partitions with low modularity. Also, the partitions have low minimal internal cluster density and a high external density relative to the overall graph density. As an example, for $\theta = 0.4$ the minimal internal density of a cluster is of the same magnitude as the edge density of the entire graph for both algorithms. Also, the external cluster edge density is approximately half that of the overall graph density, indicating that the identified clusters are not well separated. All these results indicate that the market graph lacks a strong community structure for lower thresholds. This is not really surprising as at lower thresholds even weakly correlated stocks can be connected in the graph making it more difficult to distinguish which instruments truly form clusters.

Also, for these lower thresholds the size of the largest cluster found is very large, especially for the greedy modularity approach where the largest cluster consists of nearly $2/3$ of the considered nodes. This cluster has low internal density and is strongly connected to the rest of the graph. This further supports the idea that the market graph lacks a clear community structure for lower thresholds. Figure 5 shows a partition of the giant connected component obtained by the modularity approach for $\theta = 0.5$. One can see that even though each cluster seems strongly connected, most of them are not well separated.

For higher values of the thresholds ($\theta \in [0.6, 0.7]$) the quality of the partitions increases. Partitions of these graphs display higher modularity, combined with higher minimal internal cluster density and lower external density. As an example, for $\theta = 0.7$ the minimal internal cluster density is more than three times as high as the overall graph density and the external cluster density is less than $\frac{1}{10}$ the edge density of the entire graph. Hence, clusters are both more dense and better separated compared to partitions for lower thresholds. This result was found for all the approaches. Figures 6 and 7 show the partitions of the largest connected component for $\theta = 0.7$ for both algorithms. In this case the partitions seem very similar, this is also confirmed by computing the ARI for the two partitions, $ARI = 0.9295$, further indicating a large overlap between the two partitions.

In Fig. 8 we have plotted the internal cluster density against cluster size for both approaches (with 20 partitions of each) applied on market graphs with $\theta = [0.5, 0.7]$. From Fig. 8 we can notice that the result from the two approaches becomes more similar for larger values of θ, indicating a stronger community structure for higher thresholds in the graph. Figure 9 shows the normalized cut plotted against cluster size of the corresponding partitions. Here we can notice

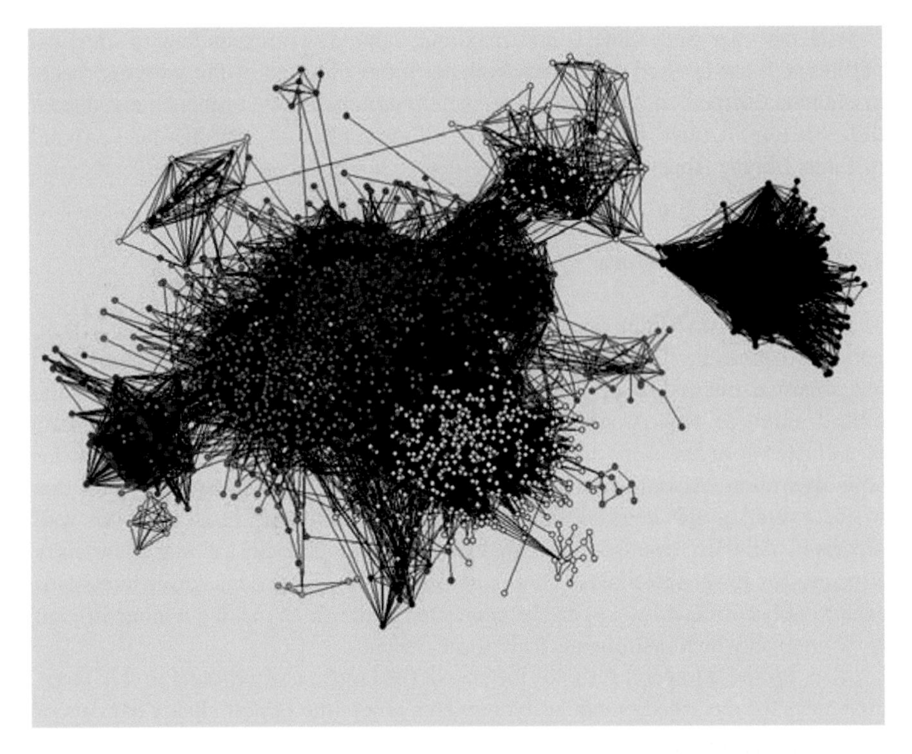

Fig. 5 Partition of the market graph ($\theta = 0.5$) into 18 clusters from modularity approach

a difference between the approaches since the modularity formulation produces partitions with smaller fluctuations in the normalized cut of clusters than the cut formulation.

Comparing the two approaches one can notice that using modularity generally gives a larger giant cluster than the method using normalized cut (when they are set to find the same number of clusters).

4.2.3 Internal Cluster Structure

Since a common way to classify instruments in portfolio management is by dividing them into industrial sectors we will compare the internal structure of the clusters with 12 industrial sectors of the market. First, the sector representation in the data and for the giant component of different market graphs can be found in Table 3 in the Appendix. It is especially interesting to note that as the threshold θ increases some sectors such as finance, basic industries, and energy increase their percentage in the largest connected component of the graph while other sectors, such as health, drastically decrease. Hence, the degree of correlation between industrial sectors in the market differs.

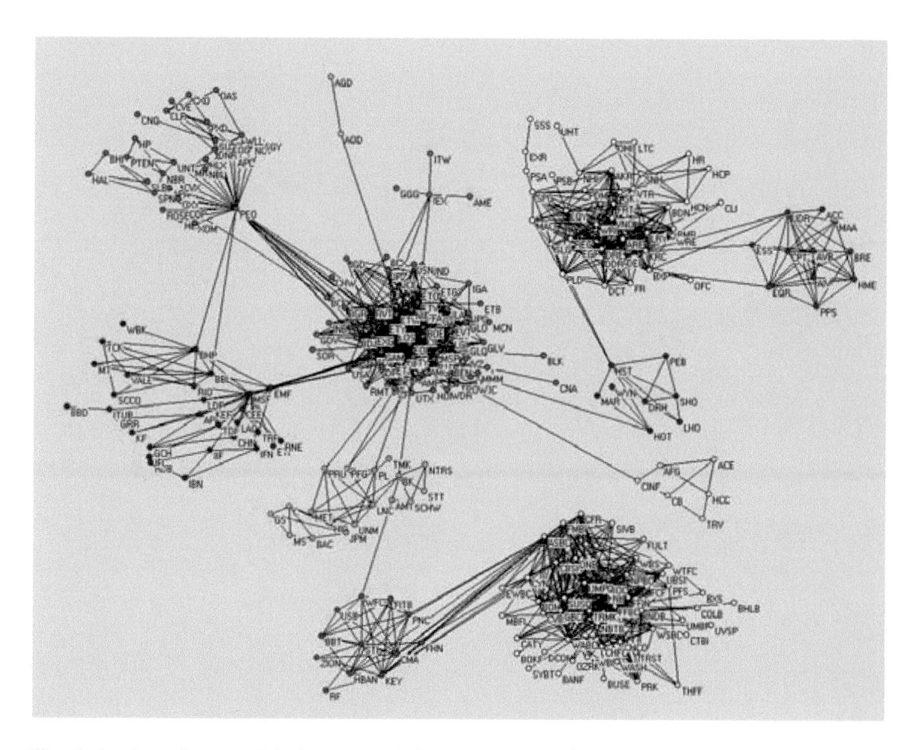

Fig. 6 Partition from modularity approach for market graph $\theta = 0.7$

To analyze the internal structure of the clusters in the graph partitions we study the percentage representation of each sector in every cluster. Table 3 in the Appendix shows the relation between industrial classification and clusters in partitions of the giant connected component of the market graph, obtained by the greedy modularity approach for different thresholds.

For lower thresholds $\theta = 0.5$ the sector overlap between clusters is relatively large. This is especially clear for the two largest clusters in the partition. In these sets all 12 industry sectors are represented. Hence, for moderate correlations there is no strong connection between clusters and industry sectors. It is also interesting to note that the financial sector stands out by existing in all of the four largest clusters, indicating that this sector is connected to many other sectors at this correlation level. This is further confirmed by the fact that when selecting the ten nodes with the highest degrees in the graph more than 50 % of these belong to the financial sector.

For higher thresholds the correlation between industrial sectors and clusters is stronger. At threshold 0.7, no cluster includes stocks from all sectors and more clusters now only consist of one sector. However, even though the correlation between cluster and sector is very strong, it is not complete, even at this high threshold level. This phenomenon introduces the possibility to use graph clusters instead of industry sectors in portfolio diversification.

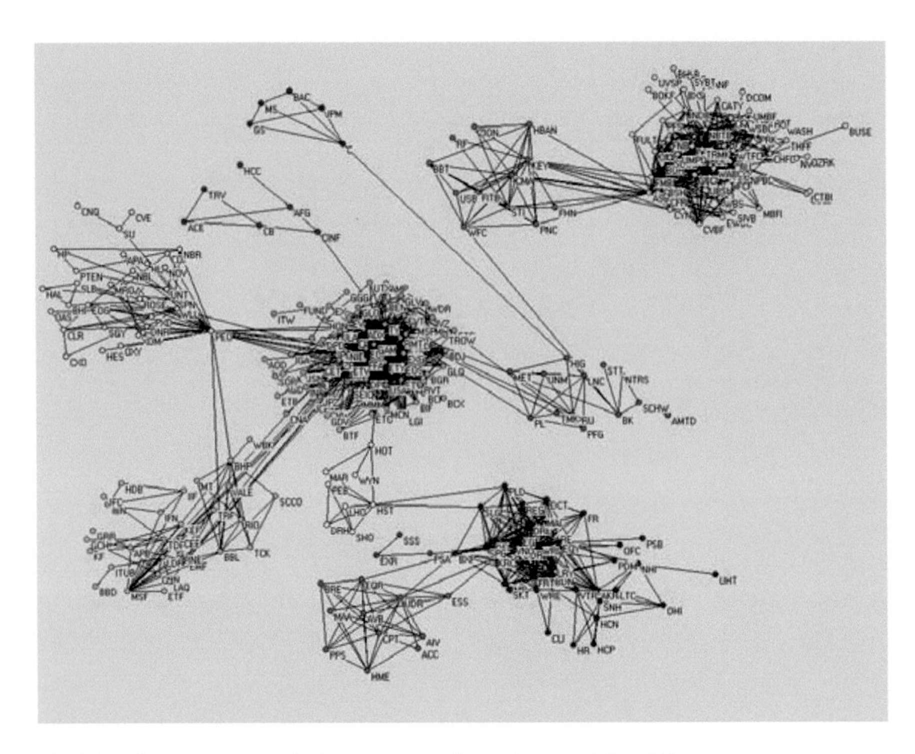

Fig. 7 Partition from normalized cut approach for market graph $\theta = 0.7$

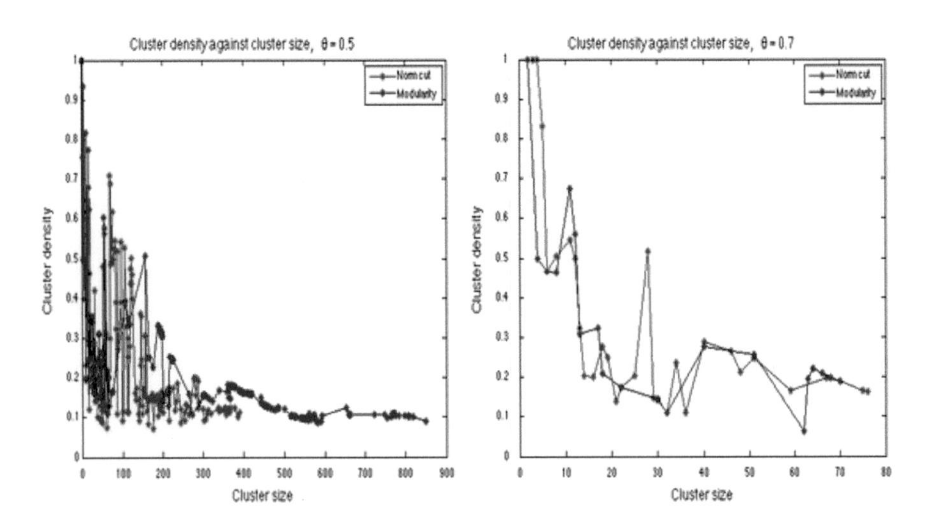

Fig. 8 Internal cluster density against cluster size for GM (*blue*) and SC (*red*) $\theta = [0.5, 0.7]$

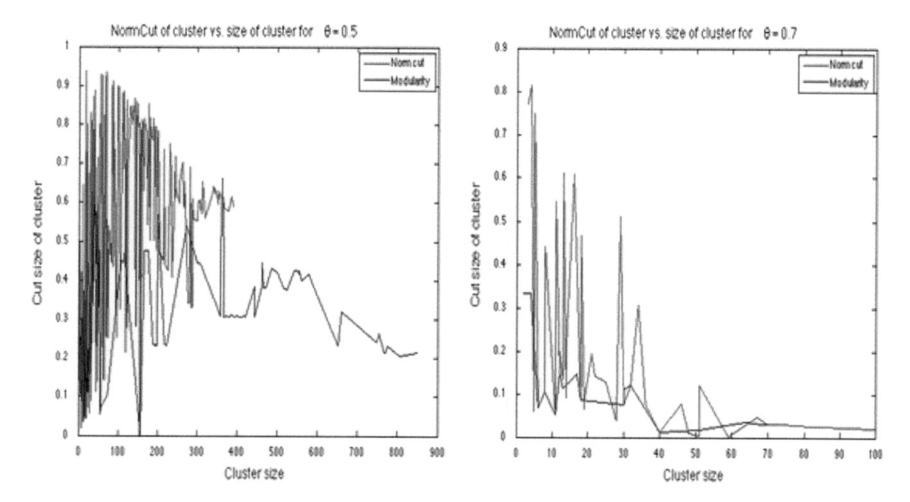

Fig. 9 Normalized cut of cluster vs. cluster size for GM (*blue*) and SC (*red*) $\theta = [0.5, 0.7]$

4.2.4 Dynamics in Partition Structure

To study possible dynamics and stability in the partition structure we divide our data of price returns into four periods, each consisting of 150 days, and with 50 days overlap between each consecutive period. For each time series we construct a market graph for $\theta = 0.7$. First, computing the giant connected (GC) component of each market graph we notice that its size varies greatly, from 253 instruments in period 3 up to 701 in period 4. Thus the market correlations in these sub-periods are quite different compared to the correlations obtained by using data covering all time series. Moreover, by considering the overlap for these different GCs we can see that it is not only the cardinality that changes but also which instruments that are present in the GC. The intersection between all the GCs is 131 indicating that the instruments composing the GCs change over different time periods. However, the edge density of the GC is almost constant over all four periods.

Using the greedy modularity approach, partitions for the different time periods were obtained (20 for each graph), the results are reported in Table 4. From these numbers it can be observed that the modularity decreases between period 1 and period 4. This together with the fact that the external cluster density increases over the same time could imply that the community structure of the graph decreases over time. Partitions of the market graph for $\theta = 0.7$ from period 1 and 4 can be found in Figs. 10 and 11, these graphs confirm the differences between the partitions.

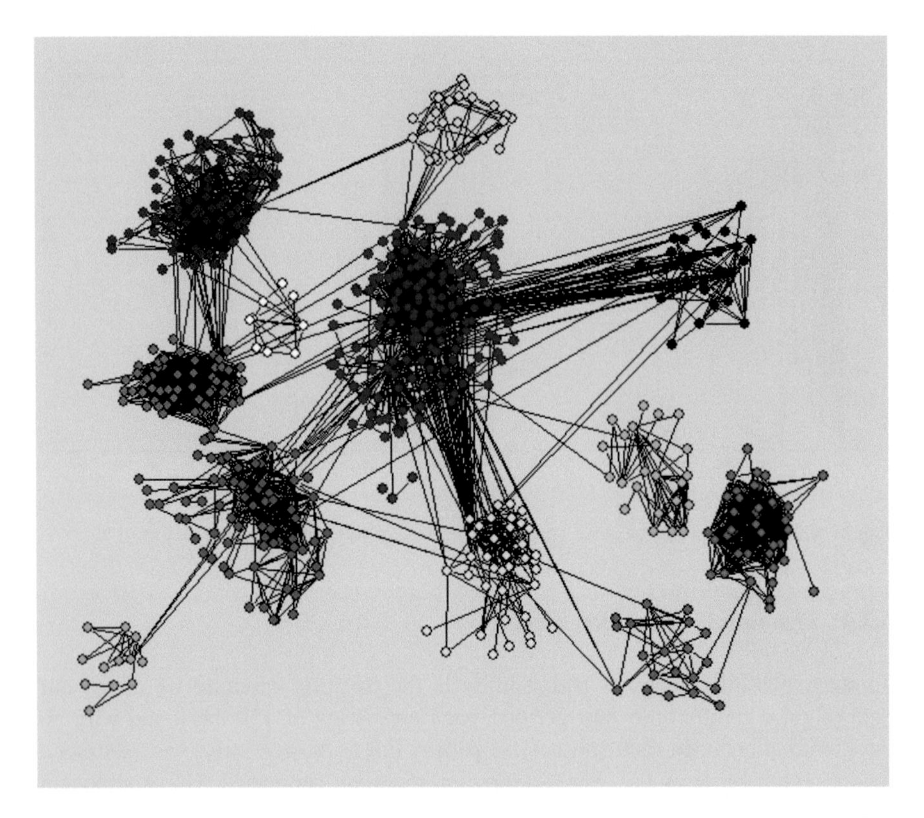

Fig. 10 Partitions of market graph with $\theta = 0.7$ for period 1 into 11 clusters

By studying the sector composition of the clusters it was found that the connection between sectors and clusters was weaker in all sub-periods compared to the entire period. Also, this pattern increased for every considered period, and in the last period most clusters consisted of several different industrial sectors. Hence, the correlation between industrial sectors and graph clusters is weaker for shorter time series.

Appendix

Number of runs is N = 50

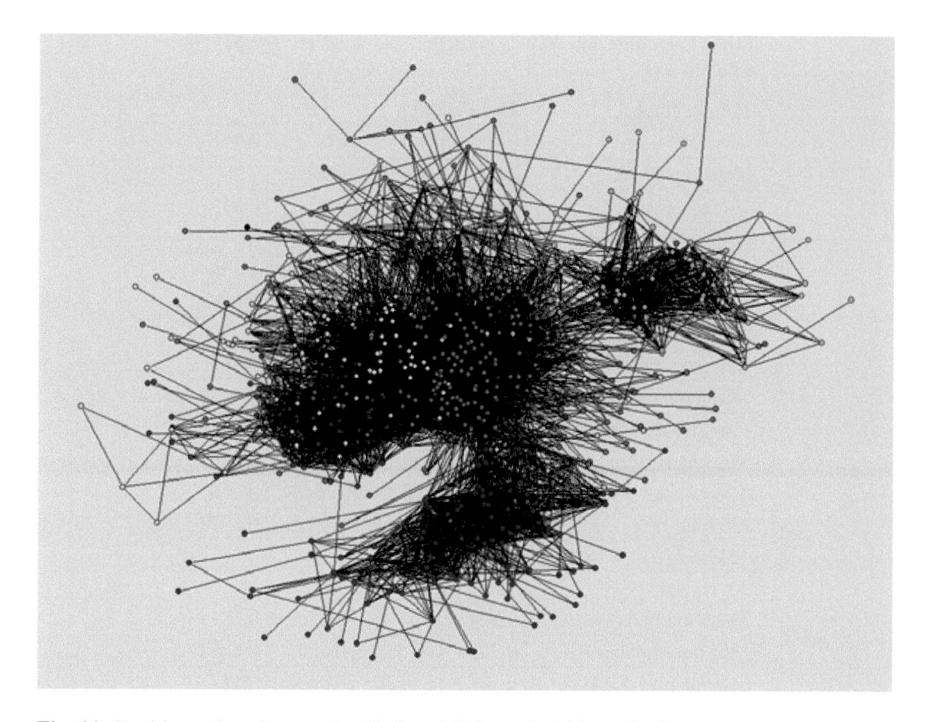

Fig. 11 Partitions of market graph with $\theta = 0.7$ for period 4 into 12 clusters

Adjusted Rand Index

ARI for different partition approaches and market graphs. The ARI was obtained by choosing three pairs of partitions, at random from each type, computing the ARI for each of them, and then taking the average of these values. (GM—greedy modularity, SNC—spectral normalized cut.)

Industrial Sectors in the Market Graph

Industrial Sectors and Clusters

Market graph $\theta = 0.5$ with industrial sectors and clusters from greedy modularity. The italic numbers represent the largest sector in every cluster.

Market graph $\theta = 0.7$ with industrial sectors and clusters from greedy modularity.

Partitions of market graph for different periods.

Table 1 Result for the algorithms on market graphs for different θ

Market graph	Threshold θ			
	0.4	0.5	0.6	0.7
Nr. nodes in GC	2964	2020	934	278
Global density (GC)	0.0928	0.0400	0.0282	0.0337
Greedy modularity				
Mean nr. clusters	14.1	20.66	13.66	11.08
Max/Min nr. clusters	[9,20]	[15 33]	[10 19]	[10 12]
Mean internal density	0.5768	0.4686	0.4562	0.3758
Max internal density	0.9968	0.9571	0.9933	0.8691
Min internal density	0.0924	0.0996	0.0714	0.1009
External cluster density	0.0400	0.0123	0.0037	0.0013
Min cluster size	2.1	2.5	2.64	3.32
Max cluster size	938.78	630.22	276.78	66.48
Modularity	0.2406	0.4353	0.6738	0.7271
Spectral Ncut, $k = 10$				
Nr. clusters	10	10	10	10
Mean internal density	0.3243	0.2765	0.3162	0.3514
Max internal density	0.7932	0.6511	0.6506	0.8282
Min internal density	0.1083	0.0526	0.0632	0.117
External cluster density	0.0689	0.0188	0.0051	0.0021
Min cluster size	13.40	11.70	9.40	5.15
Max cluster size	1086	976	280.8	71.45
Modularity	0.0490	0.1360	0.6290	0.7044
Spectral Ncut, $k = 20$				
Nr. clusters	20	20	20	20
Mean internal density	0.2865	0.3037	0.3596	0.5113
Max internal density	0.8373	0.7370	0.8168	0.9950
Min internal density	0.0925	0.0874	0.0745	0.1524
External cluster density	0.0852	0.024	0.0079	0.0043
Min cluster size	7.25	8.80	4.10	2.60
Max cluster size	484.35	351.0	157.85	54.95
Modularity	0.0417	0.2844	0.5984	0.6349
Spectral Ncut, $k = 30$				
Nr. clusters	30	30	30	30
Mean internal density	0.2576	0.3084	0.4412	0.5526
Max internal density	0.8150	0.7967	0.9837	1
Min internal density	0.0801	0.0979	0.0880	0.1599
External cluster density	0.0901	0.0274	0.0094	0.0108
Min cluster size	9.62	4.5	2.6	2
Max cluster size	257.75	227.5	127.4	43.7
Modularity	0.0489	0.2559	0.5654	0.5575

Table 2 Adjusted rand index for different algorithms and market graphs

Market graph	Threshold θ		
	0.5	0.6	0.7
ARI			
GM & GM	0.5846	0.7845	0.9593
SNC & SNC	0.5317	0.9124	0.9212
GM & SNC	0.4084	0.5900	0.8963

Table 3 Sector representation in data and market graphs giant connected component for $\theta = [0.5, 0.6, 0.7]$

Sector representation	Threshold θ			
	Data	0.5	0.6	0.7
Sector				
Basic industries	0.0755	0.1065	0.1192	0.0294
Capital goods	0.0860	0.1097	0.1205	0.0196
Consumer Dur.	0.0326	0.0368	0.0314	0.0049
Consumer Non-Dur.	0.0508	0.0387	0.0226	0
Consumer Serv.	0.1525	0.1503	0.1267	0.2745
Energy	0.0712	0.0958	0.1192	0.1569
Finance	0.1649	0.2245	0.2961	0.5049
Health	0.1128	0.0323	0.0138	0.0049
Miscellaneous	0.0315	0.0146	0.0013	0
Public Util.	0.0590	0.0729	0.0803	0
Technology	0.1379	0.0977	0.0590	0.0049
Transportation	0.0252	0.0203	0.0100	0

Table 4 Partitions for different time periods and $\theta = 0.7$ from greedy modularity approach

	Time Periods			
Market graph	P1	P2	P3	P3
Nr. nodes in GC	532	559	253	701
Global density (GC)	0.0313	0.0209	0.0273	0.0291
Greedy modularity				
Mean nr. clusters	9.1	12.5	10	13.45
Max/Min nr. clusters	[7, 11]	[11 13]	[8 12]	[10 16]
Mean internal density	0.2853	0.2614	0.3281	0.4988
Max internal density	0.7047	0.5613	1	1
Min internal density	0.0643	0.0511	0.09264	0.0727
External cluster density	0.0012	0.0012	0.0025	0.0070
Min cluster size	8.3000	8.1000	2.0	2.0
Max cluster size	134.4	120.5	62.8	239.7
Modularity	0.7483	0.7847	0.6143	0.5120

Cluster	*Size*	Bas.Ind	Cap.Go	Cons.D	Cons.N	Cons.S	Ener	Fin	Heal	Missc.	Pub.U	Tech	Trans
1	188	0.01	0	0	0	*0.61*	0.01	0.03	0	0	0.35	0	0
2	29	0.07	0	0.07	*0.59*	0	0	0	0.21	0	0.07	0	0
3	157	0	*1*	0	0	0	0	0	0	0	0	0	0
4	758	0.12	0.16	0.06	0.04	0.09	*0.18*	0.09	0.04	0.02	0.02	0.16	0.02
5	3	0	0	0	0	0	0	0	0	*1*	0	0	0
6	5	0	0	0	0	*1*	0	0	0	0	0	0	0
7	3	0	0	0	0	*0.67*	0	0	0	0.33	0	0	0
8	10	0	0	0	0	*1*	0	0	0	0	0	0	0
9	15	0	0	0	0	0	0	0	0	0	0	*1*	0
10	57	*1*	0	0	0	0	0	0	0	0	0	0	0
11	8	0	0	0	0	0	*1*	0	0	0	0	0	0
12	20	0	0	0	0	0.1	0.4	0	0	0	*0.5*	0	0
13	2	0	0	0	0	0	0	*1*	0	0	0	0	0
14	407	0.04	0.08	0.03	0.03	0.05	0.01	*0.66*	0.03	0.02	0.02	0.06	0.01
15	52	0.12	*0.37*	0.08	0.02	0.08	0	0	0.02	0	0	0	0.31
16	7	*1*	0	0	0	0	0	0	0	0	0	0	0
17	32	0	0	0	0	*0.94*	0	0.06	0	0	0	0	0
18	3	0	0	0	0	0.33	0	0	0	0	*0.67*	0	0
19	262	0.06	0.08	0	0.04	0.04	0.06	*0.33*	0.04	0.02	0.27	0.04	0.04
20	2	0	0	0	0	0	*1*	0	0	0	0	0	0

Cluster	Size	Bas.Ind	Cap.Go	Cons.D	Cons.N	Cons.S	Ener	Fin	Heal	Missc.	Pub.U	Tech	Trans
1	6	0	0	0	0	0	0	*1*	0	0	0	0	0
2	2	0	0	*1*	0	0	0	0	0	0	0	0	0
3	32	0	0	0	0	0	*1*	0	0	0	0	0	0
4	11	0	0	0	0	*1*	0	0	0	0	0	0	0
5	4	0	*0.5*	0.25	0	0	0	0	0	0	0	0.25	0
6	40	0	0	0	0	*1*	0	0	0	0	0	0	0
7	51	0	0	0	0	0	0	*1*	0	0	0	0	0
8	30	*0.5*	0	0	0	0	0.08	0.42	0	0	0	0	0
9	64	0	0.14	0	0	0	0	*0.79*	0.07	0	0	0	0
10	30	0	0	0	0	0	0	*1*	0	0	0	0	0
11	8	0	0	0	0	*1*	0	0	0	0	0	0	0

References

1. Boginski, V., Butenko, S., Pardalos, P.M.: On structural properties of the market graph. Innov. Financ. Econ. Netw. **48**, 29–35 (2003)
2. Boginski, V., Butenko, S., Pardalos, P.M.: Mining market data: a network approach. Comput. Oper. Res. **2**, 3171–3184 (2006)
3. Boginski, V., Butenko, S., Pardalos, P.M.: Statistical analysis of financial networks. Comput. Stat. Data Anal. **48**, 431–443 (2005)
4. Bollobas, B. (2nd ed.) Random Graphs, pp. 10–13. Cambridge University Press, Cambridge (2001)
5. Brandes, U., Delling, D., Gaertler, M. et al.: On modularity clustering. IEEE Trans. Knowl. Data Eng. **20**(2), 172–188 (2008)
6. Bu, T., Towsley, D.: On distinguishing between internet power law topology generators. In: Proceedings-IEEE INFOCOM, vol. 2, pp. 638–647 (2002)
7. Erdös, P., Rényi, A.: The evolution of random graphs. Magyar Tud. Akad. Mat. Kutató Int. Közl. **5**, 17–61 (1960)
8. Faloutsos, M., Faloutsos, P., Faloutsos, C.: On power-law relationships of the Internet topology. SIGCOMM Comput. Commun. Rev. **29**, 251–262 (1999)
9. Fortunato, S.: Community detection in graphs. Phys. Rev. E **486**, 75–174 (2010)
10. Huang, W., Zhuang, X., Yao, S.: Network analysis of the Chinese stock market. Phys. A **338**, 2956–2964 (2009)
11. Hubert, L., Arabie, P.: Comparing partitions. J. Classif. **2**(1), 193–218 (1985)
12. Le Martelot, E., Hankin, C.: Fast multi-scale detection of relevant communities. Comput. J. **16**, 1136–1150 (2013)
13. Luxburg, U.: A tutorial on spectral clustering. Stat. Comput. **17**(4), 395–416 (2007)
14. Matsuda, H., Ishihara, T., Hashimoto, A.: Classifying molecular sequences using a linkage graph with their pairwise similarities. Theor. Comput. Sci. **210**, 305–325 (1999)
15. Newman, M.E.: Spectral methods for community detection and graph partitioning. Phys. Rev. E **88**(4), 321–354 (2013)
16. Newman, M.E., Girvan, M.: Fast algorithm for detecting community structure in networks. Phys. Rev. E **69**(2), 18–33 (2004)
17. Newman, M.E., Girvan, M.: Spectral methods for community detection and graph partitioning. Phys. Rev. E **88**(44), 32–54 (2013)
18. Pattillio, J., Youssef, N., Butenko, S.: Clique relaxation models in social network analysis. In: Thai, M.T., Pardalos, P.M. (eds.) Handbook of Optimization in Complex Networks, pp. 143–162. Springer, New York (2012)
19. Reka, A., Albert-Laszlo, B.: Statistical mechanics of complex networks. Rev. Mod. Phys. **74**(1), 47–97 (2002)
20. Schaeffer, S.E.: Survey: graph clustering. Comput. Sci. Rev. **1**(1), 27–64 (2007)
21. Shi, J., Malik, J.: Normalized cuts and image segmentation. IEEE Trans. Pattern Anal. Mach. Intell. **22**(8), 888–905 (2000)
22. Vizgunov, A., Goldengorin, B., Kalyagin, V., Koldanov, A., Koldanov, P., Pardalos, P.M.: Network approach for the Russian stock market. Comput. Manage. Sci. **11**, 44–55 (2014)

Part III
Economics and Other Applications

A Semantic Solution for Seamless Data Exchange in Supply Networks

Elena Andreeva, Tatiana Poletaeva, Habib Abdulrab, and Eduard Babkin

Abstract This paper proposes the semantic solution for data exchange in dynamically changing supply networks. Because of the dynamic nature of supply networks, there is a necessity to manage their efficiency. The first step in this direction is proper data exchange, which leads to transparency of networks and speeds up interactions of their participants. Despite plenty of data standards, there is a lack of approaches to data modeling allowed the seamless knowledge sharing within supply networks. This paper introduces a new ontology-based data metamodel of supply networks based on organizational ontology, consistent theory for data modeling, and the ontologized standard in logistics domain. At the first part of the paper, selected approaches to data modeling are explained and justified. Then, basic information patterns applicable for modeling of supply networks are explained. Finally, we elaborate how the proposed ontological framework facilitates knowledge integration in logistics.

Keywords Semantic data modeling • Domain ontology • Knowledge management • Semantic interoperability • Supply network

1 Introduction

Nowadays the structure of supply networks is getting more complex with the increasing volume of transportation load, the number of people and organizations involved in logistical processes, stronger requirements to the quality of product distribution, and the growing volume of information flow. Therefore, businesses realize the necessity of applying information systems to facilitate the control of products within supply networks. However, seamless information exchange between

E. Andreeva (✉) • T. Poletaeva • E. Babkin
National Research University Higher School of Economics, Nizhny Novgorod, Russia
e-mail: eandreeva@hse.ru; ta.poletaeva@gmail.com; eababkin@hse.ru

H. Abdulrab
INSA de Rouen, LITIS Lab., Rouen, France
e-mail: abdulrab@insa-rouen.fr

V.A. Kalyagin et al. (eds.), *Models, Algorithms and Technologies for Network Analysis*, Springer Proceedings in Mathematics & Statistics 156,
DOI 10.1007/978-3-319-29608-1_17

information systems of supply network parties is hindered by poor semantic interoperability, in general, and data integration problems, in particular [1].

The lack of semantic interoperability of information systems comes out from the impossibility to declare the universal standard for naming and meaning of measures and characteristics of the supply network among its current and potential future parties. The storage and transmission of logistics records have been the object of many initiatives concerning standardization, to wit: Universal Business Language, UBL (OASIS project),[1] UN/CEFACT[2] logistics module, GS1[3] Logistics Interoperability Model, etc. Nevertheless, these standards are not amenable to foster semantic interoperability, because their metamodels are not smoothly integrable. In addition to the use of different standards and approaches to data modeling, developers of information systems specify data storages with the particular concepts and data patterns, which are rather based on their experience and intuition than on any theoretical foundations [2]. The following commonly encountered problem is a shining example of the problems of information exchange. The attribute "selling price" of the same products has different meanings and, therefore, different values for manufacturers and distributors. For a manufacturer the selling price of a product equals the sum of net cost and extra cost. But a distributor considers the selling price to be the sum of purchase price and some extra costs. As for other cases, the root of heterogeneity lies in the differences of the definitions given for the same concept. Aforementioned lack of semantic interoperability occurs when automated discovery of concept definition and meaning in a data model is not possible.

The widely known solution for the problem of information exchange is the use of domain ontologies. Logistics domain ontologies provide a theory to address semantic interoperability and data/standard integration between information systems of supply network parties, allow automatic analysis by means of artificial intelligence, and convey a knowledge repository. As soon as the elements of particular data models are mapped onto complete collection of logically related concepts provided by the domain ontology, the problem of information exchange is resolved. Meanwhile the mapping between a particular data model and the ontology-related data patterns is easier than the mapping between two particular data models, because the meaning of ontological data elements is unambiguously defined by their interrelations and related axioms [3]. Creation of underlying domain ontology is propagated by many standards in different domains (ISO 15926,[4] IDEAS,[5] DoDAF,[6] OMG FIBO,[7] etc.). However, there are only some attempts of creation the domain ontology in logistics [4–6].

[1]OASIS: https://www.oasis-open.org/.

[2]UN/CEFACT: http://www.unece.org/CEFACT/.

[3]GS1: http://www.gs1.org/.

[4]ISO 15926: http://15926.org/.

[5]IDEAS: http://ideasgroup.org/.

[6]DoDAF: http://dodcio.defense.gov/TodayinCIO/DoDArchitectureFramework.aspx.

[7]OMG FIBO: http://www.omg.org/.

The main drawback of known approaches in building the logistics domain ontologies is the underestimation of the importance of the underlying foundational ontology. Prevailing Entity paradigm [2] for data modeling lefts high levels of freedom in the definition of entities and their attributes. On the other hand, it provides a very weak theory about parts and wholes, types and instances, identity, dependency, and unity. That is why relational data models contain ambiguous definitions of real objects. Consequently, it is difficult to automate the mapping process and the integration of the data models based on Entity paradigm.

Our research is aimed to overcome the problem of information exchange in logistic networks by developing a logistic domain ontology based on the Object paradigm by means of the business objects reference ontology (BORO) methodology [2]. In opposition to the Entity paradigm, the Object paradigm of data modeling [2] has no superficial division of the world into entities and attributes. Everything in the world is considered as an explicit map of objects and their patterns of relationships [2], whether it is individual, class, individuals' relationship, or class of relationships. The Object paradigm, fully expressed in the BORO methodology [2], (1) provides strong logical foundations for classification and identity; (2) provides clearer and more precise representation of reality based on spatial and temporal dimensions of objects; (3) provides the ability to capture dynamic objects showing their changes over time; and (4) allows avoiding subjective and possibly erroneous ontological assumptions in modeling constructs [7]. That is why we chose this paradigm for modeling of dynamic supply networks. BORO-based formal ontology has been already designed by the IDEAS Group. IDEAS ontology was used for developing of data-modeling standards—DoDAF, MODAF, and MODEM. Despite the fact that the IDEAS-based data standards were developed to support the exchange and sharing of military enterprise architectures, IDEAS upper-level ontology is domain-independent [8]. In our work we partially use the IDEAS upper-level ontology, and extend it with the concepts related to physical flows of enterprises.

In order to build the domain-specific level of the logistics ontology, we need a comprehensive set of concepts that are used in this field. For this purpose we chose supply chain operations reference model (SCOR model) [9] as a standardized description of business operations in supply chains and the source of unambiguously defined terms that participants of supply networks operate.

Thus, in this paper we present the data metamodel built in accordance with the BORO methodology partially using the IDEAS upper-level ontology. Moreover, the proposed metamodel is built on top of the formal enterprise ontology described in detail in [10]—a core ontology for supply networks. Providing essential ontologized domain-independent knowledge about enterprises, this ontology forms a shape for the integration of heterogeneous data models in the logistics domain. The domain-specific level of the metamodel is built by applying business objects of the SCOR model to the domain-independent level. The proposed metamodel is aimed to facilitate significantly the supply chain management and simplify interaction and data exchange.

The outline of the paper is as follows. First, the theoretical background of proposed ontology is summarized in Sect. 2. Proposed logistics ontology and basic information patterns applicable for modeling of supply networks are explained in Sect. 3. Then, in order to verify the completeness and coherence of the developed ontology, we used the ontology to build the data metamodel of a particular supply process. The details of this data metamodel are described in Sect. 4. In Sect. 5 we demonstrate the semantic power of the metamodel by the knowledge that can be extracted from related data model. Moreover, in this section we elaborate on the integration of the particular logistics data metamodel with the proposed one. Finally, Sect. 6 provides conclusions and directions for further research.

2 The Ontological Engineering Approach

In our research the BORO methodology [2] is a theoretical basis for the ontology development. The Object paradigm of data modeling underlies the BORO methodology and defines some fundamental approaches to data modeling.

Firstly, according to the BORO, only the concepts of tangible objects that exist in a real world should form the ontological core. Moreover, object identity is defined as an object's extension in the universe [7]. The extensional approach provides an ontology that is well suited to deconflicting multiple identification schemes [11]. The second modeling principle is the consideration of spatio-temporal (4D) extensions of any object. This approach intrinsically represents the temporality of objects by the assumption that they just partly exist at each time instant of their life span. Together with the essential information patterns listed hereafter, this modeling principle makes the methodology perfectly suitable for designing data models of dynamically changing logistic networks.

The ontological constructs used in the Object paradigm are individuals, classes, and tuples (coupled relations) [2]. It is assumed that all these constructs represent four-dimensional objects. Individuals are four-dimensional objects that persist through time. A class is an object according to its definition as a collection of 4D-individuals. A tuple as the ordered pair of interconnected individuals or classes is also an object. Any object is a sequence of 4D-states within its lifecycle. Any state is bounded by the events that mark its beginning and the end. Since an event happens at a point of time, it has no time duration, but, in turn, it has a spatial extension. A spatial extension of an event is a time slice that is a 3D-part of an object.

Finally, in the Object paradigm C. Partridge proposed fundamental information patterns for expressing relationships between 4D-objects. He proposed using the "whole-part" pattern to assign that an object is a part of another one. Also the sequence of two states of an object can be related by means of the "before-after" pattern. The role of the "pre-condition" pattern is to link an event with the conditions required to make it happen.

The Object paradigm proposes the unique description of production processes as the set of four-dimensional objects involved into this process. Consequently, a process is also considered as a 4D-object. Thus, the extension of a simple process can be the aggregation of the extensions of an actor and a product changed by the actor. Changes of the states of constituent objects define the states of the process. The sequence of process states is unambiguously expressed by the means of the "before-after" and "pre-condition" patterns.

In accordance with the Object paradigm, the BORO methodology allows revealing business objects as well as relationships between them. Moreover, the methodology defines some conceptual modeling rules for defining new ontological concepts.

3 The Logistics Ontology

In this section we explain the developed ontology and we give the notion of invented concepts and the reasons why they were added to the ontology.

The developed ontology consists of three levels of abstraction: the Framework, the Application level, and the Operational level [2]. Because of the domain requirements, the conceptualization of physical flows of enterprises (the Application level) was created upon the formal enterprise ontology (the Framework) described in detail in [10]. The Application level contains the concepts of standard supply processes specified by the SCOR model. This ontological level is supposed to be the common core of different data metamodels of supply networks putting aside their specifics. The Operational level in turn comprises business objects related to a particular business scenario. Also, as it was mentioned in the Introduction, the notation of the IDEAS standard [8] was used for the visualization of created conceptual model.

The top level of the Framework fully corresponds to the BORO fundamental ontology. According to the extensional approach, all concepts have corresponding

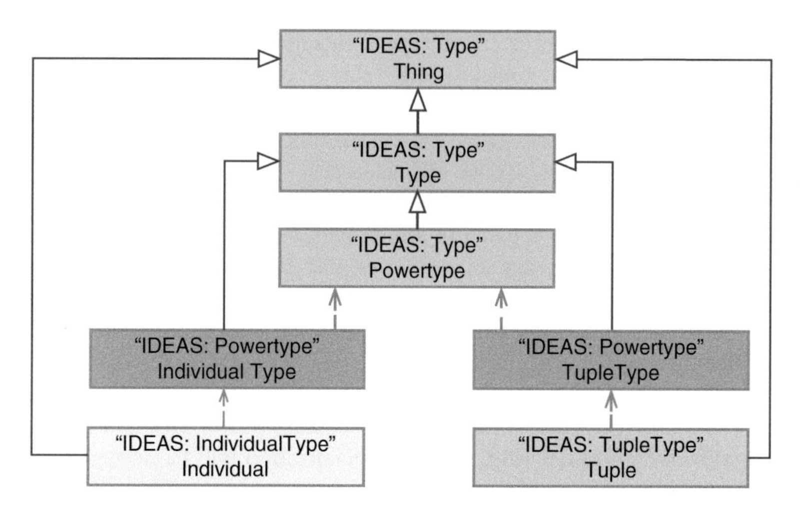

Fig. 1 The top concepts of the Framework

spatial equivalents. Thus, the top concept presented in Fig. 1 is a *Thing*—"a union of *Individual*, *Type*, and *Tuple*" [2] where an *Individual* is an object that has a spatio-temporal extent, i.e. it is tangible in our world [2], a *Tuple* is "a relationship between two or more things" [2]. Spatio-temporal extension of a tuple comprises the extensions of its placeholders. Specified set of individuals or tuples is a *Type* (class). According to the IDEAS notation, yellow rectangles depict classes of *Individual* objects, green rectangles—classes of relationships, and purple rectangles—classes of classes of *Individual* or *Tuple* objects in figures of this section. A type and its members are connected through *classification* relationships (presented by dotted brown lines in the graphical notation of the metamodel). A relationship between a type and one of its subtypes is the *specialization* relationship (presented by navy blue arrow in the graphical notation of the metamodel).

Specialization of the *Type* depends on the nature of its possible members. Thus, the *IndividualType* is the set of classes with members, which are *Individual* instances. The *TupleType* is the set of classes with members, which are *Tuple* instances. A class that has other classes as its members is called *Powertype* (a class of classes). Subclasses of the *Type* can relate to each other. Thus, the *IndividualType* and the *TupleType* are instances of the *Powertype* class.

Since the model reflects the concepts of the enterprise ontology [12], it includes several categories of objects: agents, roles, resources, processes, transactions, and facts. The concept *Agent* represents people or organizations that are authorized and can take the responsibility to participate in business processes. *Resource* concept expresses any kind of resources involved into processes, and *Product* as a subclass of *Resource* is a class of end products. *Process* is an object of *Individual* type whose extent is defined by its involvements [2]. And finally, world changes (facts) and agents' interactions about these facts (transactions) are represented in the metamodel in accordance with DEMO enterprise ontology [12] by *Transaction*, *CoordinationFact*, and *ProductionFact* concepts. Speaking in the language of BORO we can consider facts as 3D-objects, events that change the state of business processes. Besides, the data model includes classes of relationships between given concepts. These relationships allow linking objects with their parts and their states; they allow assigning process states and transactions into one sequence, relating processes and involved roles, agents, and resources.

Following BORO fundamental ontology, proposed conceptual model exploits four fundamental information patterns: whole-part, before-after, pre-condition, and participation. For assigning agents, resources, and processes with their parts and states the following concepts are used: *agentWholePart* and *agentWholeState*, *resourceWholePart* and *resourceWholeState* (Fig. 2), and *processWholePart* and *processWholeState*.

To link two successive states of an object BORO provides with *beforeAfter* pattern. The *beforeAfter* tuple class has two subclasses: *resourceStateBeforeAfter* and *processStateBeforeAfter* (Fig. 3). This specialization will be needed in further computer processing of the data model.

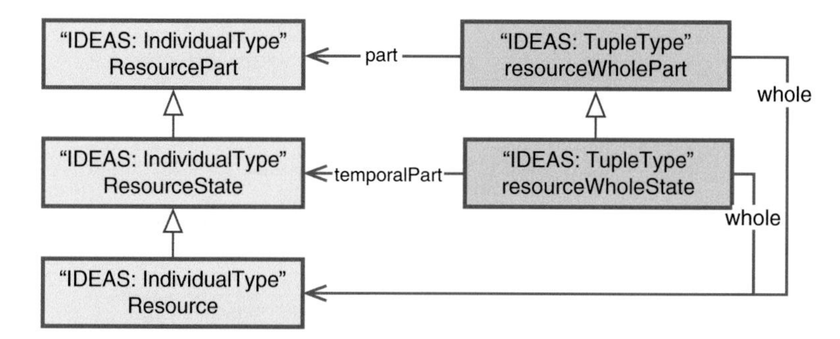

Fig. 2 *resourceWholePart* and *resourceWholeState* patterns

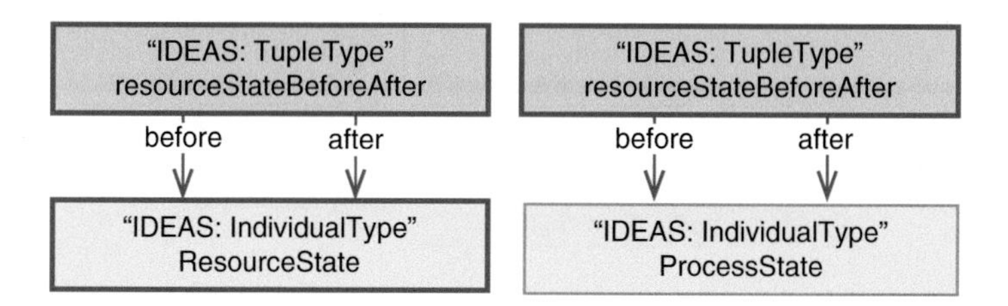

Fig. 3 *beforeAfter* pattern

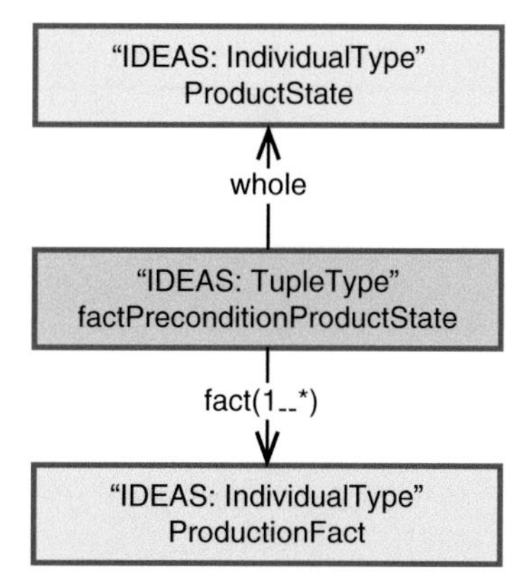

Fig. 4 *factPreconditionProductState* pattern

Links between events and their preconditions are important in the domain of supply networks. For that reason *factPreconditionProductState* concept was added to the metamodel (Fig. 4). In human language it means that a resource will not take a certain state unless a fact happens.

In the domain of supply networks a process is performed by some actors and involves some resources or products. This kind of relations is supported by means of active (*agentParticipation*) roles (Fig. 5) and passive (*passiveParticipation*) (Fig. 6). When *Agent* takes a certain role in a process, it becomes an *ActiveParticipationExtent* related to this process (Fig. 7).

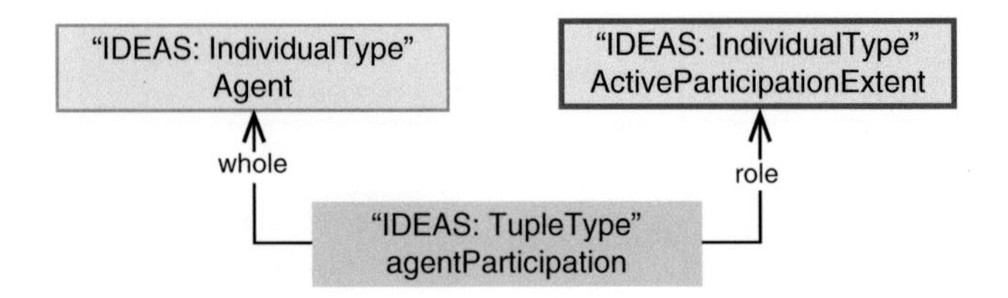

Fig. 5 *agentParticipation* pattern

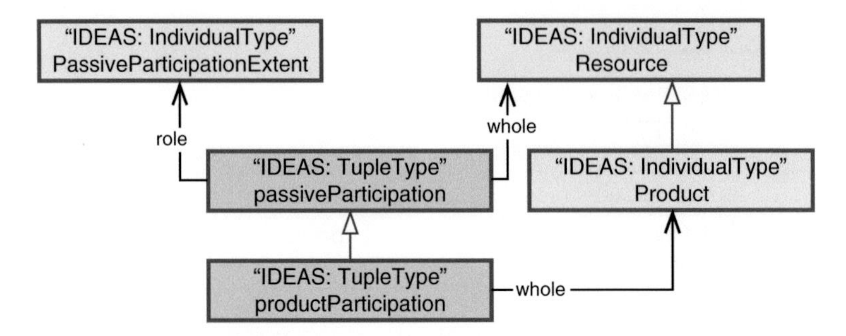

Fig. 6 *passiveParticipation* pattern

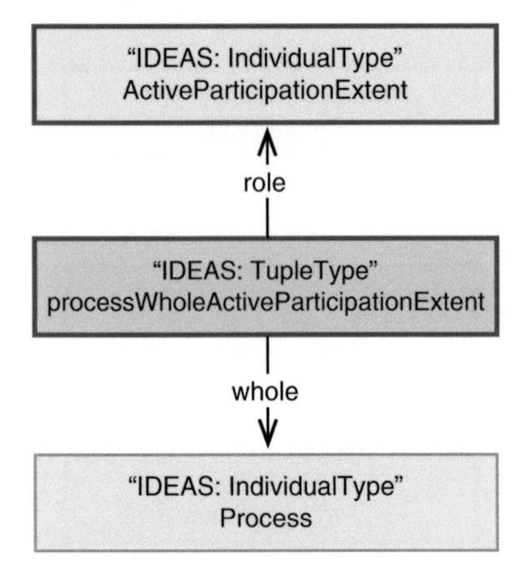

Fig. 7 *processWholeActiveParticipationExtent* pattern

But the notion "role" does not define responsible actors for certain processes. To do this we use *responsibleForProductionFact* class of relationships (Fig. 8). The responsibility for coordination facts is expressed by *responsibleForCoordination-Fact* pattern.

Following the enterprise ontology [12], we consider a process as a sequence of transactions. The model reflects essential parts of any transaction: coordination and production facts. A process is assigned with transactions through

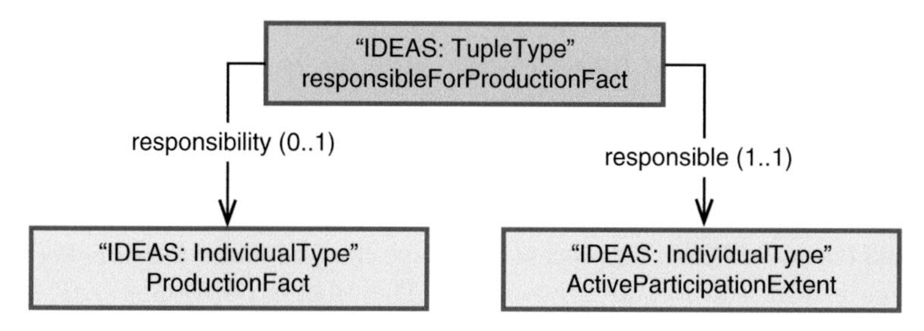

Fig. 8 *responsibleForProductionFact* pattern

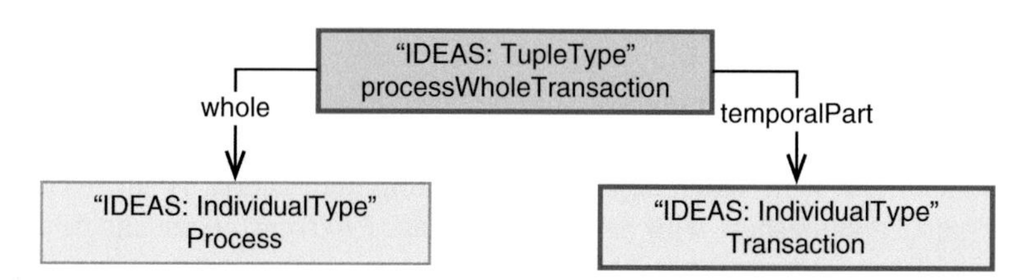

Fig. 9 *processWholeTransaction* pattern

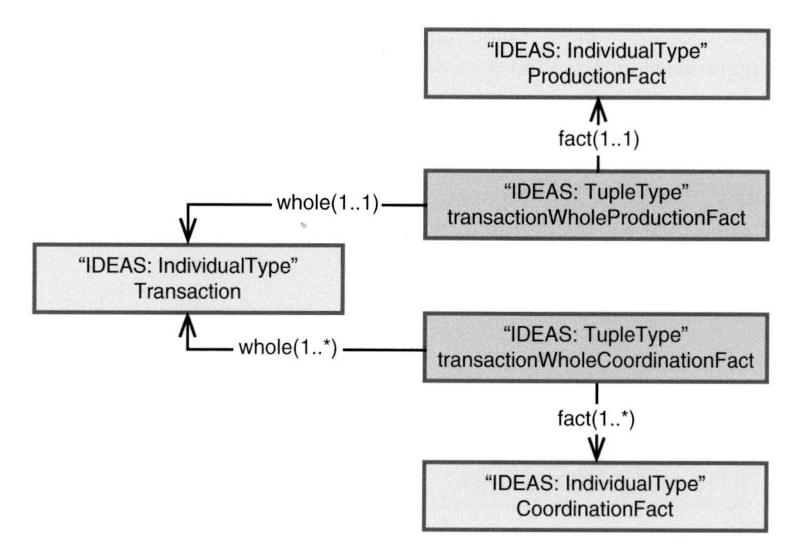

Fig. 10 *transactionWholeProductionFact* and *transactionWholeCoordinationFact* patterns

processWholeTransaction pattern, and a transaction is assigned with facts through *transactionWholeFact* pattern. Relevant diagrams are shown on Figs. 9 and 10.

The Application level is the specification and extension of the Framework in application to logistics domain. The concepts of this level express specific kinds of processes, transactions, resource states, and roles that may appear in supply networks according to the supply chain operation reference (SCOR) model [9].

To that moment Source Stocked Resource and Deliver Stocked Resource processes [9] were translated from the standard to the ontology. By analyzing the description of these processes given in the SCOR standard there were revealed ontological transactions, production, and coordination facts, actor roles, and relationships between all of these classes of objects. The process *SourceStockedResource* includes five transactions: *SchedulingDeliveries*, *Receiving*, *Verification*, *Transferring*, and *Payment*. Within the process a resource changes states: *ResourceReceived*, *ResourceVerified*, and *ResourceTransferred*. The process *DeliverStockedResource* includes nine transactions: *InventoryReservation*, *OrderConsolidation*, *LoadsBuilding*, *ShipmentsRouting*, *CarrierSelection*, *Receiving*, *Pick*, *Pack*, and *Payment*. Within this process a resource changes states: *ResourceReceived*, *ResourcePicked*, and *ResourcePacked*.

All the patterns of the Framework are reflected to the Application level. As an illustration of Application level patterns we shall consider Transferring transaction and its relations with other objects. Transferring transaction is involved into "whole-part" relationships with its production fact *PF-Transferring* and *SourceStockedResource* process (Fig. 11). Resource states as well as transactions form a sequence by means of *beforeAfter* patterns (Fig. 12). A sequence of transactions linked by preconditions (Fig. 13) forms the whole process (*SourceStockedResource* or *DeliverStockedResource*). Some transactions change resource states as a result; therefore, proposed metamodel provides an opportunity to trace resource lifecycle within a certain process. In addition, each production fact has responsible actors (*responsibleForProductionFact* pattern) (Fig. 14).

The power of the metamodel is also provided by cardinalities and additional logic rules of objects' associations. These parts of proposed metamodel were implemented in OWL and in the form of coded rules. Moreover, first order logic

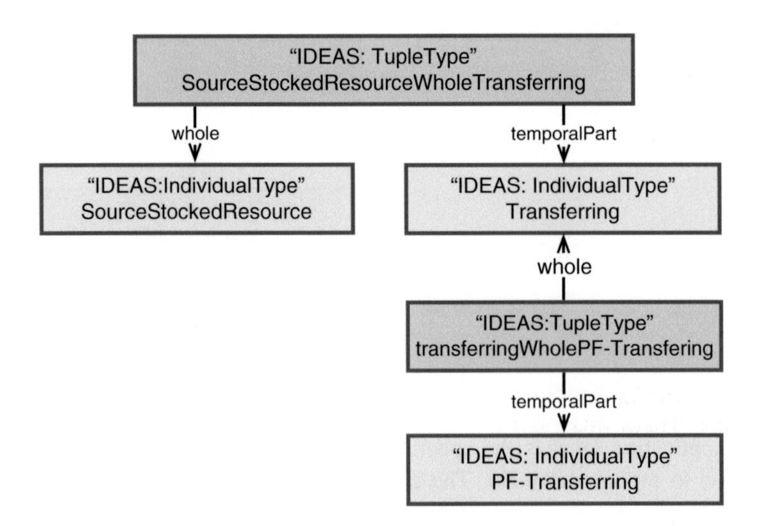

Fig. 11 *wholePart* pattern for *Transferring* transaction

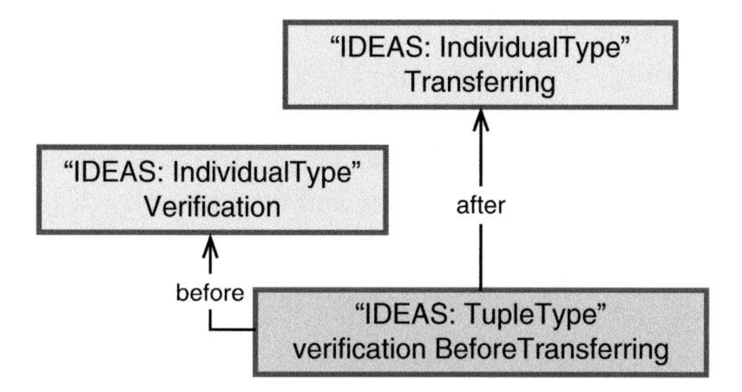

Fig. 12 *beforeAfter* pattern for *Verification* and *Transferring* transactions

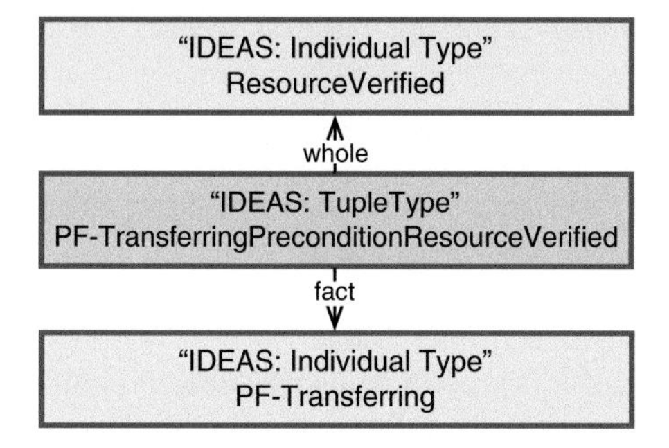

Fig. 13 *precondition* pattern for production fact of *Transferring* transaction

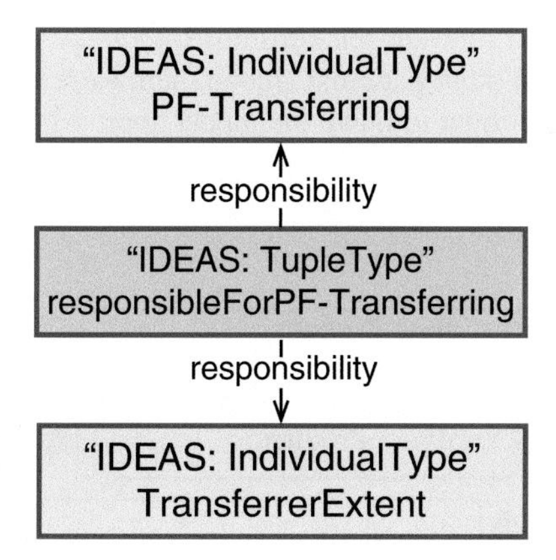

Fig. 14 *responsibleForProductionFact* pattern for production fact of *Transferring* transaction

rules were used for extracting new knowledge that is not explicitly presented in the metamodel. These rules add additional semantic power and allow building more complex queries to related data model (Sect. 5).

The Application level of created conceptual model contains 49 classes of individual objects and 71 associations (classes of relations). All created classes are the subclasses of the formal enterprise ontology [10].

At the stage of implementation (in accordance with SABiO's Development Process [13]) the ontology was presented in OWL format (dialect FULL). The OWL representation takes into account all the classes and classes of classes of objects and relations, cardinalities of object dependencies, restrictions on classes' properties, comments with the definitions of concepts, and the hierarchy of metamodel levels.

4 Case Study of Supply Network

To verify completeness and coherence of the developed metamodel we apply it to the use-case[8] that describes a possible business scenario close to supply processes of the SCOR model.

The Fig. 15 shows the sequence of subprocesses of the whole process of supply. The case includes four stages: Scheduling Product Deliveries, Verifying Product, Transferring Product, and Payment. The first three stages change product states: On Stock, Requested, Verified, and Transferred.

Apart from process stages and product states, the use-case gives information about involved agents and their roles (Table 1).

The stages of the business process being described resemble subprocesses of S1 process of the SCOR model except Receiving Product subprocess. The flexibility of the metamodel allows avoiding fixed sequence of process stages and, therefore, applying the model to various cases.

All objects above and relevant relationships (*beforeAfter, wholePart, precondition, responsibleFor*, and other patterns) among them were added to the data model and presented in OWL format to make the model appropriate for further computer processing. Ultimately, OWL representation of the Operational level contains 30 instances of individual objects and 37 instances of relationships.

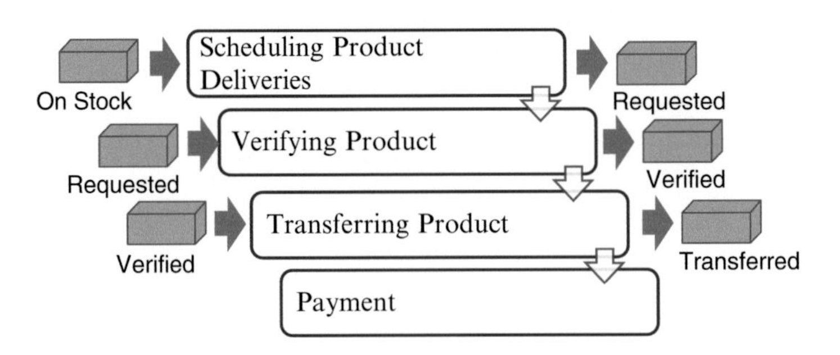

Fig. 15 The structure of the business process given in the use-case

[8]Taken from GS1 Logistics Interoperability Model (GS1 LIM), Version 1, Issue 1.0 (2007).

Table 1 Agents and their roles within the process of replenishment

Role	Agent
# Manufacturer	# Laithwaite's_ wine
# Material supplier	# Le_ Chai_ au_ Quai
# Manufacturer warehouse	# Laithwaite's_ wine_ Warehouse
# Material supplier warehouse	# Le_ Chai_ au_ Quai_ Warehouse
# Transport service provider	# HAROPA

5 Semantic Analysis with Proposed Logistics Ontology

The use-case gave us instances of created classes on the Operational level of the data model. According to the systematic approach for building ontologies (SABiO's Development Process [13]), we made a set of queries to the data model based on some competency questions. Competency questions are the most important questions from concerned parties' perspective. The amount and the complexity of queries the metamodel is able to answer shows its completeness and coherence, and its semantic power. Therefore, it is crucial to test the metamodel by means of competency questions.

The logic of queries building is the following. Each information pattern of the metamodel is an atomic part of more complex information item. In such a way, complex queries associated with complex information items comprise a set of atomic queries associated with basic information patterns (ref. to Sect. 3). This logic was implemented into software prototype designed on Apache Jena[9] platform by means of coded rules for extracting new knowledge. Table 2 contains some spread questions in the natural language. Questions 1–6 are more general while the rest questions are based on the basic Application level patterns. Evidently, the Table 2 can be extended because the set of possible queries is defined by the set of numerous combinations of information patterns. Notably, the cardinalities of associations and implemented logic rules of proposed metamodel form a "semantic glue" of complex information items.

The semantic power makes the metamodel a core for different metamodels integration. As a particular case we consider combined data metamodel of three projects of the EU 7th Framework program (FP7): EURIDICE,[10] iCargo, and[11] e-Freight.[12] These projects put joint efforts on development of the new generation of information systems in logistics, including the multilayered semantic metamodel [14].

The top level of the FP7 metamodel includes concepts: "Activity," "Event," "Role," "Actor," "StaticResource," and "MoveableResource". "Activity" denotes

[9]Jena Ontology API: http://jena.apache.org/documentation/ontology/.

[10]EURIDICE project: http://www.euridice-project.eu.

[11]i-Cargo project: http://i-cargo.eu.

[12]E-Freight project: http://www.efreightproject.eu.

Table 2 Implemented competency questions and their associated basic information patterns

No	What is/are	Information pattern
1	The role of an agent in a process?	• *processWholeActiveParticipationExtent* • *agentParticipation*
2	The resources/products involved into a process?	• *processWholePassiveParticipationExtent* • *passiveParticipation*
3	The sequence of process transactions?	• *processWholeTransactions* • set of *processStateBeforeAfter*
4	The result of a process?	• *processWholeTransaction* • set of *processStateBeforeAfter* • *transactionWholeProductionFact* • *factPreconditionProductState*
5	The detailed lifecycle of a product?	• set of *resourceStateBeforeAfter* • *productStatebeforeAfter*
6	The responsible agent for a transaction?	• *responsibleForProductionFact* • *agentParticipation*
7	The role of an agent in Source Stocked Resource process?	• *agentParticipation* • *sourceStockedResourceWhole-ActiveParticipationExtent*
8	The resources/products involved into Source Stocked Resource process?	• *sourceStockedResourceWhole-PassiveParticipationExtent* • *passiveParticipation*
9	The previous transaction for Transferring transaction?	• *verificationBeforeTransferring*
10	The result of Transferring transaction?	• *transferringWholePF-Transferring* • *PF-TransferringPrecondition-ResourceTransferred*
11	The previous state for Resource Transferred state?	• *resourceVerifiedBeforeTransferred*
12	The responsible agent for Verification transaction?	• set of *verificationWholePF-Verification* • *responsibleForPF-Verification*

an action and is connected with "Actor" via "hasProvider" and "hasConsumer" relations, "Event" means something that happens and has no time duration. "Actor" performing an "Activity" has "Role" that is expressed by means of "hasRole" relationship. And finally, "Activity" is associated with resources ("StaticResource" and "MoveableResource") by "hasStaticResource" and "hasMoveableResource" relations.

Following the meaning of enumerated concepts we integrated FP7 metamodel with our metamodel. "Activity" can be considered as a subclass of *Transaction* class, "Actor"—a subclass of *Agent*, and "Role"—a subclass of *ActiveParticipationExtent* (Fig. 16). "Event" has the meaning of *CoordinationFact*, "StaticResource" and "MoveableResource" are subclasses of *Resource* (Fig. 17). Relationships of FP7 metamodel can also be integrated as following. "hasProvider"

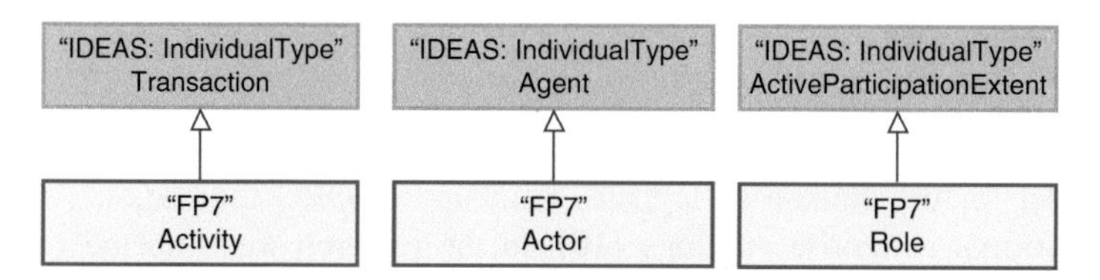

Fig. 16 "Activity," "Actor," and "Role" integrated with the metamodel

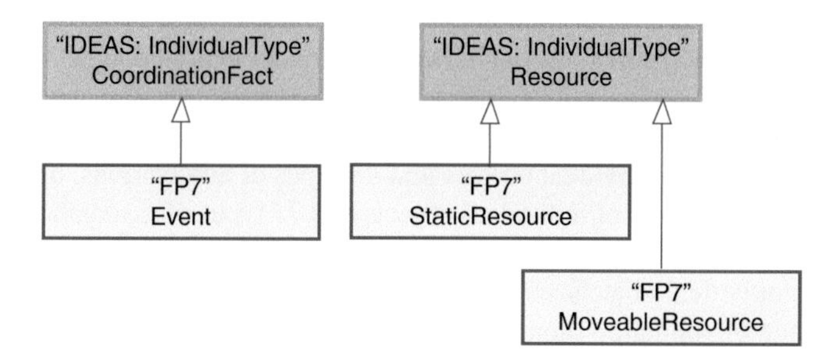

Fig. 17 "Event," "StaticResource," and "MoveableResource" integrated with the metamodel

and "hasConsumer" become subclasses of *responsibleForProductionFact* and *responsibleForCoordinationFact* classes, correspondingly, "hasRole" a subclass of *agentParticipation*, and "hasStaticResource" and "hasMoveableResource" are subclasses of *passiveParticipation* class. Thus, all top level classes and the main associations of the combined FP7 metamodel were easily integrated with the proposed metamodel. Together with other subclasses of the FP7 metamodel, obtained result of metamodels integration can be considered as a particular metamodel of the open supply network supported ontology-based data exchange.

Thus, the strong semantics of proposed logistics ontology facilitates integration of metamodels. Moreover, it helps to reveal semantic gaps in the integrated conceptual models. Following the constraints of proposed metamodel, we found that complex information patterns cannot be formed on the basis of initial FP7's data metamodel. For example, the information patterns answering questions: "What are results of an activity?," "How resources/products were transformed within a process/activity?," "What is the sequence of activities", and so on cannot be built. Moreover, some redundant associations were revealed in the initial metamodel during the integration process.

In such a way, the proposed metamodel can be a platform for further integration of different logistic standards and existing ontologies due to its semantic power. Thus, it forms the platform for semantic interoperability.

6 Conclusions

The article proposes the logistics ontology and its implementation—the data metamodel for the domain of logistics built upon BORO foundational ontology and the formal enterprise ontology [10]. In the research we extracted consistent conceptual model from the SCOR standard using the BORO methodology. On the other hand, we used the formal enterprise ontology to build semantically reach data metamodel describing the lifecycle of products and providing the transparency of supply networks. Proposed ontology-based metamodel is aimed at serving a basis for the integration of different ontologized logistic standards and particular data metamodels that are built upon the SCOR standard. The top level of the metamodel contains 38 classes of individual objects and 48 types of associations, the logistics-specific level—49 classes of individual objects and 71 types of associations. For the validation and testing of proposed metamodel we used simulated data, to wit: 30 instances of individual objects and 37 particular associations were specified.

At the stage of validation the metamodel was applied to ordinary use-case close to supply processes of the SCOR model. The metamodel was tested by means of the set of queries in order to prove its completeness and coherence and to evaluate its semantic power. For extracting new knowledge from the ontology-based data model the software prototype was implemented on Apache Jena platform. To show the suitability of proposed logistics ontology for the resolution of interoperability problem, the developed metamodel was integrated with another existing metamodel of the logistics domain. Developed metamodel helped to reveal semantic gaps in the integrated metamodel as well as allowed to perform the integration without any changes in the integrated metamodel.

In the nearest future both created logistics ontology and the data metamodel as its implementation will be reinforced by the shared core of leadership standards in logistics (GS1 and UN/CEFACT).

Acknowledgements This research is partially supported by LATNA Laboratory, NRU HSE, RF government grant 11.G34.31.0057, and by the CLASSE ("Les Corridors Logistiques: Application a la Vallee de la Seine et son Environnement") project of the Grand Research Network of in Upper Normandy (Grand Réseaux de Recherche de Haute-Normandie).

References

1. European Interoperability Framework for PAN-European eGovernment Services. European Communities (2004). ISBN 92-894-8389-X
2. Partridge, C.: Business Objects: Re-Engineering For Re-Use. Springer, Berlin (2000)
3. Partridge, C.: The role of ontology in semantic integration. In: Second International Workshop on Semantics of Enterprise Integration at OOPSLA 2002, Seattle (2002)
4. Bjeladinovic, S., Marjanovic, Z.: A comparison and integration of ontologies suitable for interoperability extension of SCOR model. In: Bogdanova, A.M., Gjorgjevikj, D. (eds.) ICT Innovations 2014. Advanced in Intelligent Systems and Computing, vol. 311, pp. 75–84. Springer, Cham (2015)

5. Zdravkovic, M., Panetto, H., Trajanovic, M., Aubry, A.: An approach for formalizing the supply chain operations. J. Enterp. Inf. Syst. **5**, 401–421 (2011)
6. Magdalenic, I., Vrdoljak, B., Schatten, M.: Mapping of core components based e-business standards into ontology. In: Sanchez-Alonso, S., Athanasiadis, I.N. (eds.) Metadata and Semantic Research. Communications in Computer and Information Science, vol. 108, pp. 95–106. Springer, Berlin (2010)
7. Al Asswad, M.M., Al-Debei, M.M., de Cesare, S., Lycett, M.: Conceptual modeling and the quality of ontologies: a comparison between an object-role modeling and the object paradigm. In: Proceedings of the 18th European Conference on Information Systems, pp. 1–12. Association for Information Systems (AIS) (2010)
8. International Defense Enterprise Architecture Specification Group: http://ideasgroup.org/
9. Supply Chain Operations Reference (SCOR) Model Version 9.0. Supply Chain Council (2008)
10. Poletaeva, T., Babkin, E., Abdulrab, H.: Ontological framework aimed to facilitate business transformations. In: Guizzardi, G., Pastor, O., Wand, Y., de Cesare, S., Gailly, F., Lycett, M., Partridge, C. (eds.) Proceedings of 1st Joint Workshop Onto.Com/ODISE on Ontologies in Conceptual Modeling and Information Systems Engineering, Rio de Janeiro, 22–25 Sept 2014, vol. 1301. CEUR-WS (2014)
11. Bailey, I., Partridge, C.: Working with extensional ontology for defence applications. In: Ontology in Intelligence Conference Proceedings, Fairfax, VA (2009)
12. Dietz, J.: Enterprise Ontology ? Theory and Methodology. Springer, Berlin/Heidelberg (2006)
13. Falbo, R. de A.: SABiO: systematic approach for building ontologies. In: Guizzardi, G., Pastor, O., Wand, Y., de Cesare, S., Gailly, F., Lycett, M., Partridge, C. (eds.) Proceedings of 1st Joint Workshop Onto.Com/ODISE on Ontologies in Conceptual Modeling and Information Systems Engineering, Rio de Janeiro, 22–25 Sept 2014, vol. 1301. CEUR-WS (2014)
14. EURIDICE Integrated Project ICT 2007-216271. Workpackage 12: Domain Knowledge Formalization, Version 2.0 (2009)

Key Borrowers Detected by the Intensities of Their Short-Range Interactions

Fuad Aleskerov, Irina Andrievskaya, and Elena Permjakova

Abstract The issue of systemic importance has received particular attention since the recent financial crisis when it came to the fore that an individual financial institution can disturb the whole financial system. Interconnectedness is considered as one of the key drivers of systemic importance. Several measures have been proposed in the literature in order to estimate the interconnectedness of financial institutions and systems. However, they do not fully take into consideration an important dimension of this characteristic: intensities of agents' interactions. This paper proposes a novel method that solves this issue. Our approach is based on the power index and centrality analysis and is employed to find a key borrower in a loan market. To illustrate the feasibility of our methodology we apply it at the European Union level and find key countries-borrowers.

Keywords Power index • Key borrower • Systemic importance • Interconnectedness • Centrality

1 Introduction

The detection of a pivotal agent has received particular attention within systemic risk analysis. Specifically, this refers to the identification of pivotal or, in other words, systemically important financial institutions or countries. According to [26],

F. Aleskerov
DeCAn Lab, Institute of Control Sciences of Russian Academy of Sciences, National Research University (NRU) Higher School of Economics, Moscow, Russia
e-mail: alesk@hse.ru

I. Andrievskaya (✉)
LIA Lab, National Research University (NRU) Higher School of Economics, Moscow, Russia
e-mail: irina.k.andrievskaia@gmail.com

E. Permjakova
National Research University (NRU) Higher School of Economics, Moscow, Russia
e-mail: hellen_113@mail.ru

V.A. Kalyagin et al. (eds.), *Models, Algorithms and Technologies for Network Analysis*, Springer Proceedings in Mathematics & Statistics 156,
DOI 10.1007/978-3-319-29608-1_18

"systemic importance is not a binary concept but can be measured along a continuum: some firms, sectors, markets, or countries can be judged to be 'more' or 'less' systemically important than others, using different criteria" (p. 3).

A common approach in the literature with respect to systemic importance identification is either to consider several indicators of systemic importance (cf. [10, 21, 26]) or to examine the contribution to and participation in systemic risk (cf. [1, 2, 14, 18, 20, 24, 30, 36, 38, 40]).

Among the possible indicators of systemic importance interconnectedness is considered as the most important (cf. [18, 22, 25, 31]) and is used along with the indicators of size, cross-jurisdictional activity, substitutability, and complexity for the global systemically important banks assessment [10].

The recent financial crisis has revealed how interconnected a financial system can be at the national and international level [6]. Connections can be direct, for example, links in the interbank market, and indirect, which arise from similar portfolio holdings.

In this paper we focus on direct links among financial institutions and systems. The calculation of the level of interconnectedness of a financial institution or system is not an easy task. The Basel committee on banking supervision (BCBS) proposes using an indicator-based approach [10]. However, an indicator-based approach does not take into consideration the links among financial institutions and, therefore, does not reveal the possible level of contagion.

An alternative approach, which is receiving an increasing attention in the literature, is to employ the network theory [22]. A network can be described as the system of nodes and links among them. The network approach has been applied to different areas in economics [34]. It has also received particular attention within systemic risk, financial stability, and contagion analysis. Within the financial system framework nodes are represented by financial institutions or systems and links can be described as the mutual exposures among them. Early theoretical study that employs the network theory within financial system stability context is presented by Allen and Gale [7], where the authors investigate the effect of interregional bank claims structure on financial contagion.[1]

For the purposes of the interconnectedness measurement, the researchers propose using such measures from the network theory as *centrality* indices. For example, [26, 39] identify several centrality measures that are most suitable for the calculation of interconnectedness within a global financial system framework. Both papers consider a network of banking sectors and use bilateral data where each banking center represents a node. The network measures used to identify the "central" banking centers are the following: degree, closeness, betweenness, intermediation (employed only in [39]), and prestige (or Bonacich centrality).[2]

[1]For detailed literature review with respect to theoretical and empirical application of network approach to financial systems see [6, 32].

[2]Details on these measures are provided in Sect. 3.

These centrality measures are widely used for the interbank market investigation (see [3] for the Norwegian money market, [12] for the US Federal Funds market, [16] for the Brazilian interbank market, [28] for the Italian interbank market, and others) and global banking network analysis [33].

However, these centrality measures, when used separately, do not fully incorporate the intensity of agents' interactions. Therefore, we contribute to the systemic risk literature elaborating on agents' interactions intensities as an important dimension of interconnectedness. We also add to the network-in-finance literature proposing a centrality measure adjusted in order to take into account a particular nature of a financial system. Our methodology can be used for detecting the most interconnected financial institutions or countries from the pivotal borrower perspective.

The methodology we develop is built on the power analysis which is widely used to determine voting power of agents employing power indices. Examples of these indices are represented by Banzhaf index [9], Coleman index [19], Johnston index [29], Penrose index [35], and Shapley–Shubik index [37].

Within systemic risk analysis there are few papers that employ power indices to find systemically important financial institutions. Specifically, Tarashev et al. [38] were the first to use the Shapley value to estimate the contribution of a bank to systemic risk. The approach was extended in [20] taking into account interbank linkages. The paper [23] employs the Shapley value in a different way. The index is used to find a pivotal bank which is a bank that makes the system losses larger than a predefined threshold.

However, these methodologies focus on systemic importance as a whole rather than estimating the level of bank interconnectedness. We, in turn, concentrate directly on the interconnectedness estimation proposing to use a preference-based power index—we call it *key borrower index*—worked out in [4] and adjusted in our paper in order to take into account the nature of a financial system. This index is of a broad application. It can be used for interbank market analysis in order to detect the most interconnected financial institution. Alternatively it can be applied at the international level in order to find the most interconnected financial centers.

To the best of our knowledge, this is the first time this index is applied to loan market analysis. We show how to use our approach in a hypothetical case and compare it with the centrality measures identified in [26, 39] as the most suitable for the network analysis of financial institutions or systems. Thereafter we apply our methodology at the international level. The detection of jurisdictions with systemically important financial sectors based on their size and interconnectedness is considered by IMF as an important task for a country's financial stability assessment [27]. We demonstrate in a real-data case the applicability of our index in order to detect the most interconnected financial systems.

Quite a few policy initiatives have already been worked out in order to reduce systemic risk and systemic importance of financial institutions and, thus, enhance financial stability (cf. Basel III). The approach we propose will hopefully help in improving macroprudential regulation.

The paper is organized as follows. In the next section, we explain the proposed methodology. In Sect. 3 we compare our approach with the centrality measures commonly used in the literature. Empirical application of our methodology is presented in Sect. 4. Section 5 concludes.

2 Methodology

We modify the preference-based power index proposed in [4, 5] in order to capture the nature of the loan market and call this modified index *key borrower index*. Our aim is to find *the most pivotal borrower*. This index is calculated for each borrower in order to determine the magnitude of his/her pivotal role in the market. The pivotal role of a borrower reflects the level of his/her *interconnectedness* in the system.

We first consider a "one pure lender, many borrowers/lenders" case and then adjust our approach for the general case "many lenders/borrowers."

Consider a group of borrowers. The *distressing* group is interpreted as a group whose default can lead to the default of a creditor (while the creditor is able to cover his/her losses from the distress of the non-distressing group). Thus, the group is distressing if the total amount of its members' borrowings is greater than or equal to a predefined quota. The quota should be determined on a case-by-case basis according to the characteristics of the financial institution and system.

The *pivotal (or key) borrower* can be defined as a borrower who makes the amount of losses critical for the lender and whose exclusion from the distressing group makes it non-distressing. The *most pivotal borrower* will be the one who becomes pivotal in more distressing groups than other pivotal borrowers.

When we consider a group of borrowers there is no sense in allowing for the intensity of connection among borrowers. The crucial thing here is the intensity of connection between a borrower and a lender.

First, we define the intensity of connection $f(i, w)$ taking into account not only the *direct* links but also the *indirect* ones between a lender and a borrower:

$$f(i, w_l) = \frac{P_{li} + P'_{li}}{\sum_j P_{li}}, \tag{1}$$

where w_l is a distressing group with respect to a lender l and a borrower i is pivotal,[3] P_{li} are total direct loans taken by a borrower i from a creditor l, and P'_{li} are total *indirect* loans taken by a borrower i from a creditor l.

[3]If the borrower is not pivotal in a distressing group, no intensities of connections are calculated for this borrower. The intensities are assumed to be zero for this borrower in this distressing group. The zero value is then taken into consideration for the calculation of the key borrower index for this borrower.

It is important to emphasize that the transferred funds represent a uniform product (money). This indirect link reflects the additional channel through which the borrower can potentially affect the lender and, therefore, cannot be ignored. However, the indirect impact of a direct borrower is limited by the loans provided by the lender to the intermediate borrower.

At this stage we consider only indirect connection of the first order between a lender and a borrower. The first order indirect connection includes loans taken by a borrower from a lender through only one intermediate borrower. For example, suppose a lender L provides the amount of P_{L1} 1000 USD to the borrower B_1, P_{L2} to the borrower B_2, and P_{L3} to the borrower B_3 (see Fig. 1). The borrower B_1, in turn, provides P_{12} to the borrower B_2, P_{14} to the borrower B_4, and P_{13} to the borrower B_3. Moreover, the borrower B_2 takes P_{42} from the borrower B_4 and provides P_{23} to B_3.

The intensity of *indirect* connection between L and B_i through B_j—p_{ji}—is calculated as

$$p_{ji} = \begin{cases} \frac{P_{ji}}{\sum_k P_{Lk}}, & P_{ji} < P_{Lj},\ i \neq j,\ k = 1, \ldots, 3 \textit{ borrowers of a Lender L} \\ \frac{P_{ji}}{\sum_k P_{Lk}}, & P_{ji} > P_{Lj},\ i \neq j,\ k = 1, \ldots, 3 \textit{ borrowers of a Lender L} \end{cases} \tag{2}$$

The direct connection between L and B_i is calculated as

$$p_{Li} = \frac{P_{Li}}{\sum_k P_{Lk}} \tag{3}$$

and the total intensity of connection between L and B_i would be

$$f_{li} = \sum_j p_{ji} + p_{Li} \tag{4}$$

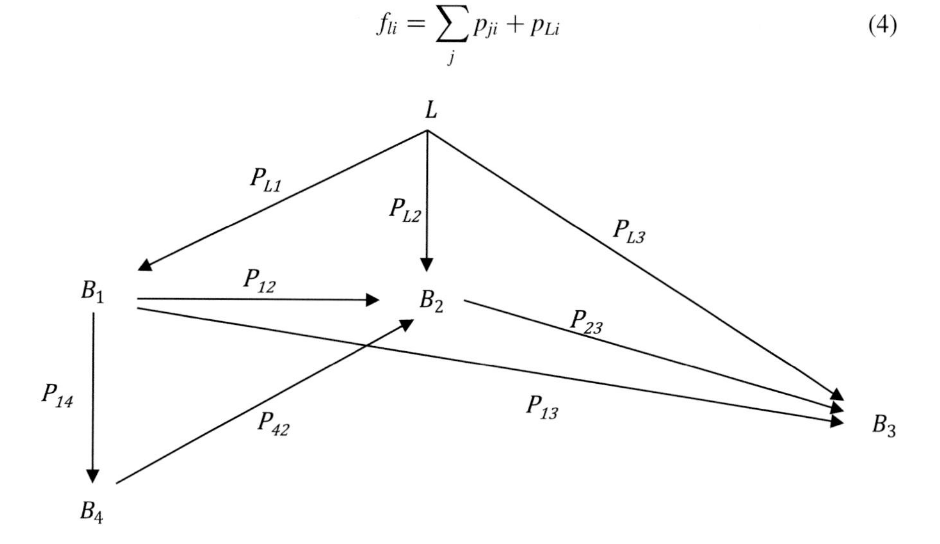

Fig. 1 Lender's connections with its borrowers

For a borrower B_2, for example, the direct connection with L includes a loan P_{L2} and the indirect connection of the first order occurs through B_1 and includes a loan P_{12} (if $P_{12} < P_{L1}$) or P_{L1} (if $P_{12} > P_{L1}$). The indirect connection of a higher order (of the second order, in this case) will be through B_4. It is important to emphasize that by including the indirect links we assume the worse-case scenario when the default of an ultimate borrower disturbs not only the lender but also the intermediate borrower. This assumption can be relaxed or adjusted (for example, by determining the relative importance of an indirect link for the stability of an intermediate borrower) depending on a loan market and on a particular problem under study.

The intensity of connection for a borrower i is calculated separately for each distressing group and then aggregated over all possible distressing groups as

$$\mathscr{X}_i = \sum_{w_l} f(i, w_l)/N_w, \tag{5}$$

where N_w—the number of borrowers in the group. The final form of the index for each borrower is calculated according to the following formula:

$$\alpha_i = \frac{\mathscr{X}_i}{\sum_j \mathscr{X}_j} \tag{6}$$

In order to demonstrate the full estimations we consider a numerical example presented in Fig. 2 below.

Based on the loan amounts presented in Fig. 2, the intensities matrix $P = (p_{ji})$, where on the diagonal line we have a direct connection between L and a borrower B_i, has the form depicted in Table 1. At the intersection of the row B_1 and column

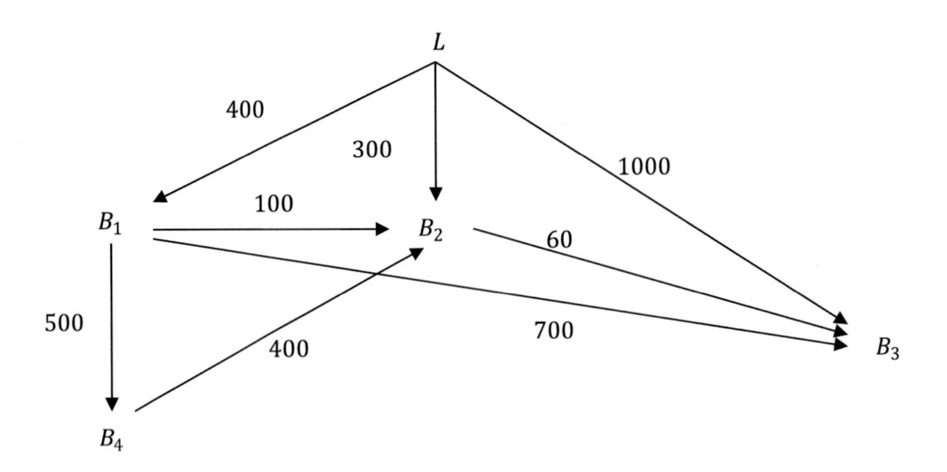

Fig. 2 Numerical example 1

Table 1 Intensities matrix

	B_1	B_2	B_3	B_4
B_1	0.24	0.00	0.00	0.00
B_2	0.06	0.18	0.00	0.00
B_3	0.24	0.04	0.59	0.00
B_4	0.24	0.00	0.00	0.00

Table 2 Distressing coalitions and pivotal borrowers

Distressing coalitions, w	Pivotal borrowers, i	$f(i, w)$
B_1B_2	B_1B_2	$f(B_1, w) = 0.12$ $f(B_2, w) = 0.12$
B_3	B_3	$f(B_3, w) = 0.59$
B_1B_3	B_3	$f(B_3, w) = 0.42$
B_2B_3	B_3	$f(B_3, w) = 0.32$
$B_1B_2B_3$		
$B_1B_2B_4$	B_1B_2	$f(B_1, w) = 0.08$ $f(B_2, w) = 0.08$
B_3B_4	B_3	$f(B_3, w) = 0.30$
$B_1B_3B_4$	B_3	$f(B_3, w) = 0.28$
$B_2B_3B_4$	B_3	$f(B_3, w) = 0.21$
$B_1B_2B_3B_4$		

Table 3 Key borrower index, q = 25%

	$\mathscr{X}_i = \sum_w f(i, w)$	Index, α_i
B_1	0.20	0.08
B_2	0.20	0.08
B_3	2.12	0.84
B_4	0.00	0.00
$\sum \mathscr{X}_i$	2.52	1.00

B_1 we have 0.24. This number is received by dividing 400 by (400 + 300 + 1000). At the intersection of the row B_2 and the column B_1 we have 0.06 which is 100 (the amount borrowed by B_2 from B_1) divided by (400 + 300 + 1000).

Let us consider the quota q being equal to 25 %. Then distressing groups and pivotal borrowers are the following (in Table 2 we omit the subscript l as we have only one lender):

Therefore, the values of the key borrower index are the following (see Table 3):

Table 3 shows that the most pivotal borrower turns out to be B_3. This borrower indeed is the most interconnected one: she borrows relatively large amounts of money from quite a few agents. B_4, in turn, is the least pivotal/interconnected in this case, which is in line with expectations as she takes money only from B_1.

For the "many lenders/borrowers" case we need to aggregate the index over all lenders. Lenders are different in terms of their lending abilities. Therefore, the aggregation of the index for each borrower over all lenders should take into account the size of each lender's total loans. The importance of a borrower for a large lender (who provides a large amount of loans in total) is not the same as its importance for a small one (who provides a small amount of loans in total). Therefore, the final value of the key borrower index has the following form:

$$\alpha_i = \sum_l \left(\frac{\mathcal{X}_i}{\sum_j \mathcal{X}_j} \times \frac{Total\_loans_l}{\sum_l Total\_loans_l} \right), \tag{7}$$

where $Total\_loans_l$ is the total amount of loans provided to all borrowers by a lender l.

When we consider a banking system, several additional points should be taken into considerations. First, loans have to be put in perspective relative to the equity position of a bank. A bank is in default if its losses are higher than or equal to its capital. Therefore, the group is "distressing" if the total amount of its members' borrowings is greater than or equal to the amount of capital of a lender. Thus, the magnitude of the quota can be specific for each bank (calculated as the ratio of total bank capital over total bank loans). Alternatively, we can use the recommendations of the Basel Committee with respect to the large exposure limits. In particular, according to [11, p. 4], "The sum of all the exposure values of a bank to a single counterparty or to a group of connected counterparties must not be higher than 25 % of the bank's available eligible capital base at all times." Within our framework, the distressing groups include, among others, those that consist of only one borrower. This situation fits the Basel Committee's recommendation with respect to the large exposure limit. When a group consists of more than one borrower, it can be considered as a group of connected counterparties in the sense that these borrowers can be prone to the common risk factors. Therefore, the 25 % limit can also be applied to this situation.

Second, the approach can be extended taking into account the bankruptcy theory according to which not all creditors lose their money when a borrower defaults (cf. [8, 13, 17]. For example, the indirect links should be included into calculations again taking into account the equity position of an intermediate bank (a bank through which the indirect connection takes place). Specifically, if the value of the indirect link is higher than or equal to the amount of intermediate bank's capital, then this link should be considered for the estimation as described above. Otherwise, this link is not included into calculations.

As a result, we obtain the value of the key borrower index for each borrower in a general case "many lenders, many borrowers." The borrower with the largest value of the index is considered as the most pivotal/interconnected one in the market.

3 Key Borrower Index vs. Centrality Measures

The aim of this section is to discuss the key borrower index in comparison with centrality measures used in the literature for the interconnectedness assessment. Centrality measures include degree centrality, indicators of closeness and betweenness, intermediation measure, and the measure of prestige. They are described in [39] and we follow the same logic.

We consider a hypothetical example with a "many lenders–many borrowers" system of interconnections. Graphical representation is shown in Fig. 3. B_i can be a lender and a borrower at the same time. Arrows indicate the direction of the money flows. For example, B_1 borrows 10,000 USD from B_3, while B_3 borrows 5000 USD from B_1.

In-degree and *out-degree* centrality measures of B_i are calculated as the number of its ingoing and outgoing links. The level of *closeness* of B_i is the inverse from the average distance from B_i to all other participants in the system. *Betweenness* of B_i is based on the probability that the path between B_j and B_k lies through B_i. The level of *betweenness* of B_i is the sum of all these probabilities over all pairs B_j and B_k. While *intermediation* is an extension of betweenness, it takes into account the value of the links and is calculated as the total probability (over all pairs B_j and B_k) that

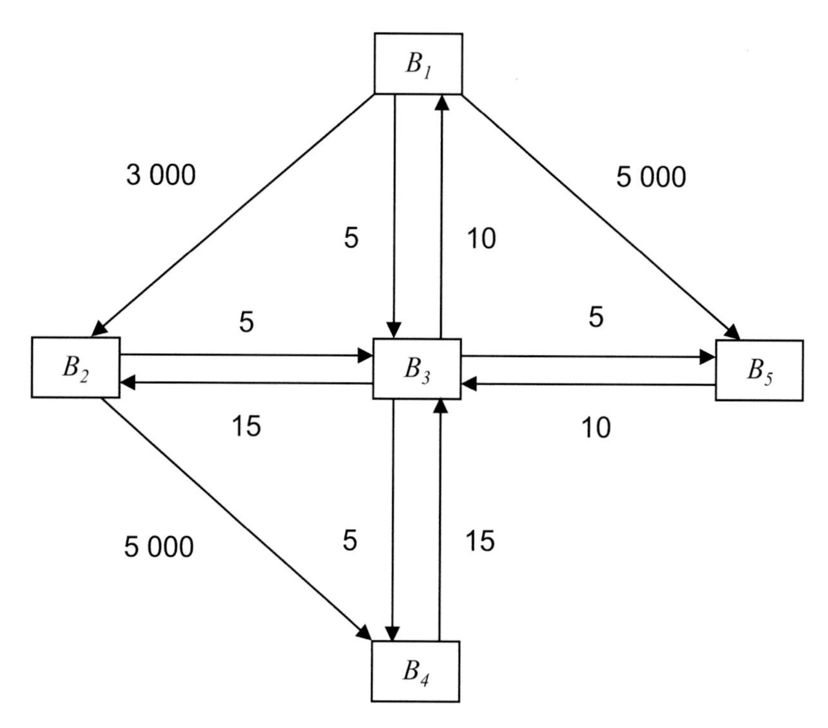

Fig. 3 Numerical example 2

Table 4 Centrality measures and the key borrower index

Indices	Borrower				
	B_1	B_2	B_3	B_4	B_5
Key borrower index[a] $\alpha_{i,\,q=0.25}$	0.00	0.23	0.00	0.38	0.38
Closeness (from a borrower perspective)	0.16	0.19	0.28	0.19	0.19
Intermediation	0.12	0.12	0.65	0.06	0.06
Betweenness	0.11	0.19	0.62	0.05	0.03
Bonacich centrality (eigenvector centrality)	0.10	0.19	0.35	0.24	0.11
In-degree	1.00	2.00	4.00	2.00	2.00
Out-degree	3.00	2.00	4.00	1.00	1.00

[a]For higher levels of the quota, B_5 gets higher scores while B_2 gets lower scores

monetary unit sent from B_j to B_k will be delivered through B_i. *Bonacich centrality*,[4] or *prestige*,[5] takes into account the scores of all the counterparties. It is received by solving the linear system $v = R'v$, where v is the vector of the importance scores of B_i and R is the matrix of relationships (in the rows we have R_{ij}, the money borrowed by B_j from B_i). The solution is represented by the eigenvector corresponding to the eigenvalue 1. This vector contains the prestige levels of each B_i.

Based on the structure of connections and the loan amounts presented in Fig. 3, the key borrower index and other centrality measures described above are the following (see Table 4 below):

As can been seen from Fig. 3, there is one participant—B_3—that takes and provides money to all other participants. The values of its borrowing and lending, though, are relatively small. Therefore, the default of B_3 would unlikely seriously affect any lender or cause contagion in the market. However, the centrality measures described above show the opposite: B_3 turns out to be the central player in the market.

This occurs due to some weaknesses of these measures. For example, the degree centrality, closeness, and betweenness measures lack information about the value of the links (borrowed/lent amounts of money). It is taken into account by the intermediation metrics. However, estimation of the intermediation level of B_i does not control for the differences in the magnitude of the links. For example, B_i transferring small amounts of money and B_j transferring large amounts of money can receive the same intermediating rating (or even B_j can receive a lower rating, which is visible from our example: B_3 is rated higher than B_2 even though B_2

[4]The logic behind this measure is well explained in [15].

[5]We estimate only the outgoing Bonacich centrality in order to rank the borrowers (not the lenders).

transfers relatively large amounts). The key borrower index solves this issue as it is calculated separately with respect to each lender and then it is aggregated using as weights the size of each lender (the share of the total amount given by a lender in total amount of money provided by all lenders). Moreover, our approach takes into account the degree centrality. For example, if a borrower takes small amounts of money but from a lot of lenders (thus, its degree centrality is high), its key borrower index will be higher compared to a borrower who takes small amounts of money from fewer lenders.

The Bonacich measure seems to be the most appropriate one. Nevertheless, it does not work well when a node is connected to other nodes with zero importance scores. In such a situation the node receives the zero score as well. For example, when B_i borrows from lenders who do not borrow from anyone else, the centrality score of B_i will be zero. In order to solve this issue, Bonacich and Lloyd [15] propose incorporating in the Bonacich centrality measure the exogenous importance. However, this exogenous importance should also be calculated somehow.

Our key borrower index, in turn, does not have this weakness. At the same time it takes into account the importance of counterparties, as the final index is computed as the sum of the indices with respect to each lender weighted by the share of this lender in total system lending.

As a result, using this simple hypothetical example, we demonstrate the applicability of the key borrower index. It incorporates the desired features of the existing centrality measures and at the same time lacks their deficiencies described above.

4 Empirical Application: Country Assessment

Now we apply our methodology to detect the most pivotal/interconnected countries-borrowers. As mentioned in Sect. 1, the identification of systemically important financial centers is necessary for a country's financial stability assessment [27]. And interconnectedness is considered by IMF as an important determinant (along with size) of systemic importance.

We estimate the level of country interconnectedness using the key borrower index. Data are taken from the Bank of International Settlements (BIS) database.[6] The sample covers countries from the European Union that have bank foreign claims and obligations (28 countries) for the 1Q2013. The claims include outstanding loans and holdings of securities as well as derivative contracts and contingent facilities. The quota is assumed to be 25 %.[7]

[6]Consolidated banking statistics, table 9D "Foreign claims by nationality of reporting banks, ultimate risk basis" http://www.bis.org/statistics/consstats.htm.

[7]For an individual banking system analysis the quota should be set specifically for each bank as 25 % of its capital following the recommendations of the Basel Committee [10]. At the country level we likewise consider the same level of the quota. The results are similar also for higher levels of the quota.

Table 5 Countries' ranking based on the key borrower index

Country	Key borrower index α_i	Country	Key borrower index α_i
United Kingdom	0.232	Romania	0.008
Germany	0.188	Portugal	0.008
France	0.102	Slovakia	0.005
Italy	0.084	Croatia	0.005
Netherlands	0.069	Hungary	0.005
Belgium	0.054	Sweden	0.003
Spain	0.046	Bulgaria	0.002
Luxembourg	0.036	Cyprus	0.002
Denmark	0.034	Slovenia	0.001
Austria	0.025	Estonia	0.000
Finland	0.024	Greece	0.000
Ireland	0.024	Latvia	0.000
Poland	0.022	Lithuania	0.000
Czech Republic	0.021	Malta	0.000

Table 5 below presents the rating of the countries-borrowers according to our key borrower index. The highest ratings belong to the United Kingdom, Germany, and France, which corresponds to the findings in [39]. These jurisdictions indeed have a broad coverage of the global financial system [26]. They have high sovereign ratings. Therefore, investments in their securities are an attractive tool for many investors, which make these countries large borrowers. This result reflects the fact that banks prefer to invest in countries with high sovereign ratings even at the cost of lower profits. Moreover, we can conclude that financial sectors of these countries could be of systemic importance and should be more closely monitored.

Malta, Lithuania, Latvia, Greece, and Estonia, in turn, have the lowest values of the key borrower index. This is also not surprising and reflects the current weak economic situation in these jurisdictions. They are relatively low-rated and are not able to attract a lot of investments and loans.

5 Conclusions

Methods and techniques for systemic risk and systemic importance analysis have been substantially advanced in terms of their complexity starting from the recent financial crisis. One of the crucial determinants of systemic importance is considered to be the interconnectedness of financial institutions or countries. However, such an important dimension of interconnectedness as the intensity of agent interaction has not been fully considered.

Our paper fills this gap in the literature. We developed a methodology in order to detect key borrowers in a loan market, the failure of whom could potentially endanger the stability of this market. The approach was based on the power index analysis and network theory. Under this approach we proposed a new type of a centrality measure—key borrower index—to take into consideration the nature of a financial system and the intensity of connections among market participants.

We carried out estimations for a hypothetical example and at the level of the European Union and demonstrated the feasibility of the proposed methodology. The empirical results based on our methodology are in line with the conclusions made by IMF. However, we avoided the calculation of a broad range of interconnectedness indicators used by IMF and at the same time we took into consideration the intensities of agent interactions. Our approach has wide-ranging applications and adds to the on-going discussion of systemic importance and macroprudential regulation.

Acknowledgements We are grateful to Professor Hasan Ersel from the Sabancı University (Turkey) for his valuable comments and to Vyacheslav Yakuba from the Institute of Control Sciences of the Russian Academy of Sciences for providing us the software. The work was supported by the International Laboratory of Decision Choice and Analysis (HSE) and by the International Laboratory for Institutional Analysis of Economic Reforms (HSE).

References

1. Acharya, V.V., Pedersen, L.H., Philippon, T., Richardson, M.: Measuring systemic risk. Federal Reserve Bank of Cleveland Working Paper No. 02 (2010)
2. Adrian, T., Brunnermeier, M.: CoVaR. Federal Reserve Bank of New York Staff Report No. 348 (2010)
3. Akram, Q.F., Christophersen, C.: Interbank overnight interest rates—gains from systemic importance. Norges Bank Working Paper No. 11 (2010)
4. Aleskerov, F.T.: Power indices taking into account agents' preferences. In: Simeone, B., Pukelsheim, F. (eds.) Mathematics and Democracy, pp. 1–18. Springer, Berlin (2006)
5. Aleskerov, F.T.: Power indices taking into account the agents' preferences to coalesce. Dokl. Math. **75**(3), 463–466 (2007)
6. Allen, F., Babus, A.: Networks in Finance. In: Kleindorfer, P., Wind, J. (eds.) Network-Based Strategies and Competencies, pp. 367–382. Wharton School Publishing, Upper Saddle River (2009)
7. Allen, F., Gale, D.: Financial contagion. J. Polit. Econ. **108**(1), 1–33 (2000)
8. Aumann, R.J., Mashler, M.: Game theoretic analysis of a bankruptcy problem from the Talmud. J. Econ. Theory **36**(2), 195–213 (1985)
9. Banzhaf, J.: Weighted voting doesn't work: a mathematical analysis. Rutgers Law Rev. **19**, 317–343 (1965)
10. Basel Committee on Banking Supervision (BCBS): Global systemically important banks: updated assessment methodology and the higher loss absorbency requirement. Consultative Document (2013)
11. Basel Committee on Banking Supervision (BCBS): Supervisory framework for measuring and controlling large exposures. Consultative Document (2014)
12. Bech, M.L., Atalay, E.: The topology of the federal funds market. Federal Reserve Bank of New York Staff Report No. 354 (2008)
13. Bergantiños, G., Lorenzo, L., Lorenzo-Freire, S.: A characterization of the proportional rule in multi-issue allocation situations. Oper. Res. Lett. **38**(1), 17–19 (2010)
14. Bluhm, M., Krahnen, J.P.: Systemic risk in an interconnected banking system with endogenous asset markets. J. Financ. Stab. **13**, 75–94 (2014)

15. Bonacich, P., Lloyd, P.: Eigenvector-like measures of centrality for asymmetric relations. Soc. Netw. **23**(3), 191–201 (2001)
16. Cajueiro, D.O., Tabak, B.M.: The role of banks in the Brazilian interbank market: does bank type matter? Phys. A: Stat. Mech. Appl. **387**(27), 6825–6836 (2008)
17. Calleja, P., Borm, P., Hendrickx, R.: Multi-issue allocation situations. Eur. J. Oper. Res. **164**(3), 730–747 (2005)
18. Chan-Lau, J.A.: The global financial crisis and its impact on the Chilean banking system. IMF Working Paper No. 108 (2010)
19. Coleman, J.S.: Control of collectivities and the power of a collectivity to act. In: Lieberman, B. (ed.) Social Choice. Gordon and Breach, London (1971)
20. Drehmann, M., Tarashev, N.: Measuring the systemic importance of interconnected banks'. Bank for International Settlements Working Paper No. 342 (2011)
21. ECB: December 2006 Financial Stability Review (2006) https://www.ecb.europa.eu/pub/pdf/other/financialstabilityreview200612en.pdf?eeb9342332f4cd3127e55b523c51c9ff
22. ECB: June 2010 Financial Stability Review (2010) https://www.ecb.europa.eu/pub/pdf/other/financialstabilityreview201006en.pdf?19bfebe20cdfd99b872c14e1abdb52e0
23. Garratt, R., Webber, L., Willison, M.: Using Shapley's asymmetric power index to measure banks' contributions to systemic risk. Bank of England Working Paper No. 468 (2012)
24. Huang, X., Zhou, H., Zhu, H.: Assessing the systemic risk of a heterogeneous portfolio of banks during the recent financial crisis. J. Financ. Stab. **8**(3), 193–205 (2012)
25. IMF/BIS/FSB: Guidance to Assess the Systemic Importance of Financial Institutions. Markets and Instruments: Initial Considerations Report to the G-20 Finance Ministers and Central Bank Governors (2009)
26. IMF: Integrating Stability Assessments Under the Financial Sector Assessment Program into Article IV Surveillance: Background Material (2010)
27. IMF: Integrating Stability Assessments Under the Financial Sector Assessment Program into Article IV Surveillance (2010)
28. Iori, G., de Masi, G., Precup, O.V., Gabbi, G., Caldarelli, G.: A network analysis of the Italian overnight money market. J. Econ. Dyn. Control. **32**(1), 259–278 (2008)
29. Johnston, R.J.: On the measurement of power: some reactions to Laver. Environ. Plan. **10**(8), 907–914 (1978)
30. Lehar, A.: Measuring systemic risk: a risk management approach. J. Bank. Financ. **29**(10), 2577–2603 (2005)
31. Leon, C., Murcia, A.: Systemic importance index for financial institutions: a principal component analysis approach. Banco de la República Working papers (Central Bank of Colombia) No. 741 (2012)
32. Martinez-Jaramillo, S., Alexandrova-Kabadjova, B., Bravo-Benitez, B., Solrzano-Margain, J.P.: An empirical study of the Mexican banking system's network and its implications for systemic risk. J. Econ. Dyn. Control. **40**, 242–265 (2014)
33. Minoiu, C., Reyes, J.A.: A network analysis of global banking: 1978–2010. J. Financ. Stab. **9**(2), 168–184 (2013)
34. Nagurney, A.: Financial and economic networks: an overview. In: Nagurney, A. (ed.) Innovations in Financial and Economic Networks, pp. 1–26. Edward Elgar Publishing, Cheltenham (2003)
35. Penrose, L.S.: The elementary statistics of majority voting. J. R. Stat. Soc. **109**, 53–57 (1946)
36. Segoviano, M.A., Goodhart, C.: Banking stability measures. IMF Working Paper No. 4 (2009)
37. Shapley, L.S., Shubik, M.: A method for evaluating the distribution of power in a committee system. Am. Polit. Sci. Rev. **48**, 787–792 (1954)
38. Tarashev, N., Borio, C., Tsatsaronis, K.: Attributing systemic risk to individual institutions. BIS Working Papers No. 308 (2010)
39. von Peter, G.: International banking centres: a network perspective. BIS Q. Rev. 33–45 (2007)
40. Zhou, C.: Are banks too big to fail? Measuring systemic importance of financial institutions. Int. J. Cent. Bank. **6**(34), 205–250 (2010)

Langmuir Solitons in Plasma with Inhomogeneous Electron Temperature and Space Stimulated Scattering on Damping Ion-Sound Waves

N.V. Aseeva, E.M. Gromov, T.V. Nasedkina, I.V. Onosova, and V.V. Tyutin

Abstract Dynamics of Langmuir solitons is considered in the framework of the extended nonlinear Schrödinger equation (NLSE), including a *pseudo-stimulated-Raman-scattering* (pseudo-SRS) term, caused by stimulated scattering on damping ion-sound waves. Also included are spatially decreasing second-order dispersion (SOD) and increasing self-phase modulation (SPM), caused by spatially decreasing electron temperature of plasma. It is shown that the wavenumber downshift of solitons, caused by the pseudo-SRS, may be compensated by an upshift provided by the decreasing SOD and increasing SPM coefficients. An analytical solution for solitons is obtained in an approximate form. Analytical and numerical results agree well.

Keywords Extended nonlinear Schrödinger equation • Soliton solution • Stimulated scattering • Damping low-frequency waves • Inhomogeneity • Second-order dispersion • Self-phase modulation analytical solutions • Numerical simulation

1 Introduction

The great interest to the dynamics of solitons is motivated by their ability to travel long distances keeping the shape and transferring the energy and information with no little loss. Soliton solutions are relevant to nonlinear models in various areas of physics which deal with the propagation of intensive wave fields in dispersive media: optical beams and pulses in fibers and spatial waveguides, electromagnetic waves in plasma, surface waves on deep water, etc. [1–7]. Recently, solitons have also drawn a great deal of interest in plasmonics [8–10].

N.V. Aseeva (✉) • E.M. Gromov • T.V. Nasedkina • I.V. Onosova • V.V. Tyutin
National Research University Higher School of Economics, 25/12 Bolshaja Pecherskaja str., Nizhny Novgorod, 603155, Russia
e-mail: naseeva@hse.ru; egromov@hse.ru; tvnasedkina@edu.hse.ru; vonosova@hse.ru; vtyutin@hse.ru

V.A. Kalyagin et al. (eds.), *Models, Algorithms and Technologies for Network Analysis*, Springer Proceedings in Mathematics & Statistics 156,
DOI 10.1007/978-3-319-29608-1_19

Dynamics of long high-frequency (HF) wave packets is described by the second-order nonlinear dispersive wave theory. The fundamental equation of the theory is the nonlinear Schrödinger equation (NLSE) [11, 12], which includes the second-order dispersion (SOD) and self-phase modulation (SPM). Soliton solutions in this case arise as a result of the balance between the dispersive stretch and nonlinear compression of wave packets.

The dynamics of narrow HF wave packets is described by the third-order nonlinear dispersive wave theory [1], which takes into account the nonlinear dispersion (self-steeping) [13], stimulated Raman scattering (SRS) [13–15], and third-order dispersion (TOD). The basic equation of the theory is the extended NLSE [15–19]. Soliton solutions in the framework of the extended NLSE with TOD and nonlinear dispersion were found in [20–27]. In [28, 29], stationary kink waves were found as solutions of the extended NLSE with SRS and nonlinear dispersion terms. This solution exists as the equilibrium between the nonlinear dispersion and SRS. For localized nonlinear wave packets (solitons), the SRS gives rise to the downshift of the soliton spectrum [13–15] and eventually to destabilization of the solitons. The use of the balance between the SRS and the slope of the gain for the stabilization of solitons in long telecom links was proposed in [30]. The compensation of the SRS by emission of linear radiation fields from the soliton's core was considered in [31]. In addition, the compensation of the SRS in inhomogeneous media was considered in several situations, *viz*., periodic SOD [32, 33], shifting zero-dispersion point of SOD [34], and dispersion-decreasing fibers [35].

Intensive short pulses of HF electromagnetic or Langmuir waves in plasmas, as well as HF surface waves in deep stratified water, suffer effective induced damping due to scattering on low-frequency (LF) waves, which, in turn, are subject to the action of viscosity. These LF modes are ion-sound waves in the plasma, and internal waves in the stratified fluid. The first model for the damping induced by the interaction with the LF waves was proposed in [36–38]. This model gives rise to an extended NLSE with the spatial-domain counterpart of the SRS term, that was called a *pseudo-SRS* one. The equation was derived from the system of the Zakharov's type equations [39] for the coupled HF and LF waves in plasmas. The pseudo-SRS leads to the self-wavenumber downshift, similar to what is well known in the temporal domain [1, 13–15, 40] and, eventually, to destabilization of the solitons. The model elaborated in [36–38] also included smooth spatial variation of the SOD, accounted for by a spatially decreasing SOD coefficient, which leads to an increase of the soliton's wavenumber, making it possible to compensate the effect of the pseudo-SRS on the soliton by the spatially inhomogeneous SOD. However, the consideration was carried out in disregard of spatial inhomogeneity of the SPM.

In this work the dynamics of Langmuir wave packets in plasma with spatial inhomogeneous electron temperature and nonlinear interaction with damping ion-sound waves is considered. Spatial inhomogeneous electron temperature in original Zakharov-type system of equations for the Langmuir and ion-sound fields leads to spatial variation of SOD Langmuir waves and ion-sound velocity. In the third-order approximation of the dispersion-wave theory, the original Zakharov-type system is reduced to an extended NLSE with pseudo-SRS and with spatially inhomogeneous coefficients of SOD and SPM terms. Spatially decreasing electron temperature gives a spatially increasing coefficient of SPM term and spatially increasing coefficient of SOD leads to an increase of the soliton's wavenumber. The balance between the pseudo-SRS and the increasing SPM and decreasing SOD, arising as result of decreasing electron temperature, leads to stabilization of the soliton's wavenumber spectrum. An analytical soliton solution is found in an approximate form.

2 The Basic Equation and Integrals Relations

We consider the evolution of slowly varying envelope $U(\xi, t)$ of the intense Langmuir wave field $U(\xi, t)exp(-i\omega_p t)$ in the plasma with space inhomogeneous electron temperature $T(\xi)$, taking into account the nonlinear interaction with damping ion-sound variations of the plasma density $n(\xi, t)$. The unidirectional propagation of the fields along coordinate ξ with taking into account HF losses of ion-sound waves is described by the system of the Zakharov's type [39]:

$$2i\frac{\partial U}{\partial t} + 3\omega_p \frac{\partial}{\partial \xi}[R^2(\xi)\frac{\partial U}{\partial \xi}] - \omega_p \frac{n}{N_0} U = 0, \qquad (1)$$

$$\frac{\partial n}{\partial t} + C_s(\xi)\frac{\partial n}{\partial \xi} - v\frac{\partial^2 n}{\partial \xi^2} = -\frac{1}{16\pi m_i}\frac{\partial(|U|^2)}{\partial \xi}, \qquad (2)$$

where $\omega_p = \sqrt{4\pi e^2 N_0/m_e}$ is the plasma frequency, $R(\xi) = \sqrt{k_B T(\xi)/(4\pi e^2 N_0)}$ is the Debye radius, k_B is the Boltzmann constant, N_0 is the unperturbed plasma density, v is the coefficient of the HF losses of ion-sound wave perturbations n, and $C_s(\xi) = \sqrt{T(\xi)/m_i}$ is the inhomogeneous velocity of ion-sound waves. Assuming space inhomogeneous electron temperature $T(\xi)$ heterogeneity is larger than the packet envelope scale heterogeneity, $D_T >> D_{|U|}$, in the third-order approximation of the theory (for Langmuir wave packets $v/\Delta << C_s$, where Δ is extension of the wave packet) the nonlinear response of media has term with small spatial antisymmetric nonlocality

$$n = -\frac{|U|^2}{16\pi m_i C_s(\xi)} - \frac{v}{16\pi m_i C_s(0)}\frac{\partial |U|^2}{\partial \xi}.$$

Antisymmetric term causes with HF losses for ion-sound waves. In this case the system (1)–(2) is reduced to model extension NLSE with small antisymmetric nonlinear response

$$2i\frac{\partial U}{\partial t} + \frac{\partial}{\partial \xi}[q(\xi)\frac{\partial U}{\partial \xi}] + 2\alpha(\xi)U|U|^2 + \mu U\frac{\partial |U|^2}{\partial \xi} = 0, \qquad (3)$$

where $q(\xi) = 3\omega_p R^2(\xi) \sim T(\xi)$ is the SOD coefficient, $\alpha(\xi) = 1/(32\pi m_i C_s(\xi)) \sim 1/\sqrt{T(\xi)}$ is the SPM coefficient, and $\mu = v/(16\pi m_i C_s(0))$. The last term in Eq. (3) describes the stimulated scattering of Langmuir fields on ion-sound space domains with HF losses. It is the spatial counterpart of SRS term in the temporal domain.

Equation (3) with zero boundary conditions at infinity, $U|_{\xi\to\pm\infty} \to 0$, gives rise to the following integral relations for field moments, which will be used below:

$$\frac{dW}{dt} \equiv \frac{d}{dt}\int_{-\infty}^{+\infty} |U|^2 d\xi = 0, \tag{4}$$

$$2\frac{d}{dt}\int_{-\infty}^{+\infty} K|U|^2 d\xi =$$

$$-\mu\int_{-\infty}^{+\infty}\left[\frac{\partial\left(|U|^2\right)}{\partial\xi}\right]^2 d\xi - \int_{-\infty}^{+\infty}\frac{dq}{d\xi}\left|\frac{dU}{d\xi}\right|^2 d\xi + \int_{-\infty}^{+\infty}\frac{d\alpha}{d\xi}|U|^4 d\xi, \tag{5}$$

$$W\frac{d\bar{\xi}}{dt} \equiv \frac{d}{dt}\int_{-\infty}^{+\infty}\xi|U|^2 d\xi = \int_{-\infty}^{+\infty} qK|U|^2 d\xi, \tag{6}$$

where the complex field is represented as $U \equiv |U|exp(i\phi)$, and $K \equiv \partial\phi/\partial\xi$ is the local wavenumber.

3 Analytical Results

For analytical consideration of the wave-packet dynamics, we assume that the scale of the inhomogeneity of electron temperature is much larger than the inhomogeneity scale of the wave-packet envelope, $D_T >> D_{|U|}$, and solution of the system (4)–(6) will be found in adiabatic sech-like approximation

$$U(\xi,t) = A(t)sech\left[\frac{\xi-\bar{\xi}(t)}{\Delta(t)}\right]exp\left[ik(t)\xi - i\int\Omega(t)dt\right], \tag{7}$$

Solution (7) contains two free parameters: center-of-mass coordinate $\bar{\xi}(t)$ and additional wavenumber $k(t)$. Using (7) the system of Eqs. (4)–(6) can be cast in the form of evolution equations for the following parameters of the wave packets:

$$2\frac{dk}{dt} = -\mu\frac{8}{15}\frac{\alpha_0}{q_0}\frac{A_0^4}{\theta^{9/2}(\bar{\xi})} - \frac{2}{3}A_0^2\alpha_0\frac{\theta'(\bar{\xi})}{\theta^3(\bar{\xi})} - q_0\theta'(\bar{\xi})k^2, \tag{8}$$

$$\frac{d\bar{\xi}}{dt} = kq_0\theta(\bar{\xi}), \tag{9}$$

where $\theta(\bar{\xi}) = T(\bar{\xi})/T_0$ is the normalized electron temperature, $T_0 = T(0)$, $\theta'(\bar{\xi}) = (d\theta/d\xi)_{\bar{\xi}}$, $q_0 = q(0)$, $\alpha_0 = \alpha(0)$, and $A_0 = A(0)$. The equilibrium state of the system (8)–(9) is achieved under the conditions

$$k = 0, 4\mu A_0^2 = -5\theta'(0)q_0. \tag{10}$$

The equilibrium is achieved for spatially decreasing electron temperature $\theta'(0) < 0$. With the help of substitutions, $\eta \equiv \bar{\xi}/D_T$, $\tau \equiv tq_0/(\sqrt{3}D_T\Delta_0)$, and $y \equiv \sqrt{3}k\Delta_0$, Eq. (9) is transformed into a scaled form,

$$2\frac{dy}{d\tau} = -\frac{2\lambda}{\theta^{9/2}} - (\frac{2}{\theta^3} + y^2)\frac{d\theta}{d\eta}, \frac{d\eta}{d\tau} = y\theta. \tag{11}$$

where $\lambda \equiv 4\mu A_0^2 D_T/5q_0$. The first integral of Eq. (11) is

$$\theta y^2 + 2\lambda \int_0^\eta \frac{d\eta'}{\theta^{9/2}} - \frac{1}{\theta^2} = const. \tag{12}$$

In order to analyze the dynamics of the wave packet with non-equilibrium parameters, we assume that electron temperature θ decreasing as linear $\theta = 1 - \eta$. The first integral in this case is:

$$(1-\eta)y^2 - y_0^2 + \frac{4\lambda}{7}\left(\frac{1}{(1-\eta)^{7/2}} - 1\right) - \frac{1}{(1-\eta)^2} + 1 = 0. \tag{13}$$

In Fig. 1, this relation between variables y and η is plotted for initial condition $y_0 = 0$ with different λ.

4 Numerical Results

We consider numerically the initial-value problem for the dynamics of the wave packet, $U(\xi, t = 0) = \operatorname{sech}\xi$, in the framework of Eq. (3) for $q(\xi) \equiv 1 - \xi/10$ and $\alpha(\xi) \equiv 1/\sqrt{1-\xi/10}$ and different values of μ. The analytically predicted equilibrium value of pseudo-SRS for the initial pulse is $\mu_* = 1/8$. In direct simulations, the initial pulse for $\mu = 1/8$ is transformed into a stationary localized distribution (the solid curve in Fig. 2) with zero wavenumber.

Variation of parameter μ leads to variation of the soliton's parameters (wavenumber and amplitude). Spatial distributions of $|U|$ at different moments of time for $\mu = 1.2/8 \equiv 1.2\mu_*$ are shown in Fig. 3.

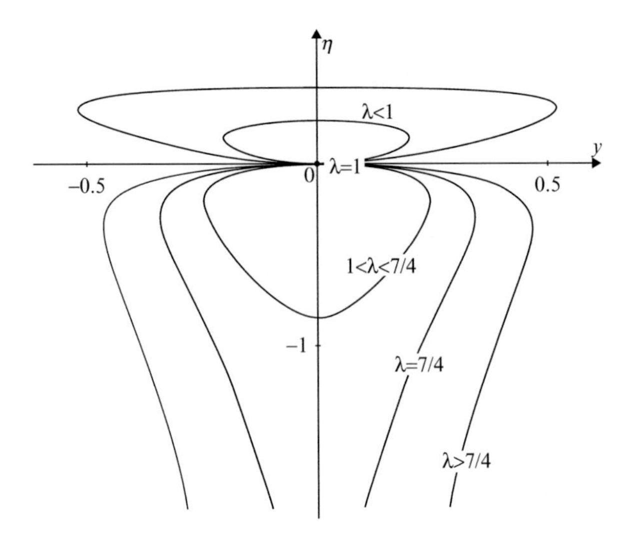

Fig. 1 First integral (13) in the plane (y, η) for initial condition $y_0 = 0$ with different values of constant λ

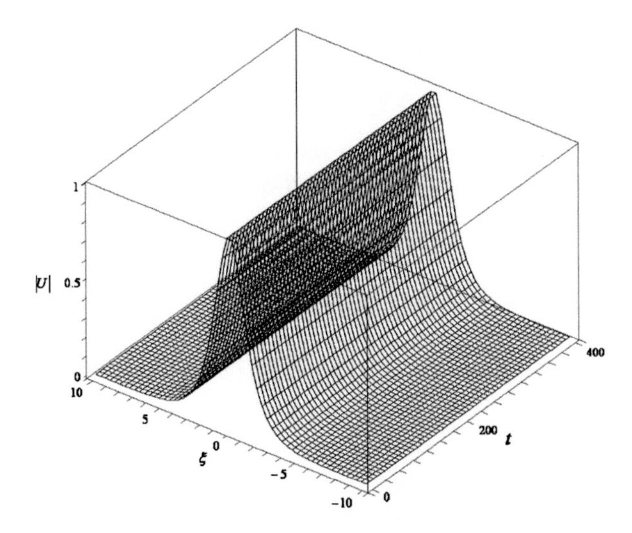

Fig. 2 The numerical result of the soliton's envelope, $|U(\xi)|$, for $\mu = 1/8 \equiv \mu_*$

In Fig. 4 numerical (solid curves) and analytical (dotted curves) results for the local wavenumber at the point of the maximum of the wave-packet's envelope as functions of t for $q(\xi) \equiv 1 - \xi/10$ and $\alpha(\xi) \equiv 1/\sqrt{1 - \xi/10}$ and different values of μ are shown.

For $\mu = 1/8 \equiv \mu_*$, the local wavenumber at the soliton's center does not vary. It corresponds to the exact equilibrium between the pseudo-SRS and the inhomogeneous SOD and SPM. For $\mu \neq 1/8$, the analytical and numerical results are seen to agree well. A similar picture is observed at other values of the parameters.

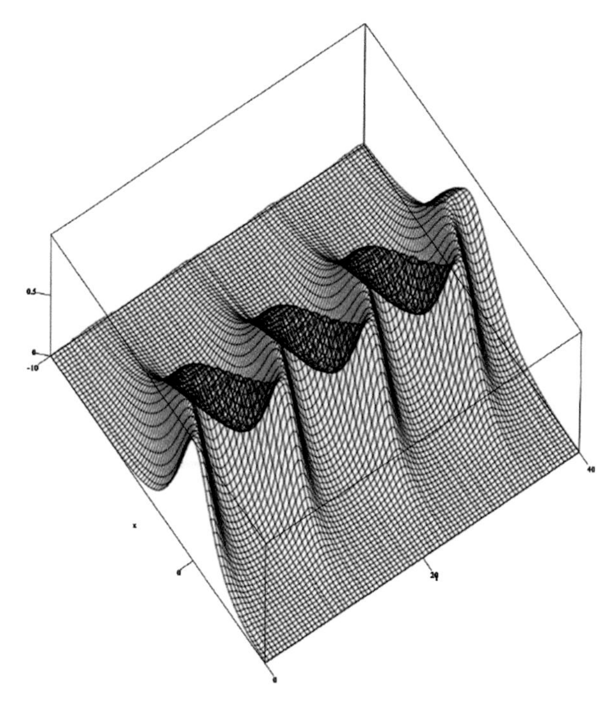

Fig. 3 Numerical results for space–time distributions of $|U(\xi, t)|$ for $\mu = 1.2/8 \equiv 1.2\mu_*$

5 Conclusion

In the work the soliton dynamics is studied in the framework of the extended inhomogeneous NLSE, which includes the pseudo-SRS effect (diffusive interaction with LF waves) and inhomogeneous SOD and SPM, corresponding to linear decreasing electron temperature profile. The results were obtained by means of numerical and analytical methods. The solitons exist due to the balance between the self-wavenumber downshift, caused by the pseudo-SRS, and the upshift induced by the inhomogeneous SOD and SPM. The analytical soliton solution is found in an approximate analytical form.

In this work the soliton dynamics was considered in the model neglecting the nonlinear dispersion and third-order linear dispersion. The compensation of the pseudo-SRS in inhomogeneous media which takes into regard these higher-order terms will be considered elsewhere.

Acknowledgements This work was supported by the Russian Foundation for Basic Research projects No 15-02-01919. The article was prepared within the framework of the Academic Fund Program at the National Research University Higher School of Economics (HSE) in 2015 (grant No 15-09-0240).

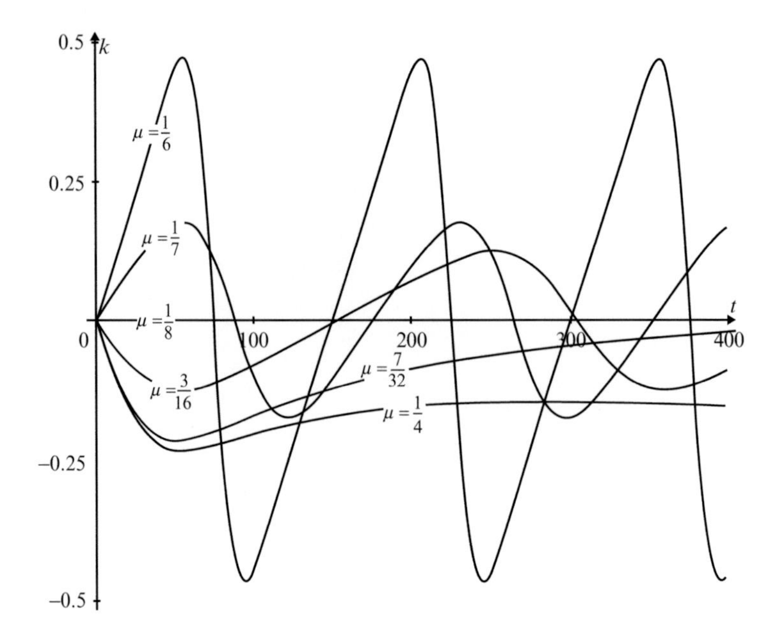

Fig. 4 Numerical (*solid curves*) and analytical (*dotted curves*) results for the local wavenumber at the point of the maximum of the wave-packet envelope versus t for $q(\xi) \equiv 1 - \xi/10$ and $\alpha(\xi) = 1/\sqrt{1 - \xi/10}$ and different μ

References

1. Infeld, E., Rowlands, G.: Nonlinear Waves, Solitons, and Chaos. Cambridge University Press, Cambridge (2000)
2. Agrawal, G.P.: Nonlinear Fiber Optic. Academic Press, San Diego (2001)
3. Yang, Y.: Solitons in Field Theory and Nonlinear Analysis. Springer, New York (2001)
4. Kivshar, Y.S., Agraval, G.P.: Optical Solitons: From Fibers to Photonic Crystals. Academic, San Diego (2003)
5. Dickey, L.A.: Soliton Equation and Hamiltonian Systems. World Scientific, New York (2005)
6. Malomed, B.A.: Soliton Management in Periodic Systems. Springer, New York (2006)
7. Dauxois, T., Peyrard, M.: Physics of Solitons. Cambridge University Press, Cambridge (2006)
8. Sich, M., Krizhanovskii, D.N., Skolnick, M.S., Gorbach, A.V., Hartley, R., Skryabin, D.V., Cerda-Méndez, E.A., Biermann, K., Hey, R., Santos, P.V.: Nat. Photonics **6**, 50–55 (2012)
9. Kauranen, M., Zayats, A.V.: Nat. Photonics **6**, 737–748 (2012)
10. Cerda-Ménde, E.A., Sarkar, D., Krizhanovskii, D.N., Gavrilov, S.S., Biermann, K., Skolnick, M.S., Santos, P.V.: Phys. Rev. Lett. **111**, 146401 (2013)
11. Zakharov, V.E., Shabat, A.B.: Sov. Phys. - JETP **34**, 62–69 (1972)
12. Hasegawa, A., Tappert, F.: Appl. Phys. Lett. **23**, 142–144 (1973)
13. Mitschke, F.M., Mollenauer, L.F.: Opt. Lett. **11**, 659–661 (1986)
14. Gordon, J.P.: Opt. Lett. **11**, 662–664 (1986)
15. Kodama, Y.: J. Stat. Phys. **39**, 597–614 (1985)
16. Kodama, Y., Hasegawa, A.: IEEE J. Quantum Electron. **23**, 510–524 (1987)
17. Zaspel, C.E.: Phys. Rev. Lett. **82**, 723–726 (1999)
18. Hong, B., Lu, D.: Int. J. Nonlinear Sci. **7**, 360–367 (2009)
19. Karpman, V.I.: Eur. Phys. J. B **39**, 341–350 (2004)

20. Gromov, E.M., Talanov, V.I.: J. Exp. Theor. Phys. **83** 73–79 (1996)
21. Gromov, E.M., Talanov, V.I.: Chaos **10**, 551–558 (2000)
22. Gromov, E.M., Piskunova, L.V., Tyutin, V.V.: Phys. Lett. A **256**, 153–158 (1999)
23. Obregon, M.A., Stepanyants, Yu.A.: Phys. Lett. A **249**, 315–323 (1998)
24. Scalora, M., Syrchin, M., Akozbek, N., Poliakov, E.Y., D'Aguanno, G., Mattiucci, N., Bloemer, M.J., Zheltikov, A.M.: Phys. Rev. Lett. **95**, 013902 (2005)
25. Wen, S.C., Wang, Y., Su, W., Xiang, Y., Fu, X., Fan, D.: Phys. Rev. E **73**, 036617 (2006)
26. Marklund, M., Shukla, P.K., Stenflo, L.: Phys. Rev. E **73**, 037601 (2006)
27. Tsitsas, N.L., Rompotis, N., Kourakis, I., Kevrekidis, P.G., Frantzeskakis, D.J.: Phys. Rev. E **79**, 037601 (2009)
28. Kivshar, Y.S.: Phys. Rev. A **42**, 1757–1761 (1990)
29. Kivshar, Y.S., Malomed, B.A.: Opt. Lett. **18**, 485–487 (1993)
30. Malomed, B.A., Tasgal, R.S.: J. Opt. Soc. Am. B **15**, 162–170 (1998)
31. Biancalama, F., Skrybin, D.V., Yulin, A.V.: Phys. Rev. E **70**, 011615 (2004)
32. Essiambre, R.-J., Agraval, G.P.: J. Opt. Soc. Am. B **14**, 314–322 (1997)
33. Essiambre, R.-J., Agrawal, G.P.: J. Opt. Soc. Am. B **14**, 323–330 (1997)
34. Andrianov, A., Muraviev, S., Kim, A., Sysoliatin, A.: Laser Phys. **17**, 1296–1302 (2007)
35. Chernikov, S., Dianov, E., Richardson, D., Payne, D.: Opt. Lett. **18**, 476–478 (1993)
36. Gromov, E.M., Malomed, B.A.: J. Plasma Phys. **79**, 1057–1062 (2013)
37. Gromov, E.M., Malomed, B.A.: Opt. Commun. **320**, 88–93 (2014)
38. Aseeva, N.V., Gromov, E.M., Tyutin, V.V.: Radiophys. Quantum Electron. **56**, 157–166 (2013)
39. Zakharov, V.E.: Collapse of Langmuir waves. Sov. Phys. JETP **35**, 908–914 (1972)
40. Oliviera, J.R., Moura, M.A.: Phys. Rev. E **57**, 4751–4755 (1998)
41. Blit, R., Malomed, B.A.: Phys. Rev. A **86**, 043841 (2012)

Equilibria in Networks with Production and Knowledge Externalities

Vladimir Matveenko and Alexei Korolev

Abstract We consider a game equilibrium in a network in each node of which an economy is described by the simple two-period model of endogenous growth with production and knowledge externalities. Each node of the network obtains an externality produced by the sum of knowledge in neighbor nodes. Uniqueness of the inner equilibrium is proved. Three ways of behavior of each agent are distinguished: active, passive, and hyperactive. Behavior of agents in dependence on received externalities is studied. It is shown that the equilibrium depends on the network structure. We study the role of passive agents and, in particular, possibilities of connection of components of active agents through components of passive agents. Changes of the equilibrium under changes in the network structure are studied. It is shown that appearance of a new link, as a rule, leads to decrease of knowledge in all nodes, but sometimes knowledge in some nodes increases. A notion of type of node is introduced and classification of networks based on this notion is provided. It is shown that the inner equilibrium depends not on the size of network but on its structure in terms of the types of nodes, and in similar networks of different size agents of the same type behave in similar way.

Keywords Network • Structure of network • Network game • Nash equilibrium • Externality • Network formation

---

V. Matveenko (✉) • A. Korolev
National Research University Higher School of Economics, 16 Soyuza Pechatnikov Street, St. Petersburg 190121, Russia
e-mail: vmatveenko@hse.ru; danitschi@gmail.com

V.A. Kalyagin et al. (eds.), *Models, Algorithms and Technologies for Network Analysis*, Springer Proceedings in Mathematics & Statistics 156,
DOI 10.1007/978-3-319-29608-1_20

1 Introduction

Actions of agents/actors[1] in a network are largely determined by the actions of their neighbors or the information obtained from them. Models which do not take the mutual dependence of agents into account become non-adequate, though in many cases recommendations concerning economic and social policies are still based on them.

Multi-agent networks are a natural object for studying interrelations in social and economic systems. Network economics and network games theory consider questions of networks formation, spreading (diffusion) of information in networks, positive and negative externalities, complementarity and substitutability of activities (see reviews [4, 6, 7]).

In the modern world mutual dependence includes, first of all, exchange of information as well as other multiple externalities. *Externalities*, i.e. influence of other agents, which does not go through the price mechanism, possess properties of public goods and are not fully paid. In particular, the so-called "Jacobian" positive externalities [8] are directly related to complementarity of agents' activities. Positive externalities, and among them externalities of knowledge and human capital, spring up both in processes of production [9, 13] and consumption [1], and it is important to account for them in economic and sociologic analysis, forecasting and mechanism design.

In case of *complementarity* (and, correspondingly, supermodularity) a marginal effect of the agent's effort depends positively on efforts of other agents [3, 10–12, 14]. The agent is interested in increase of his/her efforts if his/her neighbors in the network create enough externalities. The more efforts neighbors do, the more efforts will be done by the agent. Vice versa, in case of *substitutability* (submodularity), if other agents increase their efforts, then the efforts of the agent can become unessential, and he/she may rely on other agents [5, 7]; thus, the so-called free-rider problem arises [2].

In game theory a branch related to analysis of the role of positive externalities in networks has appeared, but attention there is devoted not to production externalities but mostly to consumption externalities connected with distribution of efforts. A typical model is as follows (e.g., [2]). Each of n agents in a network can make some efforts. Efforts by neighbors produce externalities which allow to increase the agent's utility. The efficient, from the point of view of social welfare, distribution of efforts and the Nash equilibrium essentially depend on the network structure.

In the present paper we continue the line of research of Nash equilibria in networks in presence of positive externalities, but our work contains several principally new elements in comparison to previous research.

[1]The term "agent" is used in economics, while the term "actor" is used in management, sociology, and politology. We speak further about "agents" despite results of our work may have applications in analysis of economic as well as social and political relations.

Firstly, we study production but not consumption externalities; efforts in our model have meaning of investments, in particular, investments into creation of knowledge. The presence of production block allows us to interpret concepts of complementarity (supermodularity) and substitutability (submodularity) as, correspondingly, absence and presence of productivity. We carry out comparative analysis of these concepts within the same model.

Secondly, our model for the first time in the network games literature uses the notion of the "Jacobian" production externality in definition of the concept of equilibrium. The essence of this notion is that any agent makes his/her decision staying in a particular environment which depends on actions by the agent himself/herself and by his/her neighbors. When making his/her decision, the agent considers the state of the environment as exogenous; this means that the agent does not take into account that his/her actions can directly influence the state of the environment.

As a simplest example imagine a game equilibrium in a collective of smokers and non-smokers. When the agent decides whether to quit smoking, he/she is located in a particular environment, which is defined by the fact that he/she smokes.

The third novation of our work is the use of dynamic approach. Essentially, our model is a generalization of the simple two-period model of endogenous growth and knowledge externalities due to Romer [13].

We show that equilibria depend on the network structure, and explain presence of three ways of behavior of agents: passive, active, and hyperactive.

We introduce a notion of type of node and propose an algorithm of subdividing the set of nodes. We provide a classification of networks by use of the types of nodes and show the role of this classification in characterizing equilibria in classes of networks which possess different sizes but similar structure of types of nodes.

Also we study consequences of adding a new link and connecting networks. A general theorem is proved which formulates conditions under which the sum of knowledge decreases after addition of a new link.

The paper is organized in the following way. In Sect. 2 the model is described. The uniqueness of the inner equilibrium is proved, if it exists. A theorem is proved, which serves further as a basic tool for comparison of utilities. In Sect. 3 behavior of the agent in dependence on received pure externality is analyzed. Section 4 is devoted to pure corner equilibria. In Sect. 5 equilibria in equidegree networks are studied. In Sect. 6 possibilities of attaching of a node with passive agent to an equidegree network with active agents, in such a way that the behavior of the agents does not change, are considered. In Sect. 7 possibilities of connection of equidegree components of active agents through nodes with passive agents are studied. In Sect. 8 a notion of type of node is introduced and an algorithm of subdivision is described. In Sects. 9 and 10, correspondingly, inner and corner equilibria for networks with two types of nodes are studied. In Sect. 11 a network with three types of nodes is considered. In Sect. 12 conditions are found under which addition of a new link leads to a decrease in the sum of knowledge. In Sect. 13 connection of full networks and stars is considered. Section 14 studies addition of a new link to cycles. Section 15 concludes.

2 Model

We consider a network with n nodes $i = 1, 2, \ldots, n$. Let $\mathbf{M}$ be the adjacency matrix: elements M_{ij} and M_{ji} of this matrix are equal to 1 if nodes i and j are connected by a link and equal to 0 in the opposite case. We set $M_{ii} = 0$ for all i.

In each of the nodes there is an agent, whose preferences at two periods of time, 1 and 2, are described by a utility function $U(c_1^i, c_2^i)$, where c_1^i, c_2^i are consumptions of the final good in node i in periods 1 and 2. The function U is assumed to be twice continuously differentiable, increasing and concave in each argument (at least in one of them—strictly concave).

In period 1 each agent is endowed by volume e of final good. This quantity may be used for consumption in period 1, and for investment into knowledge: $e = c_1^i + k_i$. There is a research technology which produces knowledge one-to-one from the invested good.

For an agent (index i is omitted now for notational simplicity), let k be his/her investment into knowledge. Let $\tilde{K}$ be externality which is the sum of investments of his/her close neighbors, and $K = k + \tilde{K}$ be his/her *environment*. Thus, the environment is the sum of investments in the neighboring nodes and in the node itself. The vector of environments of the agents $\mathbf{K} = (K_1, K_2, \ldots, K_n)^T$ can be calculated by use of the adjacency matrix in the following way:

$$\mathbf{K} = (\mathbf{M} + \mathbf{I})\mathbf{k},$$

where $\mathbf{I}$ is the unit matrix of order n, $\mathbf{k} = (k_1, k_2, \ldots, k_n)^T$, T is the sign of transposition.

The knowledge is used in production of final good for consumption in period 2. Production of good in the node is described by a production function $F(k, K)$ depending on the state of knowledge (investment) k and the environment K. The production function $F(k, K)$ is assumed to increase in each of its arguments and be concave (may be not strictly) in k for each environment K.

The concept of externality, developed in [9, 13], means that at the moment of decision making the agent takes the environment K as exogenously given, i.e. does not account for a possibility of its change in result of his/her choice of investment k. Correspondingly, the agent solves the following optimization problem $P(K)$:

$$U(c_1, c_2) \xrightarrow[c_1,\, c_2,\, k]{} \max$$
$$\begin{cases} c_1 \leq e - k, \\ c_2 \leq F(k, K), \\ c_1 \geq 0,\ c_2 \geq 0,\ k \geq 0. \end{cases}$$

The first two constraints of problem $P(K)$ at the optimum point are, evidently, satisfied as equalities. Substituting these constraints into the objective function, one can define new function (payoff function):

$$V(k, K) = U(e - k, F(k, K)).$$

Solution of problem $P(K)$ is one-to-one defined by k which maximizes the function $V(k, K)$ under constraint $k \in [0, e]$ given environment K. Function V is, evidently, strictly concave in k and, consequently, has a unique stationary point k^s, to the left of which it increases in k, and to the right—decreases. The stationary point k^s satisfies the equation

$$\partial V(k, K)/\partial k = 0, \tag{1}$$

where D_1 denotes derivative with respect to the first argument. If $k^s \in (0, e)$, then it is the optimal solution k of problem $P(K)$; this solution will be referred as *inner solution*. If $k^s < 0$, then the optimal solution is $k = 0$; and if $k^s > e$, then the solution is $k = e$. In these cases the solution will be referred as *corner solution*.

Let us consider a game in which the players are the agents $i = 1, 2, \ldots, n$. Feasible strategies of each player i are his/her investments $k_i \in [0, e]$. The payoff of the player is his/her utility $V(k_i, K_i)$. If profile $(k_1, k_2, \ldots, k_n)$ defines a consistent set of environments and optimal solutions of the players, this profile is referred as *Nash equilibrium with externalities*. If all k_i are inner solutions, then the equilibrium $(k_1, k_2, \ldots, k_n)$ will be referred as *inner equilibrium*. In the opposite case it will be referred as *corner equilibrium*. It is clear that the inner Nash equilibrium with externalities (if it exists under given values of parameters) is defined by the system of equations

$$\partial V(k_i, K_i)/\partial k_i = 0, i = 1, 2, \ldots, n. \tag{2}$$

We will choose a particular form of the utility function and production function which allows to study the structure of equilibria in dependence on parameters.

Let the utility function U have the quadratic form:

$$U(c_1, c_2) = c_1(e - ac_1) + dc_2, \tag{3}$$

where $0 < a < 1/2,\ d > 0$. Here a is a satiation coefficient.

Let the production function have the form

$$F(k, K) = gkK,$$

where $g > 0$. Notice that, by the meaning of parameters b and B, their increase promotes investments of agents. We will use notation $b = dg$. It will be assumed that

$$a < b. \tag{4}$$

Remark 2.1. Under our assumptions, the utility function defined by (3), evidently, strictly increases in both arguments and is concave. We could use instead a strictly concave function by applying the following concave transformation:

$$U(c_1, c_2) = \frac{[c_1(e - ac_1) + dc_2]^{1-\sigma}}{1 - \sigma},$$

where $0 < \sigma < 1$, σ is a coefficient of relative risk aversion. The points of maximum for both functions do coincide, thus the problem $P(K)$ in our case also has a unique solution which is guaranteed by the following lemma.

Lemma 2.1. *The payoff function $V(k_i, K_i)$ for the i-th node, considered under given environment K_i, as a function of k_i on the whole real axis, has a unique strict global maximum. The system of equations (2) takes the form:*

$$(b - 2a)\mathbf{k} + b\mathbf{M}\mathbf{k} = \bar{\mathbf{e}}, \tag{5}$$

where

$$\bar{\mathbf{e}} = \begin{pmatrix} e(1-2a) \\ e(1-2a) \\ \dots \\ e(1-2a) \end{pmatrix}.$$

Proof.

$$\begin{aligned} V(k_i, K_i) &= (e - k_i)(e - a(e - k_i)) + bk_iK_i \\ &= e^2(1-a) - k_ie(1-2a) - ak_i^2 + bk_iK_i, \\ D_1V(k_i, K_i) &= e(2a-1) - 2ak_i + bK_i, \end{aligned} \tag{6}$$

thus, the system of equations (2) takes the form (5). The second derivative of the function $V(k_i, K_i)$ with respect to the first argument in any point is $-2a < 0$. □

Theorem 2.1. *If $b \neq 2a$, then the system of equations (5) has a unique solution.*

Proof. The matrix of system (5) is

$$\mathbf{T} = \begin{pmatrix} b-2a & a_{12} & \dots & a_{1n} \\ a_{21} & b-2a & \dots & a_{2n} \\ \dots & \dots & \dots & \dots \\ a_{n1} & a_{n2} & \dots & b-2a \end{pmatrix},$$

where $a_{ij} = bM_{ij}$ under $i \neq j$. By dividing the elements of the matrix $\mathbf{T}$ by b, we receive the matrix

$$\tilde{\mathbf{T}} = \begin{pmatrix} \alpha & M_{12} & \dots & M_{1n} \\ M_{21} & \alpha & \dots & M_{2n} \\ \dots & \dots & \dots & \dots \\ M_{n1} & M_{n2} & \dots & \alpha \end{pmatrix},$$

where, because of (4), the diagonal elements satisfy condition $0 < |\alpha| < 1$. To prove the theorem it is sufficient to check the non-singularity of the matrix $\tilde{\mathbf{T}}$.

The determinant of the matrix is

$$\alpha^n + a_2\alpha^{n-2} + a_3\alpha^{n-3} + \cdots + a_{n-1}\alpha + a_n = 0, \tag{7}$$

where all coefficients $a_2, a_3, \ldots, a_n$ are integer. Let m be the highest degree of the variable α under which the coefficient of the polynomial (7), a_{n-m}, differs from zero. If $a_n \neq 0$, then $m = 0$. In the opposite case we reduce the polynomial to obtain

$$\alpha^{n-m} + a_2\alpha^{n-m-2} + a_3\alpha^{n-m-3} + \cdots + a_{n-m-1}\alpha + a_{n-m}. \tag{8}$$

Let $\alpha_1, \alpha_2, \ldots, \alpha_{n-m}$ be roots of the polynomial (8). Then

$$\alpha_1 + \alpha_2 + \cdots + \alpha_{n-m} = 0, \tag{9}$$

$$\alpha_1\alpha_2 + \alpha_1\alpha_3 + \cdots + \alpha_{n-m-1}\alpha_{n-m} = a_2, \tag{10}$$

$$\alpha_1\alpha_2\alpha_3 + \alpha_1\alpha_2\alpha_4 + \cdots + \alpha_{n-m-2}\alpha_{n-m-1}\alpha_{n-m} = -a_3, \tag{11}$$

$$\alpha_1\alpha_2\alpha_3\alpha_4 + \alpha_1\alpha_2\alpha_3\alpha_5 + \cdots + \alpha_{n-m-3}\alpha_{n-m-2}\alpha_{n-m-1}\alpha_{n-m} = a_4, \tag{12}$$

$$\ldots$$

$$\alpha_1\alpha_2 \ldots \alpha_{n-m} = (-1)^{n-m}a_n. \tag{13}$$

Let us show that no one of the roots can satisfy condition

$$0 < |\alpha_i| < 1.$$

Assume the opposite: let, e.g., $0 < |\alpha_1| < 1$. Then (9) implies that there is another root, e.g., α_2, which is not integer. Then the product $\alpha_1\alpha_2$ is also noninteger, and it follows from (10) that there is another noninteger product, e.g., $\alpha_2\alpha_3$. Then it follows from (11) or (12) that there is one more noninteger product of three or even four roots. Continuing this process further, we see that the product of all roots $\alpha_1\alpha_2 \ldots \alpha_{n-k}$ cannot be integer; this contradicts to (13). This absurd proves that no one of the roots of the polynomial (7) can satisfy condition $0 < |\alpha| < 1$. Hence, for any feasible values of parameters of the model, the matrix $\tilde{\mathbf{T}}$ is non-singular. □

Remark 2.2. If $b = 2a$, then the equilibrium exists only if for each agent $\tilde{K} = e(1 - 2a)$. We do not consider such artifact as far as the adjacency matrix is not obliged to be non-singular.

Corollary 2.1. *If, for given values of parameters, inner equilibrium exists, then it is unique.*

It follows from (5), that the stationary solution k_i^s for agent i is defined by

$$k_i^s = \frac{e(2a-1) + b\tilde{K}_i}{2a-b}, \tag{14}$$

where $\tilde{K}_i$ is the pure externality received by the agent. In the inner equilibrium, $k_i^* = k_i^s; i = 1, \ldots, n$.

Remark 2.3. In the theory of network games a notion of *strategic complementarity* is used. If increase of neighbors' investments leads to increase of investments by the agent himself/herself, then one says that the strategic complementarity takes place. If increase of neighbors' investments leads to decrease of the agent's investments, then one says that the *strategic substitutability* takes place. From formula (14) it becomes clear that if $b < 2a$, then the strategic complementarity takes place, and if $b > 2a$, then the strategic substitutability takes place. In our model of production, these inequalities indicate whether productivity is relatively high or low.

Definition 2.1. If $b > 2a$, we say that the *productivity is present*. In contrary case, if $b < 2a$, we say, that the *productivity is absent*.

Remark 2.4. Since it is assumed that $a < 1/2$, the inequality $b > 1$ implies presence of productivity, and, correspondingly, absence of productivity implies $b \leq 1$.

We will prove a general theorem, that will serve an instrument for utilities comparison.

Theorem 2.2. *Let W^* and W^{**} be two networks in inner equilibria; i is a node of W^*, and j is a node of W^{**}; k_i^*, K_i^*, U_i^* are, correspondingly, optimal investment, environment, and utility of agent i; k_j^{**}, K_j^{**}, U_j^{**} are corresponding values for agent j. Then:*

(1) if $K_i^* < K_j^{**}$, then $U_i^* < U_j^{**}$;
(2) if $K_i^* \leq K_j^{**}$, then $U_i^* \leq U_j^{**}$; and
(3) if $K_i^* = K_j^{**}$, then $U_i^* = U_j^{**}$. If k_i^* is a corner solution, $k_i^* = 0$, and $k_j^{**} > 0$, then

$$U_i^* = U(e, 0) < U_j^{**}.$$

If k_i^* or k_j^{**} (or both) are corner solutions, $k_i^* = e$ or $k_j^{**} = e$, then the statements 1)–3) are also true.

Proof. Let be $K_i^* < K_j^{**}$ $(K_i^* \leq K_j^{**})$. Since function $V(k_j, K_j^{**})$ reaches its maximum at point k_j^{**}, we have $V(k_i^*, K_j^{**}) \leq V(k_j^{**}, K_j^{**})$. Because $\partial V(k, K)/\partial K > 0$ for any $k \neq 0$ and K, we obtain that $V(k_i^*, K_i^*) < V(k_i^*, K_j^{**})$ (respectively, $V(k_i^*, K_i^*) \leq V(k_i^*, K_j^{**})$). Thus, $V(k_i^*, K_i^*) < V(k_i^*, K_j^{**}) \leq V(k_j^{**}, K_j^{**})$ (respectively, $V(k_i^*, K_i^*) \leq V(k_i^*, K_j^{**}) \leq V(k_j^{**}, K_j^{**})$). Hence, $U_i^* < U_j^{**}$ (respectively, $U_i^* \leq U_j^{**}$).

Combining previous results, we see that if $K_i^* = K_j^{**}$, then $U_i^* = U_j^{**}$. Obviously

$$U_j^{**} = V(k_j^{**}, K_j^{**}) > V(0, K_j^{**}) = V(0, K_i^*) = U_i^*,$$

since, as far as $k = 0$, function V does not depend on K. □

The last statement of the theorem is quite obvious.

Remark 2.5. If $k_i^* \neq k_j^{**}$, then as it is seen from the proof, in case 2) in the theorem, we have $U_i^* < U_j^{**}$ (in power of strict convexity of V in its first argument).

3 The Agent's Behavior

Let us introduce the following terminology.

Definition 3.1. If the agent makes zero investments into knowledge, $k = 0$, we will say that the agent is *passive*. If he/she makes investments $0 < k < e$, he/she is *active*. If the agent makes maximal possible investments, e (and, correspondingly, does not consume at period 1), we will say that he/she is *hyperactive*.

The following lemma describes necessary and sufficient conditions of different ways of behavior by the agent in dependence on the size of the pure externality $\tilde{K}$ which he/she receives. Index i is omitted.

Lemma 3.1. *The necessary and sufficient conditions for various types of behavior by the agent are the following.*

(1) Under absence of productivity:

The agent is passive if

$$\tilde{K} \leq \frac{e(1-2a)}{b}.$$

The agent is active if

$$\frac{e(1-2a)}{b} < \tilde{K} < \frac{e(1-b)}{b}.$$

The agent is hyperactive if

$$\tilde{K} \geq \frac{e(1-b)}{b}.$$

(2) Under presence of productivity:
The agent is passive if

$$\tilde{K} \geq \frac{e(1-2a)}{b}.$$

The agent is active if

$$\frac{e(1-b)}{b} < \tilde{K} < \frac{e(1-2a)}{b}.$$

The agent is hyperactive if

$$\tilde{K} \leq \frac{e(1-b)}{b}.$$

Proof. It follows from (6) that

$$k^s = \frac{e(2a-1) + b\tilde{K}}{2a-b}. \tag{15}$$

If $k^s \leq 0$, then the agent makes no investments, $k = 0$. If $0 < k^s < e$, the solution is in the inner point $k = k^s$. If $k^s \geq e$, the agent makes maximal possible investment, $k = e$. Writing these conditions in detail, we receive the inequalities listed in the formulation of the lemma. □

Proposition 3.1. *Under presence of productivity, if $b > 1/2$, then each agent, who has a hyperactive neighbor, is not hyperactive, moreover, if $b + 2a \geq 1$ (which implies $b > 1/2$), he/she is even passive.*

Proof. If agent i has a neighbor, who is hyperactive, then i obtains pure externality $\tilde{K} \geq e$. Hence, if $b > 1/2$, then $\tilde{K} \geq e > e(1-b)/b$, and, by Lemma 3.1, agent i is not hyperactive. Moreover, let $b + 2a \geq 1$, i.e. $(1-2a)/b \leq 1$. Then $\tilde{K} \geq e \geq e(1-2a)/b$ and, by Lemma 3.1, the agent is passive. □

Proposition 3.2. *Under absence of productivity, if $b \geq 1/2$, then each agent, who has a hyperactive neighbor, is hyperactive.*

Proof. If the agent has a hyperactive neighbor, then, as in the proof of Proposition 3.1, $\tilde{K} \geq e$. Hence, if $b \geq 1/2$, then $\tilde{K} \geq e \geq e(1-b)/b$. By Lemma 3.1, the agent is hyperactive. □

Proposition 3.3. *Under presence of productivity, an agent, who stays in an isolated node, or for whom all neighbors are passive, is active if $b > 1$, and hyperactive if $b \leq 1$.*

Proof. Since $\tilde{K} = 0$, Lemma 3.1 implies that the agent is active under $1 - b < 0$; and hyperactive under $1 - b \geq 0$. □

Proposition 3.4. *Under absence of productivity, the agent, who stays in an isolated node, or for whom all neighbors are passive, is passive.*

Proof. Since $\tilde{K} = 0$, by Lemma 3.1, the agent is passive. □

4 Pure Corner Equilibria

Definition 4.1. *Pure corner equilibrium* is such equilibrium for which knowledge in each node is equal either to 0 or to e, i.e. each agent is either passive or hyperactive.

Proposition 4.1. *Under absence of productivity, the situation when all agents are passive is equilibrium.*

Proof. For each node i, if $\tilde{K}_i = 0$, then, by Lemma 3.1, the agent is passive. □

Proposition 4.2. *For a connected network with more than one node, let μ be the smallest degree (the number of neighbors) and $\hat{\mu}$ the biggest degree. Under absence of productivity, the situation when all agents are hyperactive is equilibrium iff $b \geq 1/(\mu + 1)$. Under presence of productivity, the situation when all agents are hyperactive is equilibrium iff $b \leq 1/(\hat{\mu} + 1)$.*

Proof. By Lemma 3.1, the agent who stays in a node with the smallest degree is hyperactive iff $\mu e = \tilde{K} \geq e(1-b)/b$; this is equivalent to condition $b \geq 1/(\mu+1)$. If the latter condition is fulfilled, then the agents in all other nodes are all hyperactive.

Similarly, the agent who stays in a node with the biggest degree is hyperactive under presence of productivity iff $\hat{\mu} e = \tilde{K} \leq e(1 - b)/b$; the latter is equivalent to condition $b \leq 1/(\hat{\mu} + 1)$. Under this condition, the agents in all other nodes are all hyperactive. □

Definition 4.2. A network is referred as *equidegree network* if each node has the same degree m, where $m \geq 1$.

Corollary 4.1. *In equidegree network, if $b \geq 1/(1 + m)$ under absence of productivity, or if $b \leq 1/(m + 1)$ under presence of productivity, then the situation when all agents are hyperactive is equilibrium.*

Proof. In equidegree network: $\mu = \hat{\mu} = m$. □

Corollary 4.2. *In full network consisting of n nodes, equilibrium in which all agents are hyperactive is possible under absence of productivity iff $b \geq 1/n$, and under presence of productivity iff $b \leq 1/n$.*

Proof. In full network: $\mu = \hat{\mu} = n - 1$. □

Propositions 3.1 and 3.2 imply the following fact.

Corollary 4.3. *In connected network under absence of productivity, if $b \geq 1/2$, then a situation when all agents are hyperactive is equilibrium; moreover, it is* *a unique possible equilibrium in which at least one agent is hyperactive. If in a network there is at least one link, under presence of productivity, if* $b > 1/2$*, then the situation in which all agents are hyperactive is not equilibrium.*

Proposition 4.3. *Under absence of productivity, in each network there is equilibrium in which all agents are passive. Under presence of productivity, such equilibrium is impossible.*

Proof. This follows directly from Propositions 3.3 and 3.4. □

Theorem 4.1. *Under presence of productivity, let* r *be a natural number such that* $r \leq n$*,* $br \leq 1$*, and* $br + 2a \geq 1$*. In a full network,* C_n^r *pure corner equilibria are possible, in each of which* r *nodes are hyperactive and other* $n-r$ *nodes are passive.*

Proof. A node, for which not more than $r - 1$ neighbors are hyperactive and other neighbors are passive, receives pure externality $\tilde{K} = (r - 1)e$. By Lemma 3.1, such node is hyperactive iff $r \leq 1/b$. Similarly, a node, for which more than $r - 1$ neighbors are hyperactive and other neighbors are passive, is itself passive iff $r \geq (1 - 2a)/b$. These inequalities are equivalent, correspondingly, to $br \leq 1$ and to $br + 2a \geq 1$. □

Remark 4.1. Besides pure corner equilibria which are listed in Corollary 4.2, Proposition 4.3, and Theorem 4.1, there may exist corner equilibria and a unique inner equilibrium.

Example 4.1. In a full network with four nodes, under presence of productivity and under $2b \leq 1$, $2b + 2a \geq 1$, there is an equilibrium with two hyperactive and two passive agents. Because of symmetry, any two nodes can be hyperactive, and others passive; and in all $C_4^2 = 6$ purely corner equilibria exist.

Under $b \leq 1$, $b + 2a \geq 1$ in the same network there is an equilibrium with one hyperactive and three passive nodes; in all $C_4^1 = 4$ such equilibria exist.

Under $3b \leq 1$, $3b + 2a \geq 1$ in the same network there is an equilibrium with three hyperactive and one passive nodes. In all there are $C_4^3 = 4$ such equilibria.

In case when $3b \leq 1$, $b + 2a \geq 1$ all these pure corner equilibria are possible; they are $6 + 4 + 4 = 14$ in all.

Besides these 14 pure corner equilibria, in this network there are also other corner equilibria and a unique inner equilibrium.

5 Equilibria in Equidegree Networks

For an equidegree network, an equilibrium in which all agents have the same level of knowledge (i.e., make the same investments) will be referred as *symmetric*. For a symmetric equilibrium, Eq. (15) (under $\tilde{K} = mk$) implies

$$k^s = \frac{e(1 - 2a)}{b(m + 1) - 2a}. \tag{16}$$

If $b > 1/(m+1)$, then $k = k^s$, i.e. the equilibrium is inner: the agents are active. If $2a/(m+1) < b \leq 1/(m+1)$, then $k = e$, i.e. the equilibrium is corner: the agents are hyperactive.

Remark 5.1. Thus, the condition of existence of an inner equilibrium in an equidegree network is $b > 1/(m+1)$ (remind also permanent constraints $a < 1/2$, $a < b$).

5.1 Examples of Equidegree Networks

1. Cycle. In this case $m = 2$. According to (16), investment by an agent does not depend on the size of cycle.
2. Full network. In this case $m = n - 1$, where n is the number of nodes in the network. This case is similar to [13]. According to (16), knowledge in nodes declines with increase in the size of network. The sum of knowledge $nk^s = e(1-2a)/(b-2a/n)$ also declines and converges to $e(1-2a)/b$.
3. Chain with two nodes. It is the case of $m = 1$ (and also a case of full network).
4. Networks in which each node has degree $m = 3$ (see Fig. 1).

Remark 5.2. Equation (16) is also true for isolated node, under $m = 0$. It is seen from (16) that in an isolated node:

(1) under $b < 2a$ the agent is passive;
(2) under $2a < b \leq 1$ the agent is hyperactive; and
(3) under $b > 1$ the agent is active and

$$k = \frac{e(1-2a)}{b-2a}.$$

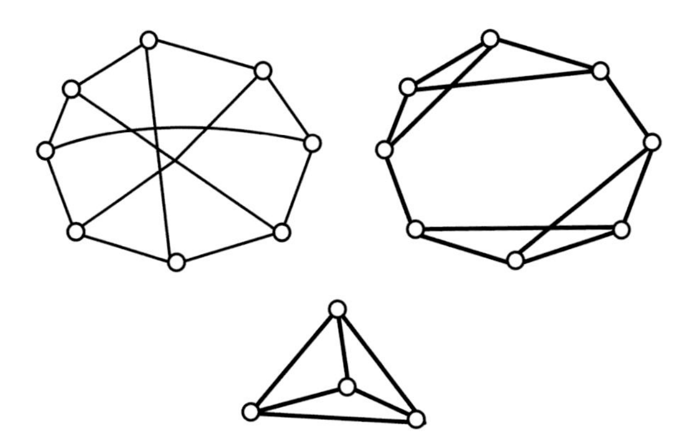

Fig. 1 Examples of equidegree networks with degree $m = 3$

Generally, (16) allows to study dependence of knowledge, consumption, and utility on the degree of nodes, m, of the equidegree network. Knowledge k decreases with respect to degree. Consumptions at the first and the second periods of time are, correspondingly,

$$c_1 = e - k = e\frac{(m+1)b - 1}{(m+1)b - 2a},$$

$$c_2 = F(k, (1+m)k) = g(1+m)k^2 = \frac{g(m+1)e^2(1-2a)^2}{[(m+1)b - 2a]^2}.$$

One can check that c_1 increases and c_2 decreases with respect to degree.

Proposition 5.1. *For inner equilibria in equidegree networks, utility decreases with respect to degree and converges to* $U(e, 0)$.

Proof. The utility function turns into

$$U = e^2\frac{[(m+1)b - 1][(m+1)b(1-a) - a] + b(m+1)(1-2a)^2}{[(m+1)b - 2a]^2}.$$

Differentiating U with respect to $x = 1 + m$ (as if x is continuous) we obtain

$$\frac{dU}{dx} = -\frac{2ab(bx - 2a)(2a-1)^2}{(x^2b^2 - 4abx + 4a^2)^2}.$$

Under $x \geq 2$, i.e. under $m \geq 1$, the inequality $bx - 2a > 0$ is fulfilled, hence $dU/dx < 0$; thus, utility in each node decreases with respect to degree. Under $m \to +\infty$ we have $c_1 \to e$, $c_2 \to 0$ and, because of continuity, the limit of the utility is equal to $U(e, 0)$. □

This result corresponds to intuition: in big social and economic systems utility can be high because of diversity, but in a system consisting of similar agents, the world, probably, loses its utility under very high degrees of nodes if there is no diversity.

6 Adding a Node with Passive Agent into an Equidegree Network

We have seen that, in definite areas of parameters, equilibrium in equidegree network can be rather simple: all agents are hyperactive, or all are active. At the same time, passive agents do not create externalities, i.e. do not influence the environments of other agents. This means that under some conditions an equilibrium is possible which consists of components with either active or hyperactive agents and of groups of passive agents which connect these components.

Below in Sect. 7 we consider connection of equidegree networks through nodes with passive agents. As a preliminary, in this section we study a possibility of adding a node with passive agent to equidegree network.

Proposition 6.1. *Let a node with passive agent be connected by use of l links to an equidegree network with degree $m > 0$, which is initially in inner equilibrium. Necessary and sufficient condition of existence of such equilibrium, in which the adjoined agent remains passive and the active agents remain in the previous inner equilibrium, is the following:*

under absence of productivity,

$$l \leq m + 1 - \frac{2a}{b}; \tag{17}$$

under presence of productivity,

$$l \geq m + 1. \tag{18}$$

Proof. For the newly adjoined node the pure externality is equal to

$$\tilde{K} = \frac{le(1-2a)}{b(m+1)-2a}.$$

By Lemma 3.1, the adjoined agent can stay passive in equilibrium iff

$$b < 2a; \quad \frac{l(1-2a)e}{b(m+1)-2a} \leq \frac{(1-2a)e}{b} \tag{19}$$

or

$$b > 2a; \quad \frac{l(1-2a)e}{b(m+1)-2a} \geq \frac{(1-2a)e}{b} \tag{20}$$

Conditions (19) and (20) are equivalent, correspondingly, to (17) and (20). □

The meaning of (17) is that, under absence of productivity, the adjoined agent can stay passive only as long as he/she is not sufficiently connected with active agents. Staying passive, he/she does not influence the initial equilibrium of active agents. But if the number of his/her neighbors becomes sufficiently big, the adjoined agent receives so big externality that he/she is not able to hold the indifferent behavior in equilibrium. When he/she starts investing, an absolutely different inner equilibrium appears in the network.

Vice versa, under presence of productivity, the agent can preserve his/her indifferent behavior only until he/she has sufficiently big number of active neighbors.

Corollary 6.1. *In case of $m = l$, adjunction of a passive agent in such a way that each of the agents preserves his/her behavior in equilibrium is impossible.*

Corollary 6.2. *Under absence of productivity, if an equidegree network with degree $m \geq 2$ is in inner equilibrium, then a node with passive agent can be adjoined by one link in such a way that each of the agents preserves his/her behavior in equilibrium.*

Proposition 6.2. *A node with passive agent can be adjoined to an isolated node, in such a way that each of the agents preserves his/her behavior in equilibrium, in the following three cases:*

(1) $b < 2a$ (in this case the agent in the isolated node is passive);
(2) $2a < b \leq 1$, $b + 2a \geq 1$ (in this case the agent in the isolated node is hyperactive); and
(3) $b > 1$ (in this case the agent in the isolated node is active and $k = e(1-2a)/(b-2a)$).

Proof. If $b < 2a$, then there are no externalities and nothing changes after adjunction. Thus, we have equilibrium of two passive agents.

If $2a < b \leq 1$, then the adjoined agent receives externality $\tilde{K} = e$ and, by Lemma 3.1, remains passive under $b \geq 1 - 2a$.

If $b > 1$ (this is a condition for the isolated agent to be active), then the adjoined agent receives externality $\tilde{K} = e(1-2a)/(b-2a)$. By Lemma 3.1, the agent will remain passive if $b - 2a \leq b$, but this inequality is certainly true. □

6.1 Examples of Adjunction of a Passive Agent to an Equidegree Network

Example 6.1. In the chain of 3 nodes 1–2–3, under $b > 1/2$, equilibrium with $k_1 = 0$, $k_2 = k_3 = (1-2a)e/[2(b-a)]$ is impossible, by virtue of Corollary 6.1.

Example 6.2. If initially there is a chain of two active agents, 2–3, and passive agent 1 establishes links to both of them, i.e. $l = 2$, $m = 1$, then, by Proposition 6.1, there is equilibrium, in which all three agents maintain their initial behavior.

Example 6.3. Let a passive agent establish $l = 4$ links with agents in equidegree network with degree $m = 3$, which is in inner equilibrium. The initial equilibrium exists only if $b > 1/4$ (see Remark 5.1). If, moreover, productivity takes place ($b > 2a$), then, by Proposition 6.1, there exists equilibrium in which all agents maintain their initial behavior.

Under absence of productivity, similarly to adjunction of one node with passive agent, any connected network with passive agents can be added.

7 Connection of Equidegree Networks Through Nodes with Passive Agents

In this section we consider connection of two equidegree networks being initially in inner equilibrium. We wonder is it possible to construct a new network from such blocks, connecting them by components of passive agents in such a way that in the new network there exists an equilibrium, in which all the agents behave in the same way as before unification.

Proposition 7.1. *Under $2a/(m-1) \leq b < 2a$ (what implies $m \geq 3$), two equidegree networks with the same degree m, being initially in inner equilibrium, can be connected through a node with passive agent in such a way that all agents maintain their initial behavior in equilibrium. Under $m = 2$ (case of cycles) such equilibrium is impossible. Under $m = 1$ (case of active dyads, when $b > 1/2$) such equilibrium is possible under presence of productivity ($b > 2a$). Under $m = 0$ (case of active isolated nodes, when $b > 1$) such connection is always possible.*

Proof. The connecting passive node receives externality from two active nodes, $l = 2$. Under absence of productivity, condition (6.1) takes the form $2a(m-1) \leq b$. Condition (6.2) takes the form $2 \geq m+1$ and is fulfilled under $m = 1$. Let $m = 0$, $b > 1$, and, hence, $b > 2a$. Initially the inner equilibrium in two isolated nodes was $k = e(1-2a)/(b-2a)$, hence the connecting node receives externality

$$\tilde{K} = \frac{2e(1-2a)}{b-2a}.$$

By Lemma 3.1, it remains passive under

$$\frac{2}{b-2a} > \frac{1}{b},$$

and the latter inequality is certainly fulfilled. □

Proposition 7.2. *Under $m \geq 2$ and the absence of productivity, two equidegree networks with the same degree m, being initially in inner equilibrium, can be connected by a chain of two or more passive nodes in such a way that behavior of the agents does not change in equilibrium. Such connection is impossible if $m = 1$. In case of $m = 0$ (isolated nodes) such connection is possible under $b > 1$ through a chain of two passive nodes but is not possible through chains of three or more passive nodes.*

Proof. Statements for $m \geq 2$ and $m = 1$ follow from Proposition 6.1 and Corollary 6.1. Statement for $m = 0$ follows from Proposition 6.2. If two active agents are connected by a chain of three or more passive nodes, then, by Proposition 3.3, the agent, who has no active neighbors, could not stay passive in equilibrium under presence of productivity. □

Under the same conditions, there exists a "cycle," consisting of equidegree networks connected by chains of nodes with passive agents. Components of active agents in such cycle alternate with components of passive agents.

8 Types of Nodes

Definition 8.1. Let the set of nodes $1, 2, \ldots, n$ be decomposed into disjoint classes in such a way that any nodes belonging the same class have the same numbers of neighbors from each class. The classes will be referred as *types* of nodes. Type j is characterized by vector $\mathbf{l}(j) = (l_1(j), l_2(j), \ldots, l_k(j))$, where $l_i(j)$ is the number of neighbors in class i for each node of class j.

Let us describe an algorithm of subdivision of the set of nodes of network into types. Let s be a current number of subsets of subdivision. Initially $s = 1$.

Iteration of the algorithm. Consider nodes of the first subset. If all of them have the same numbers of neighbors in each subset $1, 2, \ldots, s$, then the first subset is not changed. In the opposite case, we divide the first subset into new subsets in such a way that all nodes of each new subset have the same vector of numbers of neighbors in subsets.

We proceed in precisely the same way with the second, the third, …, the s-th subset. If on the present iteration the number of subsets s has not changed, then the algorithm finishes its work. If s has increased, then the new iteration is executed.

The number of subsets s does not decrease in process of the algorithm. Since s is bounded from above by the number of nodes in the network, the algorithm converges.

It is clear that the algorithm divides the set of nodes into the minimal possible number of classes.

8.1 Example

Let us apply the algorithm to the network depicted in Fig. 2. Initially $s = 1$, all nodes are from the same one set (Fig. 2).

After the first iteration we obtain the division depicted in Fig. 3.

Then, on the first step of the second iteration, we obtain the division depicted in Fig. 4.

On the second step of the second iteration we obtain the division shown in Fig. 5.

On the third iteration nothing changes, and the algorithm stops. We have obtained a subdivision of the set of nodes of the network into four types which are characterized by the vectors of numbers of neighbors: $\mathbf{l}(1) = (1, 2, 1, 0)$, $\mathbf{l}(2) = (1, 0, 1, 1)$, $\mathbf{l}(3) = (1, 2, 0, 1)$, and $\mathbf{l}(4) = (0, 2, 1, 0)$.

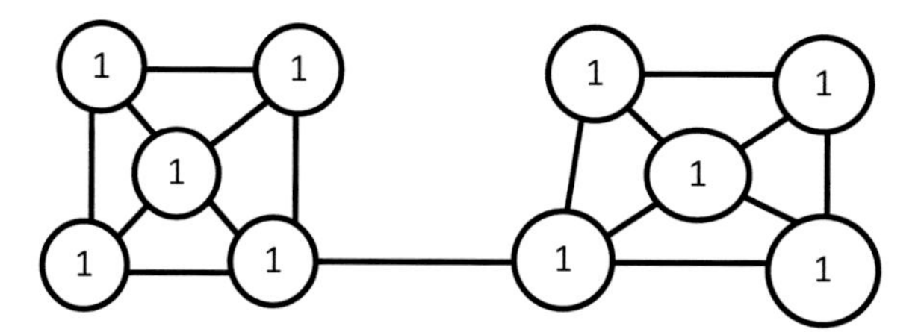

Fig. 2 Start of the algorithm: $s = 1$

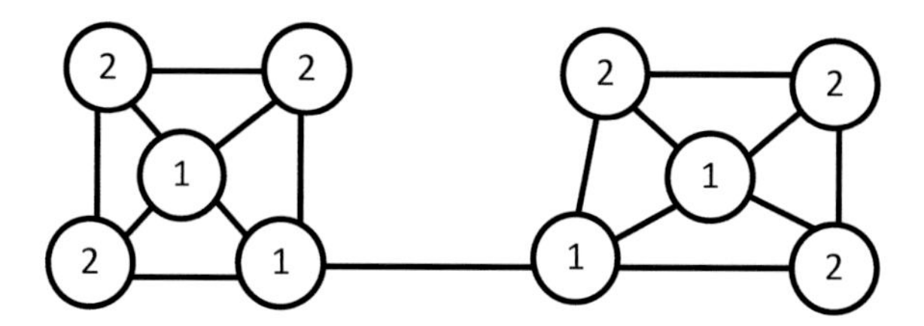

Fig. 3 Result of the first iteration: $s = 2$

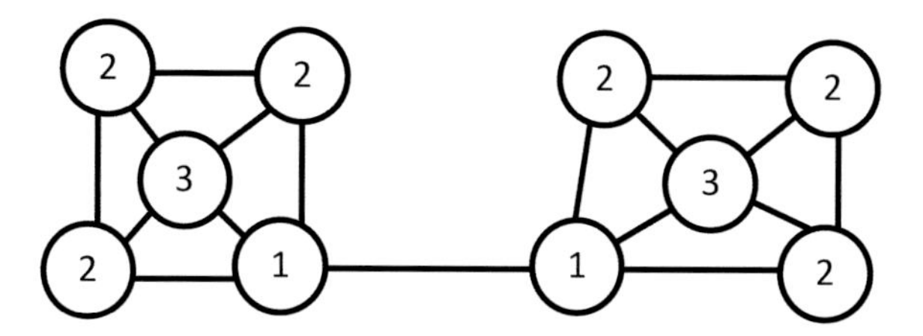

Fig. 4 The first step of the second iteration: $s = 3$

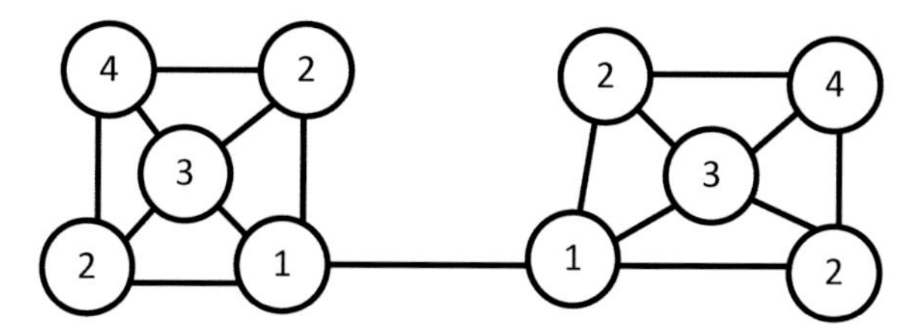

Fig. 5 The second step of the second iteration: $s = 4$

Definition 8.2. Let us call *symmetric* such equilibrium in which agents of the same type make the same investments.

Remark 8.1. Inner equilibrium is always symmetric. It follows from the uniqueness of solution of the system of Equations (5) and symmetry of this system with respect to types.

Later on k_i will denote investment in any node of type i.

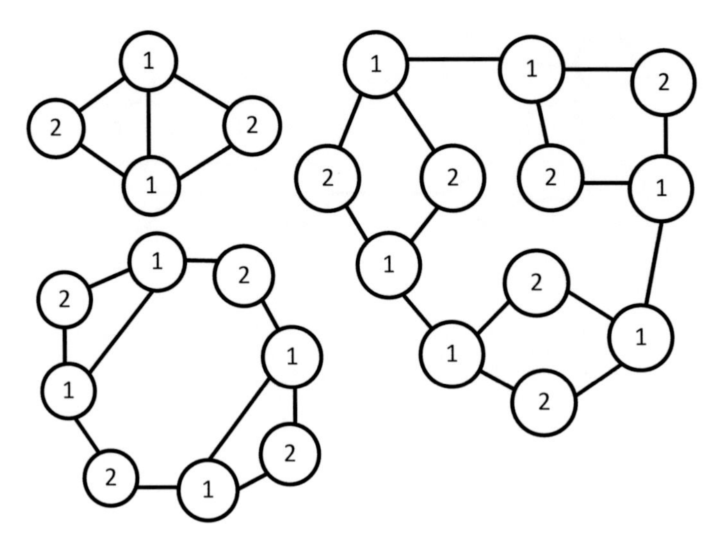

Fig. 6 Networks with "coinciding" inner equilibria

Remark 8.2. If two networks have the same number of types of nodes, S, and these types are characterized by the same set of vectors $\mathbf{l}(1), \mathbf{l}(2), \ldots, \mathbf{l}(S)$, then the inner equilibria in these networks do coincide, in the sense that agents in the nodes of the same type behave in the same way.

Figure 6 provides an example of three networks which possess the same types of nodes characterized by vectors $\mathbf{l}(1) = (1, 2)$ and $\mathbf{l}(2) = (0, 2)$. Correspondingly, these networks have the same inner equilibria, despite these networks have different sizes.

9 Inner Equilibria in Networks with Two Types of Nodes

Let a network have two types of nodes which are characterized by vectors $\mathbf{l}(1) = (s_1, s_2)$ and $\mathbf{l}(2) = (t_1, t_2)$. Here s_i is the number of links connecting a node of type 1 with nodes of type i; t_i is the number of links connecting a node of type 2 with nodes of type i; $i = 1, 2$. Then (5) implies the following system of equations:

$$\begin{cases} (b - 2a + s_1 b)k_1 + s_2 bk_2 = e(1 - 2a), \\ t_1 bk_1 + (b - 2a + t_2 b)k_2 = e(1 - 2a). \end{cases} \tag{21}$$

Its solution is the pair

$$k_1^s = \frac{e(1-2a)[b-2a+(t_2-s_2)b]}{(b-2a)^2+(s_1+t_2)(b-2a)b+(s_1t_2-t_1s_2)b^2}. \tag{22}$$

$$k_2^s = \frac{e(1-2a)[b-2a+(s_1-t_1)b]}{(b-2a)^2+(s_1+t_2)(b-2a)b+(s_1t_2-t_1s_2)b^2}. \tag{23}$$

If $0 < k_1^s < e$, $0 < k_2^s < e$, then the stationary values k_1^s, k_2^s define the inner equilibrium in the network.

9.1 Special Cases of the Network with Two Types of Nodes

1. Chain of four nodes: 2–1–1–2. Types 1 and 2 are characterized by vectors $\mathbf{l}(1) = (1, 1)$ and $\mathbf{l}(2) = (1, 0)$. Formulas (22) and (23) take the form:

$$k_1 = \frac{2ae(1-2a)}{6ab-4a^2-b^2}, \tag{24}$$

$$k_2 = \frac{e(1-2a)(2a-b)}{6ab-4a^2-b^2}. \tag{25}$$

   Inequalities $0 < k_i < e$, $i = 1, 2$ are fulfilled under absence of productivity ($b < 2a$).
2. A generalization of the previous case is a fan, i.e. a chain of two nodes, to each of which a bundle of ν hanging nodes is adjoined. The types are characterized by vectors $\mathbf{l}(1) = (1, \nu)$ and $\mathbf{l}(2) = (1, 0)$. In Sect. 13 we will consider the fan as a result of connection of two stars.
3. Star of order ν, i.e. a network, in which a central node of type 1 has ν peripheral neighbors of type 2. The types are characterized by vectors $\mathbf{l}(1) = (0, \nu)$ and $\mathbf{l}(2) = (1, 0)$. Equations (22) and (23) turn into

$$k_1 = \frac{e(1-2a)[(\nu-1)b+2a]}{\nu b^2-(b-2a)^2}, \tag{26}$$

$$k_2 = \frac{2ea(1-2a)}{\nu b^2-(b-2a)^2}, \tag{27}$$

   The pair k_1, k_2 defines inner equilibrium if $0 < k_i < e$, $i = 1, 2$, i.e. if

$$\begin{cases} \nu b^2-(b-2a)^2 > 0, \\ \nu b^2-(b-2a)^2 > (1-2a)[(\nu-1)b+2a], \\ \nu b^2-(b-2a)^2 > 2a(1-2a). \end{cases}$$

Evidently, this system of inequalities is equivalent to the second of them,

$$(\nu - 1)b^2 + [2a(\nu + 1) - (\nu - 1)]b - 2a > 0. \tag{28}$$

This inequality is fulfilled iff

$$b > \frac{-2a(\nu + 1) + \nu - 1 + \sqrt{[2a(\nu + 1) - (\nu - 1)]^2 + 8a(\nu - 1)}}{2(\nu - 1)}.$$

We see that under big ν inequality (28) is true if $b^2 + 2ab - b > 0$, which is equivalent to $b + 2a > 1$. It is also easily seen that the left-hand side of (28) increases in ν. Hence, if (28) is fulfilled for $\nu = 2$, it is fulfilled for all ν. This implies that if $b + 2a > 1$ and

$$b > \frac{-6a + 1 + \sqrt{36a^2 - 4a + 1}}{2},$$

then formulas (26) and (27) define inner equilibrium for all $\nu \geq 2$.

Proposition 9.1. *In a star, if the number of peripheral nodes ν increases, then knowledge and utility in the central node decrease under absence of productivity, but increase under presence of productivity. Knowledge and utility in each peripheral node always decrease.*

Proof. Derivative of k_1 in ν, if ν is considered as a continuous parameter, is

$$\frac{2bae(1 - 2a)(b - 2a)}{[(b - 2a)^2 - \nu b^2]^2}.$$

Hence, knowledge in the central node decreases in ν if $b < 2a$, and increases if $b > 2a$. It is directly seen from (27) that k_2 decreases in ν.

Environment for the central node is

$$K_1 = \frac{e(1 - 2a)[\nu(b + 2a) - (b - 2a)]}{\nu b^2 - (b - 2a)^2}.$$

Derivative of K_1 is

$$\frac{4a^2e(1 - 2a)(b - 2a)}{[\nu b^2 - (b - 2a)^2]^2}.$$

Theorem 2.2 implies that utility in the central node decreases in ν if $b < 2a$, and increases if $b > 2a$. Environment for any peripheral node is

$$K_2 = \frac{e(1 - 2a)[(\nu - 1)b + 4a]}{\nu b^2 - (b - 2a)^2}.$$

Derivative of K_2 is

$$\frac{-4a^2be(1-2a)}{[\nu b^2-(b-2a)^2]^2}<0.$$

Hence, by Theorem 2.2, utility in a peripheral node decreases in ν. □

Remark 9.1. When the order of the star, ν, increases, the sum of knowledge in the peripheral nodes decreases and under $\nu \to \infty$ converges to $2ae(1-2a)/b^2$, while knowledge in each separate peripheral node converges to 0. Knowledge in the central node converges to $e(1-2a)/b$.

Remark 9.2. If $\nu = 2$, then the star turns into a chain of three nodes.

4. Cycle of k nodes ($k \geq 3$), to each of which a bundle of ν "hanging" nodes is added. Equilibria in this network will be studied in Sect. 10 below.

5. Network shown in Fig. 7. The types of nodes are characterized by vectors $\mathbf{l}(1) = (2,2)$ and $\mathbf{l}(2) = (2,0)$. Equations (22) and (23) turn into

$$k_1 = \frac{e(1-2a)(b+2a)}{b^2+8ba-4a^2}, \tag{29}$$

$$k_2 = \frac{e(1-2a)(2a-b)}{b^2+8ba-4a^2}. \tag{30}$$

Positivity $k_i > 0$, $i = 1,2$ is equivalent to the absence of productivity ($b < 2a$) and fulfillment of the inequality

$$b^2+8ba-4a^2>0.$$

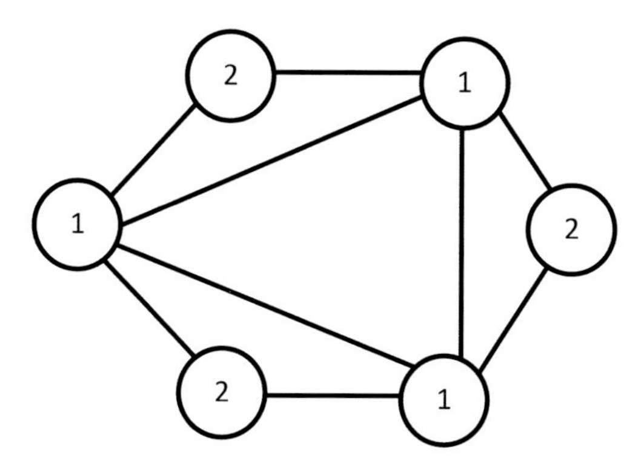

Fig. 7 Network with types characterized by vectors $\mathbf{l}(1) = (2,2)$ and $\mathbf{l}(2) = (2,0)$

Conditions $k_i < e$, $i = 1, 2$ are then equivalent to

$$b^2 + 8ba - 4a^2 > (1 - 2a)(b + 2a),$$
$$b^2 + 8ba - 4a^2 > (1 - 2a)(2a - b).$$

The system of the latter three inequalities is equivalent to the second of them, which can be written in the form

$$b^2 + 10ab - b - 2a > 0.$$

Ultimately, we obtain necessary and sufficient condition of inner equilibrium:

$$\frac{-10a + 1 + \sqrt{(10a - 1)^2 + 8a}}{2} < b < 2a.$$

Let us compare levels of knowledge and utility for the network in Fig. 7 and for the full network.

Proposition 9.2. *If the network of the type depicted in Fig. 7 is completed to become the full network, then, under absence of productivity, knowledge and utility in nodes of type 1 decrease, while knowledge and utility in nodes of types 2 increase.*

Proof. Comparing k_1 and k_2 with knowledge in a node of the full network, $k = e(1 - 2a)/((n - 1)b - 2a)$, we see that $k_1 > k$, $k_2 < k$. Comparing environments

$$K_1 = \frac{e(1 - 2a)(b + 10a)}{b^2 + 8ab - 4a^2}, \quad K_2 = \frac{e(1 - 2a)(b + 6a)}{b^2 + 8ab - 4a^2}$$

of the nodes of the initial network with environment of a node of the full network,

$$K = \frac{(n - 1)e(1 - 2a)}{(n - 1)b - 2a},$$

we see that $K_1 > K$, $K_2 < K$. Theorem 2.2 provides the needed result. □

10 Corner Equilibria in Networks with Two Types of Nodes

In this section we will study symmetric corner equilibria. The network of type depicted in Fig. 8 can be considered in two ways: as a result of addition of ν previously isolated nodes to each node of the cycle of order n, or as a result of conjunction of n stars, each of them with ν peripheral nodes, into one cycle. The types are characterized by vectors $\mathbf{l}(1) = (2, \nu)$ and $\mathbf{l}(2) = (1, 0)$.

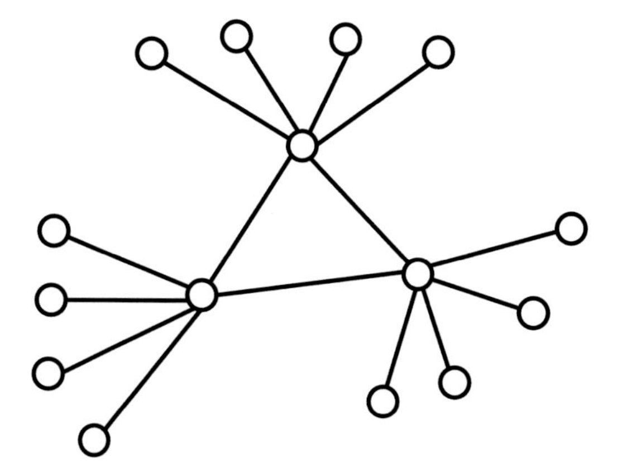

Fig. 8 Cycle of stars

Proposition 10.1. *In the network depicted in Fig. 8 inner equilibrium is impossible.*
Under presence of productivity, the corner equilibrium

$$k_1 = \frac{e(1-2a)}{3b-2a}, \quad k_2 = 0,$$

is possible if $b > 1/3$, *and the pure corner equilibrium*

$$k_1 = e, \quad k_2 = 0,$$

is possible if $b \leq 1/3$, $2a + \nu b > 1$.

Besides, under absence of productivity, the pure corner equilibrium $k_1 = k_2 = 0$ *is possible.*

Under presence of productivity, the corner equilibrium

$$k_1 = 0, \quad k_2 = \frac{e(1-2a)}{b-2a},$$

is possible if $b > 1$, *and the pure corner equilibrium*

$$k_1 = 0, \quad k_2 = e,$$

is possible if $b \leq 1$, $2a + \nu b > 1$.

In each case, increase in ν *does not influence knowledge and utilities.*

Proof. Equations (22) and (23) turn into

$$k_1^s = \frac{e(1-2a)(b-2a-\nu b)}{(b-2a)(3b-2a)-\nu b^2},$$

$$k_2^s = \frac{2e(1-2a)(b-a)}{(b-2a)(3b-2a)-\nu b^2}.$$

We see that the numerator of the expression for k_1^s is negative, and the numerator of the expression for k_2^s is positive, hence, independently on the sign of the denominator, k_1^s and k_2^s have different signs. Hence, inner equilibrium is impossible.

Let $b < 2a$, $b > 1/3$. If $k_2 = 0$, then $k_1 = e(1-2a)/(3b-2)$. But if $k_1 = e(1-2a)/(3b-2a)$, then $k_2^s = e(1-2a)(2b-2a)/(b-2a) < 0$ and, hence, $k_2 = 0$.

Let $b < 2a$, $b \leq 1/3$, $2a + \nu b > 1$. If $k_2 = 0$, then $k_1 = e$. If $k_1 = e$, then $k_2^s = e(1-2a-\nu b)/(3b-2a) < 0$ and, correspondingly, $k_2 = 0$.

Let $b > 2a$, $b > 1$. If $k_1 = 0$, then $k_2 = e(1-2a)/(b-2a)$. If $k_2 = e(1-2a)/(b-2a)$, then $k_1^s = e(1-2a)[(1-\nu)b-2a]/(3b-2a)(b-2a) < 0$ and, hence, $k_1 = 0$.

Let $b > 2a$, $b \leq 1$, $2a + \nu b > 1$. If $k_1 = 0$, then $k_2 = e$. If $k_2 = e$, then $k_1^s = e(1-2a-\nu b)/(3b-2a) < 0$ and, hence, $k_1 = 0$. □

Proposition 10.2. *Let centers of several stars, each with ν peripheral nodes, being initially in inner equilibrium, be unified into one cycle. Under absence of productivity and $b > 1/3$, knowledge in each node in equilibrium declines, and, moreover, each peripheral node becomes passive. Utility in each node declines.*

Under $b > 1$ (which implies presence of productivity), each central node becomes passive, while knowledge in each periphery node decreases if $\nu = 1$ and increases if $\nu \geq 2$. Utility in each central node decreases and in each periphery node increases.

Proof. According to (26) and (27), before unification, the equilibrium level of knowledge in each central node was

$$k_1^* = \frac{e(1-2a)[(\nu-1)b+2a]}{\nu b^2-(b-2a)^2},$$

and in each peripheral node

$$k_2^* = \frac{2ea(1-2a)}{\nu b^2-(b-2a)^2}.$$

The environments were

$$K_1^* = \frac{e(1-2a)[(\nu-1)b+2a+2\nu a]}{(\nu-1)b^2+4ab-4a^2},$$

$$K_2^* = \frac{e(1-2a)[(\nu-1)b+4a]}{(\nu-1)b^2+4ab-4a^2}.$$

Under absence of productivity and $b > 1/3$, after unification, the level of knowledge in each central node becomes

$$k_1^{**} = \frac{e(1-2a)}{3b-2a},$$

and in each peripheral node

$$k_2^{**} = 0.$$

The environment in each central node becomes

$$K_1^{**} = \frac{3e(1-2a)}{3b-2a}.$$

We see that $k_1^* > k_1^{**}$, $K_1^* > K_1^{**}$. According to Theorem 2.2, utility in each central node decreases. Evidently, utility in each peripheral node also decreases, since the node becomes passive.

If $b > 1$ (which implies presence of productivity), after unification, the level of knowledge in each central node becomes

$$k_1^{**} = 0,$$

by Proposition 10.1; and in each peripheral node:

$$k_2^{**} = \frac{e(1-2a)}{b-2a}.$$

The environment in each peripheral node becomes

$$K_2^{**} = \frac{e(1-2a)}{b-2a}.$$

We see that $k_2^* > k_2^{**}$ if $\nu = 1$, $k_2^* < k_2^{**}$ if $\nu \geq 2$, $K_2^* < K_2^{**}$. By Theorem 2.2, utility in each peripheral node increases. In each central node, utility decreases. □

Remark 10.1. Strategic complementarity is observed when the stars are unified under absence of productivity. After unification, investments in central nodes decrease, and in peripheral nodes it is not profitable to make investments. As a result, the knowledge in the central nodes is the same as if there are no peripheral nodes at all. Strategic substitutability takes place under presence of productivity. After unification, investments of peripheral nodes increase if $\nu \geq 2$ and it is not profitable for central nodes to make investment. As a result, in periphery nodes the knowledge is the same as if there are no central nodes.

11 Chains with Three Types of Nodes

Let us consider chain with five nodes of 3 types: 3–2–1–2–3.

Proposition 11.1. *If* $a < b < (\sqrt{5}-1)a$ *and parameter* a *is sufficiently close to* $1/2$*, then the following inner equilibrium exists:*

$$k_1 = \frac{2a^2 e(1-2a)}{(2a-b)(b^2+2ab-2a^2)}, \tag{31}$$

$$k_2 = \frac{e(1-2a)(b+2a)}{2(b^2+2ab-2a^2)}, \tag{32}$$

$$k_3 = \frac{e(1-2a)(4a^2-2ab-b^2)}{2(2a-b)(b^2+2ab-2a^2)}. \tag{33}$$

Proof. System of equations (5) for the chain is

$$\begin{cases} (b-2a)k_1 + 2bk_2 = e(1-2a), \\ bk_1 + (b-2a)k_2 + bk_3 = e(1-2a), \\ bk_2 + (b-2a)k_3 = e(1-2a). \end{cases}$$

A solution can be found by use of Cramer's formulas:

$$\Delta = \begin{vmatrix} b-2a & 2b & 0 \\ b & b-2a & b \\ 0 & b & b-2a \end{vmatrix} = -2(b-2a)(b^2+2ab-2a^2),$$

$$\Delta^1 = \begin{vmatrix} e(1-2a) & 2b & 0 \\ e(1-2a) & b-2a & b \\ e(1-2a) & b & b-2a \end{vmatrix} = 4a^2 e(1-2a),$$

$$\Delta^2 = \begin{vmatrix} b-2a & e(1-2a) & 0 \\ b & e(1-2a) & b \\ 0 & e(1-2a) & b-2a \end{vmatrix} = -e(1-2a)(b-2a)(b+2a),$$

$$\Delta^3 = \begin{vmatrix} b-2a & 2b & e(1-2a) \\ b & b-2a & e(1-2a) \\ 0 & b & e(1-2a) \end{vmatrix} = -e(1-2a)(b^2+2ab-4a^2),$$

from which Eqs. (31)–(33) follow. The quadratic trinomial $b^2+2ab-2a^2$ has roots $b_{1,2} = (-1 \pm \sqrt{3})a$; and the roots of quadratic trinomial $b^2+2ab-4a^2$ are $b_{1,2} = (-1 \pm \sqrt{5})a$.

Closeness of a to $1/2$ implies $k_i < e$, $i = 1,2,3$. Thus, $0 < k_i < e$, $i = 1,2,3$ are fulfilled, i.e. the equilibria are inner. □

Some corner equilibria for the chain can be found by considering this chain as a result of adjunction of passive agents to a chain of a smaller size.

Proposition 11.2. *If $(\sqrt{5}-1)a \le b < 2a$, then passive agents can be adjoined to the ends of the chain of three nodes being in inner equilibrium, in such a way that in equilibrium behavior of each of the five agents does not change. If $b < (\sqrt{5}-1)a$ or $b > 2a$, then such connection is impossible.*

Proof. Inner equilibrium in a network of three nodes was considered in Sect. 9, case 3. By Lemma 3.1, under absence of productivity, the adjoined agent is passive iff

$$\frac{2ae(1-2a)}{b^2+4ab-4a^2} \le \frac{e(1-2a)}{b},$$

which is equivalent to

$$(\sqrt{5}-1)a \le b < 2a.$$

Under presence of productivity, the adjoined agent is passive iff

$$\frac{2ae(1-2a)}{b^2+4ab-4a^2} \ge \frac{e(1-2a)}{b},$$

which is impossible. □

Similarly one or two passive agents can join the chain of four active agents, connecting to one or both of its end agents: 2–1–1–2–3, 3–2–1–1–2, or 3–2–1–1–2–3.

Proposition 11.3. *If $b < 2a$ (absence of productivity) or $b \ge (3+\sqrt{5})a$, then passive agent can be adjoined to an end of the chain of four nodes which was in inner equilibrium, in such a way that each of the five agents does not change behavior in equilibrium. If $2a < b < (3+\sqrt{5})a$, then such connection is impossible.*

Proof. Inner equilibrium for the chain of four nodes was considered in Sect. 9, case 1. For the quadratic trinomial $b^2-6ab+4a^2$ in the denominator of expressions (24) and (25), the roots are

$$b_{1,2} = (3 \pm \sqrt{5})a.$$

By Lemma 3.1, under absence of productivity, the adjoined agent is passive, since condition

$$\frac{e(1-2a)(b-2a)}{b^2-6ab+4a^2} \le \frac{e(1-2a)}{b}.$$

is fulfilled. Under presence of productivity, the adjoined agent is passive iff

$$\frac{e(1-2a)(b-2a)}{b^2-6ab+4a^2} \geq \frac{e(1-2a)}{b},$$

i.e. under $b > (3+\sqrt{5})a$. Hence, under $b < 2a$ or $b \geq (3+\sqrt{5})a$ such adjunction is possible, and under $2a < b < (3+\sqrt{5})a$ it is not possible. □

Proposition 11.4. *In the chain 3–2–1–1–2, if* $b > 1$*, then the equilibrium*

$$k_1 = k_3 = 0, \quad k_2 = \frac{e(1-2a)}{b-2a}$$

is possible. If $2a < b \leq 1$, $b + 2a \geq 1$*, then the pure corner equilibrium*

$$k_1 = k_3 = 0, \quad k_2 = e$$

is possible.

Proof. Existence of the first of the equilibria is equivalent to a possibility of adjunction of a passive agent to an isolated active agent (Proposition 6.2) and connection of isolated active agents through a passive agent (Proposition 7.2). Conditions of passivity of the agents of types 1 and 3 are, correspondingly,

$$2e \geq \frac{e(1-2a)}{b},$$

$$e \geq \frac{e(1-2a)}{b}.$$

Evidently, these inequalities are fulfilled. □

Proposition 11.5. *In the chain of five nodes 3–2–1–1–2, if* $2a < b \leq 1/2$ *and* $2b + 2a \geq 1$*, then the following pure corner equilibrium is possible:*

$$k_1 = 0, \quad k_2 = k_3 = e.$$

If $2a < b \leq 1/3$ *or* $1/2 \leq b < 2a$*, then the following pure corner equilibrium is possible:*

$$k_1 = k_2 = k_3 = e.$$

Proof. Let $2a < b \leq 1/2$ and $2b+2a \geq 1$. Assume that $k_1 = 0, k_2 = k_3 = e$. Then, by Lemma 3.1, the necessary and sufficient condition of passivity of the agents in nodes of type 1 is

$$e \geq \frac{e(1-2a)}{b}, \tag{34}$$

while the necessary and sufficient condition of hyperactivity of the agents in nodes 2 and 3 is

$$e \leq \frac{e(1-b)}{b}. \tag{35}$$

Conditions (34) and (35) are, evidently, fulfilled. The second statement follows from Proposition 4.2. □

Similarly, a chain of six nodes 3–2–1–1–2–3 can be considered.

Proposition 11.6. *In the chain of six nodes 3–2–1–1–2–3, under presence of productivity, if parameter a is sufficiently close to 1/2, then the following inner equilibrium exists:*

$$k_1 = \frac{2ae(1-2a)(b-2a)}{b^3 + 6ab^2 - 16a^2b + 8a^3}, \tag{36}$$

$$k_2 = \frac{e(1-2a)(b^2 + 2ab - 4a^2)}{b^3 + 6ab^2 - 16a^2b + 8a^3}, \tag{37}$$

$$k_3 = \frac{4ae(1-2a)(b-a)}{b^3 + 6ab^2 - 16a^2b + 8a^3}. \tag{38}$$

Proof. System of equations (5) for this chain is

$$\begin{cases} (2b-2a)k_1 + bk_2 = e(1-2a), \\ bk_1 + (b-2a)k_2 + bk_3 = e(1-2a), \\ bk_2 + (b-2a)k_3 = e(1-2a). \end{cases}$$

A solution can be found by use of the Cramer's formulas:

$$\Delta = \begin{vmatrix} 2b-2a & b & 0 \\ b & b-2a & b \\ 0 & b & b-2a \end{vmatrix} = -2(b-2a)(b^2+2ab-2a^2) = 16a^2b - b^3 - 6ab^2 - 8a^3.$$

We see that under presence of productivity $\Delta < 0$.

$$\Delta^1 = \begin{vmatrix} e(1-2a) & b & 0 \\ e(1-2a) & b-2a & b \\ e(1-2a) & b & b-2a \end{vmatrix} = -2ae(1-2a)(b-2a).$$

We have $\Delta^1 < 0$ under presence of productivity.

$$\Delta^2 = \begin{vmatrix} 2b-2a & e(1-2a) & 0 \\ b & e(1-2a) & b \\ 0 & e(1-2a) & b-2a \end{vmatrix} = e(1-2a)(4a^2 - b^2 - 2ab).$$

The roots of the quadratic trinomial $4a^2 - b^2 - 2ab$ are $a(\pm\sqrt{5} - 1)$, hence $\Delta^2 < 0$ under presence of productivity.

$$\Delta^3 = \begin{vmatrix} 2b - 2a & b & e(1-2a) \\ b & b - 2a & e(1-2a) \\ 0 & b & e(1-2a) \end{vmatrix} = -4ae(1-2a)(b-a) < 0$$

if $b > a$. Thus, presence of productivity implies $k_i > 0$, $i = 1, 2, 3$; and closeness of parameter a to $1/2$ implies $k_i < e$, $i = 1, 2, 3$; hence, (k_1, k_2, k_3) is inner equilibrium. □

Proposition 11.7. *If* $(3 + \sqrt{5})a \leq b < 2a$, *then it is possible to adjoint nodes with passive agents to the ends of a chain 2–1–1–2 of four nodes being in inner equilibrium, in such a way that the agents do not change their behavior in equilibrium. If* $2a < b < (3 + \sqrt{5})a$, *then such adjunction is impossible.*

Proof. The same proof as for Proposition 11.3. □

Proposition 11.8. *In the network of six nodes 3–2–1–1–2–3, if* $b > 1$ *(which implies presence of productivity), then the following equilibrium is possible:*

$$k_1 = k_3 = 0, \quad k_2 = \frac{e(1-2a)}{b-2a}.$$

If $2a < b \leq 1$, $b + 2a \geq 1$, *then the following equilibrium is possible:*

$$k_1 = k_3 = 0, \quad k_2 = e.$$

Proof. Let $b > 1$, $b > 2a$, $k_2 = e(1-2a)/(b-2a)$. By Lemma 3.1, the necessary and sufficient condition of passivity of agents of types 1 and 3 is

$$\frac{e(1-2a)}{b-2a} \geq \frac{e(1-2a)}{b},$$

which is, evidently, fulfilled. If $2a < b \leq 1$, $k_2 = e$, $b+2a \geq 1$, then, by Lemma 3.1, the necessary and sufficient condition of passivity of agents of types 1 and 3 is

$$e \geq \frac{e(1-2a)}{b},$$

which is also, evidently, fulfilled. □

Proposition 11.9. *In the network of six nodes 3–2–1–1–2–3 if* $2a < b \leq 1/2$ *and* $2b + 2a \geq 1$, *then the following equilibrium is possible:*

$$k_1 = 0, \quad k_2 = k_3 = e.$$

If $2a < b \leq 1/3$ *or* $1/2 \leq b < 2a$*, then the following equilibrium is possible:*

$$k_1 = k_2 = k_3 = e.$$

Proof. The same proof as for Proposition 11.6. □

12 Rise of New Links

Let us see what is going on when in equilibrium a new link in a network appears. We will provide conditions under which the sum of knowledge in the whole network decreases.

Theorem 12.1. *Let* W *be a network, and* W' *be the network obtained from* W *by addition of a new link between nodes* i *and* j*. The network* W' *is assumed to be connected. Let* $(k_1, k_2, \dots, k_n)$ *be an inner equilibrium in* W*, and* $(k'_1, k'_2, \dots, k'_n)$ *an inner equilibrium in* W'*. Then* $\sum_{i=1}^{n} k'_i < \sum_{i=1}^{n} k_i$.

Proof. Let $\mathbf{M}$ be the adjacency matrix of W, and $\mathbf{M'}$—the adjacency matrix of W'. The system of equations (5) for network W can be written as

$$b(\mathbf{M} + \mathbf{I} - 2a\mathbf{I})\mathbf{X} = \bar{\mathbf{e}}, \tag{39}$$

and for network W' as

$$b(\mathbf{M'} + \mathbf{I} - 2a\mathbf{I})\mathbf{X} = \bar{\mathbf{e}}, \tag{40}$$

where $\mathbf{I}$ is the identity matrix, $\mathbf{X} = (x_1, x_2, \dots, x_n)^T$ is the column of variables, $\mathbf{M'} = \mathbf{M} + \mathbf{D_{ij}}$, $\bar{\mathbf{e}} = (e(1-2a), e(1-2a), \dots, e(1-2a))^T$, $\mathbf{D_{ij}}$ is the $n \times n$-matrix, in which units stay in intersection of i-th row and j-th column and in intersection of j-th row and i-th column, and all other elements are zeros.

Let $\mathbf{k} = (k_1, k_2, \dots, k_n)^T$ and $\mathbf{k'} = (k'_1, k'_2, \dots, k'_n)^T$ be solutions of systems (39) and (40), correspondingly, satisfying conditions of the theorem; in particular, $\mathbf{k}$ and $\mathbf{k'}$ are strictly positive:

$$b(\mathbf{M} + \mathbf{I} - 2a\mathbf{I})\mathbf{k} = \bar{\mathbf{e}}, \tag{41}$$

$$b(\mathbf{M'} + \mathbf{I} - 2a\mathbf{I} + \mathbf{D_{ij}})\mathbf{k'} = \bar{\mathbf{e}}. \tag{42}$$

Subtracting (42) from (41), we obtain

$$b(\mathbf{M} + \mathbf{I} - 2a\mathbf{I})(\mathbf{k} - \mathbf{k'}) = A\mathbf{G_{ij}}, \tag{43}$$

where $\mathbf{G_{ij}}$ is a column, the i-th element of which is k'_j, and the j-th element is k'_i. Multiplying (41) from the left by the matrix-row $\mathbf{k^T}$, we receive

$$\mathbf{k^T}b(\mathbf{M}+\mathbf{I}-2a\mathbf{I})\mathbf{k} = e(1-2a)\sum_{i=1}^{n} k_i. \tag{44}$$

Using the symmetry of the matrix $b(\mathbf{M}+\mathbf{I}-2a\mathbf{I})$ and transposing Eq. (41), we have

$$\mathbf{k^T}b(\mathbf{M}+\mathbf{I}-2a\mathbf{I}) = \bar{\mathbf{e}}^\mathbf{T}. \tag{45}$$

At the same time, from (43) and (45), we obtain

$$\mathbf{k^T}b(\mathbf{M}+\mathbf{I}-2a\mathbf{I})\mathbf{k} = \mathbf{k^T}[b(\mathbf{M}+\mathbf{I}-2a\mathbf{I})\mathbf{k'} + b\mathbf{G_{ij}}]$$

$$= \bar{\mathbf{e}}^\mathbf{T}\mathbf{k'} + \mathbf{k^T}b\mathbf{G_{ij}} = e(1-2a)\sum_{i=1}^{n} k'_i - b(k_i k'_j + k_j k'_i). \tag{46}$$

Comparing (44) with (46), we see that

$$e(1-2a)\sum_{i=1}^{n} k'_i = e(1-2a)\sum_{i=1}^{n} k_i - b(k_i k'_j + k_j k'_i).$$

Since the values $\mathbf{k}$ and $\mathbf{k'}$ are strictly positive, we receive

$$\sum_{i=1}^{n} k'_i < \sum_{i=1}^{n} k_i.$$

□

13 Connections of Full Networks and of Stars

Let us consider the case when two full networks, each with $n \geq 2$ nodes, adjoint by a new link.

Proposition 13.1. *If two full networks, each with $n \geq 2$ nodes, being initially in inner equilibrium, adjoint, then inner equilibrium in the new network exists under following conditions: absence of productivity ($b < 2a$),*

$$a > \frac{2(n-1)}{(n+1)^2}, \tag{47}$$

$$\frac{(n+1)a - \sqrt{(n+1)^2a^2 - 2(n-1)a}}{n-1} < b < \frac{(n+1)a + \sqrt{(n+1)^2a^2 - 2(n-1)a}}{n-1}.$$

The inner equilibrium (where "ports" are nodes of type 1) is

$$k_1 = \frac{2ae(1-2a)}{2(n+1)ab - (n-1)b^2 - 4a^2}, \tag{48}$$

$$k_2 = \frac{e(1-2a)(2a-b)}{2(n+1)ab - (n-1)b^2 - 4a^2}. \tag{49}$$

The level of knowledge in the "ports" in equilibrium after junction is higher than in other nodes. In comparison with the equilibrium before junction, knowledge in ports decreases if $n = 2$, and increases if $n \geq 3$. In all other nodes, knowledge decreases for any $n \geq 2$.

Proof. The types of nodes in the new network are characterized by vectors $\mathbf{l}(1) = (1, n-1)$ and $\mathbf{l}(2) = (1, n-2)$. From (22) and (23), the stationary levels of knowledge are k_1 and k_2 given by (48) and (49). Conditions of inner equilibrium $0 < k_i < e$ are reduced to

$$2a - b > 0$$

$$2a(1-2a) < 2(n+1)ab - (n-1)b^2 - 4a^2.$$

The latter inequality is

$$(n-1)b^2 - 2(n+1)ab + 2a < 0. \tag{50}$$

Discriminant of (50) is positive under (47). In this case b has to lie between the roots, and we come to the conditions listed above.

The level of knowledge in "ports" is above knowledge in other nodes.

Let us see how the levels of knowledge in the nodes change after junction. Comparing k_1 with

$$k = \frac{e(1-2a)}{nb - 2a},$$

we see that $k_1 < k$ if $n = 2$ and $k_1 > k$ if $k \geq 3$. Comparing k_2 with k, we see that $k_2 < k$.

Comparing environment in any "ports" after unification

$$K_1 = \frac{e(1-2a)[4a + (n-1)(2a-b)]}{2(n+1)ab - (n-1)b^2 - 4a^2}$$

and before unification

$$K = \frac{ne(1-2a)}{nb-2a},$$

we see that $K_1 < K$ if $n = 2$, and $K_1 > K$ if $n \geq 3$. Comparing environment in other nodes after unification

$$K_2 = \frac{e(1-2a)[2a+(n-2)(2a-b)]}{2(n+1)ab-(n-1)b^2-4a^2}$$

with environment before unification, K, we see that $K_2 < K$. Applying Theorem 2.2, we obtain the needed statement. □

Proposition 13.2. *Let centers of two stars of order* ν, *which were initially in inner equilibrium, be connected. An inner equilibrium in the united network is possible only under absence of productivity and for parameter a sufficiently close to* $1/2$. *In this equilibrium the levels of knowledge in nodes of type 1 (centers of the stars) and of type 2 (peripheral nodes) are, correspondingly,*

$$k_1 = \frac{e(1-2a)[2a+(\nu-1)b]}{(\nu-2)b^2+6ab-4a^2}, \tag{51}$$

$$k_2 = \frac{e(1-2a)(2a-b)}{(\nu-2)b^2+6ab-4a^2}. \tag{52}$$

The level of knowledge in the central nodes is higher than in the peripheral ones. The levels of knowledge and utilities in all nodes in equilibrium after unification are smaller than before unification. Knowledge and utility in both types of nodes decrease in ν.

Proof. The types are characterized by vectors $\mathbf{l}(1) = (1, \nu,)$ and $\mathbf{l}(2) = (1, 0)$; from (22) and (23) we find stationary levels of knowledge (51) and (52). Under absence of productivity and a close to $1/2$, there will be $0 < k_i < e$, $i = 1, 2$; this is inner equilibrium. In this case, $k_1 > k_2$.

Let us see how the levels of knowledge and utilities have changed after the unification. Comparing k_1 and k_2 with the levels of knowledge in the center of star before unification

$$k_1^0 = \frac{e(1-2a)[2a+(\nu-1)b]}{(\nu-1)b^2+4ab-4a^2},$$

$$k_2^0 = \frac{2ae(1-2a)}{(\nu-1)b^2+4ab-4a^2},$$

we see that $k_1 < k_1^0$, $k_2 < k_2^0$.

Comparing the environments in the nodes after unification

$$K_1 = \frac{e(1-2a)[(3\nu-2)b-2(\nu-2)a]}{(\nu-2)b^2+6ab-4a^2},$$
$$K_2 = \frac{e(1-2a)[4a+(\nu-2)b]}{(\nu-2)b^2+6ab-4a^2},$$

and before unification

$$K_1^0 = \frac{e(1-2a)[(\nu-1)b+2(\nu+1)a]}{(\nu-1)b^2+4ab-4a^2},$$
$$K_2^0 = \frac{e(1-2a)[(\nu-1)b+4a]}{(\nu-1)b^2+4ab-4a^2},$$

we see that $K_1 < K_1^0$, $K_2 < K_2^0$. It is easy to check that k_1, K_1, k_2, K_2 decrease in ν. Theorem 2.2 leads to the needed statement. □

14 Rise of a New Link in a Cycle

Let a new link connect two nodes of a cycle. We will limit ourselves by the cases when only two types of nodes appear. In particular, let us consider the cases of cycles with $n = 4$, $n = 6$, and $n = 8$ nodes and new link connecting opposite nodes of the cycle.

Proposition 14.1. *1. After adding the new link into cycle of four nodes being in inner equilibrium, under condition*[2]

$$b > \frac{\sqrt{64a^2+1}-(8a-1)}{4}, \tag{53}$$

the following inner equilibrium exists:

$$k_1 = \frac{e(1-2a)(b+2a)}{2(b^2+3ba-2a^2)}, \tag{54}$$

$$k_2 = \frac{ae(1-2a)}{b^2+3ba-2a^2}. \tag{55}$$

[2]Since $b > a$, the necessary and sufficient condition (53) follows from simple sufficient condition $a > 3/10$.

Knowledge and utility in the nodes of type 1 ("bridges") decrease under absence of productivity. Knowledge and utility in the nodes of type 2 decrease under any values of parameters.

2. *After adding the new link into cycle of six nodes, being in inner equilibrium, under conditions*

$$2a - \sqrt{4a^2 - a} < b < 2a, \; a > \frac{1}{4},$$

the following inner equilibrium exists:

$$k_1 = \frac{ae(1-2a)}{-b^2 + 4ba - 2a^2}, \tag{56}$$

$$k_2 = \frac{e(1-2a)(2a-b)}{2(-b^2 + 4ba - 2a^2)}, \tag{57}$$

The level of knowledge and utility in the nodes of type 1 ("bridges") increase after adjunction, while the level of knowledge and utility in the nodes of type 2 decrease.

Proof. 1. The types of nodes are characterized by vectors $\mathbf{l}(1) = (1, 2)$ and $\mathbf{l}(2) = (2, 0)$, and (22) and (23) imply (54) and (55). Conditions of inner equilibrium $0 < k_i < e, i = 1, 2$ can be reduced to

$$(1 - 2a)(b + 2a) < 2(b^2 + 3ba - 2a^2).$$

The solution is described by (53). Comparing k_1 with the level of knowledge before adding the new link, $k = e(1 - 2a)/(3b - 2a)$, we see that $k_1 < k$ if $b < 2a$, and $k_1 > k$ if $b > 2a$. We see also that $k_2 < k$ independently on the value of b. Comparing environments after adding the new link,

$$K_1 = \frac{e(1-2a)[2(b+2a) + 4a]}{2b^2 + 6ab - 4a^2},$$

$$K_2 = \frac{e(1-2a)[2(b+2a) + 2a]}{2b^2 + 6ab - 4a^2},$$

with environment in a node before adding the additional link

$$K = \frac{3e(1-2a)}{3b - 2a},$$

we see that $K_1 < K$ if $b < 2a$, $K_1 > K$ if $b > 2a$, $K_2 < K$ for any b. Theorem 2.2 gives the needed result.

2. The types of nodes are characterized by vectors $\mathbf{l}(1) = (1, 2)$ and $\mathbf{l}(2) = (1, 1)$, and (22) and (23) turn into (56) and (57). Conditions of inner equilibrium $0 < k_i < e, i = 1, 2$ can be reduced to the inequality

$$2a(1-2a) < -2b^2 + 8ab - 4a^2$$

which implies

$$2a - \sqrt{4a^2 - a} < b < 2a + \sqrt{4a^2 - a}.$$

This leads to conditions of inner equilibrium listed in the formulation of the proposition. Comparing k_1 and k_2 with knowledge in a node before adding the link $k = e(1-2a)/(3b-2a)$, we see that $k_1 > k$, $k_2 < k$. Comparing the environments in the nodes after adding the new link,

$$K_1 = \frac{e(1-2a)[4a + 2(2a-b)]}{-2b^2 + 8ab - 4a^2},$$

$$K_2 = \frac{e(1-2a)[2a + 2(2a-b)]}{-2b^2 + 8ab - 4a^2},$$

with the environment in a node before the adding,

$$K = \frac{3e(1-2a)}{3b-2a},$$

we see that $K < K_1$, $K_2 < K$. Hence, by Theorem 2.2, utility in the nodes of type 1 increased, while utility in nodes of type 2 decreased after appearance of the new link. □

After adding a new link connecting two opposite nodes in a cycle of eight nodes, we receive a network with three types of nodes.

Proposition 14.2. *If in a cycle of eight nodes a new link appeared which connects two opposite nodes, then inner equilibrium in the new network is impossible.*

Proof. The system of equations for the network under consideration has the form

$$\begin{cases} (2b-2a)k_1 + 2bk_2 = e(1-2a), \\ bk_1 + (b-2a)k_2 + bk_3 = e(1-2a), \\ 2bk_2 + (b-2a)k_3 = e(1-2a). \end{cases} \quad (58)$$

The solution can be found by use of the Cramer's formulas:

$$\Delta = \begin{vmatrix} 2b-2a & 2b & 0 \\ b & b-2a & b \\ 0 & 2b & b-2a \end{vmatrix} = 2(8a^2b - 4a^3 - 2b^3 - ab^2),$$

$$\Delta^1 = \begin{vmatrix} e(1-2a) & b & 0 \\ e(1-2a) & b-2a & b \\ e(1-2a) & 2b & b-2a \end{vmatrix} = e(1-2a)(b-2a)(b+2a),$$

$$\Delta^2 = \begin{vmatrix} 2b-2a & e(1-2a) & 0 \\ b & e(1-2a) & b \\ 0 & e(1-2a) & b-2a \end{vmatrix} = e(1-2a)(b^2+2ab-4a^2),$$

$$\Delta^3 = \begin{vmatrix} 2b-2a & 2b & e(1-2a) \\ b & b-2a & e(1-2a) \\ 0 & 2b & e(1-2a) \end{vmatrix} = 2e(1-2a)(2a^2-ab-b^2),$$

$$k_1 = \frac{e(1-2a)(2a-b)(b+2a)}{2(8a^2b-4a^3-ab^2-2b^3)}, \tag{59}$$

$$k_2 = \frac{e(1-2a)(b^2+2ab-4a^2)}{2(8a^2b-4a^3-ab^2-2b^3)}, \tag{60}$$

$$k_3 = \frac{2e(1-2a)(2a^2-ab-b^2)}{2(8a^2b-4a^3-ab^2-2b^3)}, \tag{61}$$

We see that, under $b > a$, expressions (59)–(61) cannot be positive simultaneously, hence, inner equilibrium is impossible.

15 Conclusions

Our model describes situations in which agents in a network make investments of some resource (such as money or time) on the first stage (period 1 in the model), and obtain a gain on the second stage (period 2). Such situations are typical in life of families, communities, firms, countries, international organizations, etc. Thus, the model can have numerous applications in analysis of equilibria in various economic, social, and political systems.

In framework of the model, we consider questions which concern relations between network structure, incentives, behavior of the agents, and the equilibrium state of economic or social system in terms of welfare of the agents.

We introduce new concepts and develop techniques which can be used in such kind of analysis. We provide some results of studying the model, among them are results describing consequences of appearance of new links in networks and of adjunctions of components. We introduce the concept of types of nodes, propose classification of networks based on this concept, describe an algorithm of subdivision of networks into types, and demonstrate the role of types in characterizing the inner equilibria.

Interesting questions for further research, which are not touched in the present paper, are relations between different possible concepts of equilibrium, and dynamics of formation of new equilibrium after adjunction of components or after rise of new links.

Acknowledgements The research was partially supported by Russian Foundation for Basic Research (projects 14-01-00448 and 14-06-00253).

References

1. Azariadis, C., Chen, B.-L., Lu, C.-H., Wang, Y.-C.: A two-sector model of endogenous growth with leisure externalities. J. Econ. Theory **148**, 843–857 (2013)
2. Bramoullé, Y., Kranton, R.: Public goods in networks. J. Econ. Theory **135**, 478–494 (2007)
3. Bulow, J., Geanakoplos, J., Klemperer, P.: Multimarket oligopoly: strategic substitutes and complements. J. Polit. Econ. **93**(3), 488–511 (1985)
4. Galeotti, A., Goyal, S., Jackson, M.O., Vega-Redondo, F., Yariv, L.: Network games. Rev. Econ. Stud. **77**, 218–244 (2010)
5. Grossman, G., Maggi, G.: Diversity and trade. Am. Econ. Rev. **90**, 1255–1275 (2000)
6. Jackson, M.O.: Social and Economic Networks. Princeton University Press, Princeton (2008)
7. Jackson, M.O., Zenou, Y.: Games on networks. In: Young P., Zamir S. (eds.) Handbook of Game Theory, vol. 4, pp. 95–164. Elsevier Science, North-Holland (2015)
8. Jacobs, J.: The Economy of Cities. Random House, New York (1969)
9. Lucas, R. E.: On the mechanics of economic development. J. Monetary Econ. **22**, 3–42 (1988)
10. Martemyanov, Y.P., Matveenko, V.D.: On the dependence of the growth rate on the elasticity of substitution in a network. Int. J. Process Manag. Benchmark. **4**(4), 475–492 (2014)
11. Milgrom, P., Roberts, J.: The economics of modern manufacturing: technology, strategy, and organisation. Am. Econ. Rev. **80**, 511–518 (1990)
12. Milgrom, P., Roberts, J.: Complementarities and systems: understanding Japanese economic organisation. Estud. Econ. **9**, 3–42 (1994)
13. Romer, P. M.: Increasing returns and long-run growth. J. Polit. Econ. **94**, 1002–1037 (1986)
14. Topkis, D.M.: Supermodularity and Complementarity. Princeton University Press, Princeton (1998)

Index

V.A. Kalyagin et al. (eds.), *Models, Algorithms and Technologies for Network Analysis*, Springer Proceedings in Mathematics & Statistics 156,
DOI 10.1007/978-3-319-29608-1

W